Manual for Detailing Reinforced Concrete Structures to EC2

Detailing is an essential part of the design process. This thorough reference guide for the design of reinforced concrete structures is largely based on Eurocode 2 (EC2), plus other European design standards such as Eurocode 8 (EC8), where appropriate.

With its large format, double-page spread layout, this book systematically details 213 structural elements. These have been carefully selected by José Calavera to cover relevant elements used in practice. Each element is presented with a whole-page annotated model along with commentary and recommendations for the element concerned, as well as a summary of the appropriate Eurocode legislation with reference to further standards and literature. The book also comes with a CD-ROM containing AutoCAD files of all of the models, which can be directly developed and adapted for specific designs.

Its accessible and practical format makes the book an ideal handbook for professional engineers working with reinforced concrete, as well as for students who are training to become designers of concrete structures.

José Calavera is Honorary President of the Technical Institute of Materials and Construction (INTEMAC – Instituto Técnico de Materiales y Construcciones) and Emeritus Professor, School of Civil Engineering, Polytechnic University of Madrid.

Manual for Detailing Reinforced Concrete Structures to EC2

José Calavera

Spon Press
an imprint of Taylor & Francis

LONDON AND NEW YORK

First published 2012
by Spon Press
2 Park Square, Milton Park, Abingdon, Oxon OX14 4RN

Simultaneously published in the USA and Canada
by Spon Press
711 Third Avenue, New York, NY 10017

Spon Press is an imprint of the Taylor & Francis Group, an informa business

British Library Cataloguing in Publication Data
A catalogue record for this book is available from the British Library

Library of Congress Cataloging-in-Publication Data
Calavera Ruiz, José.
  Manual for detailing reinforced concrete structures to EC2 / José Calavera.
     p. cm.
  Includes bibliographical references and index.
  1. Reinforced concrete construction—Details.   2. Reinforced concrete construction—
  Standards—Europe.   I. Title.
  TA683.28.C35 2012
  624.1'8341—dc22   2011009904

ISBN: 978-0-415-66348-9

Typeset in Helvetica by RefineCatch Ltd, Bungay, Suffolk

Printed and bound in Great Britain by the MPG Books Group

Contents

List of tables

Foreword

The aim of this book is to present a fairly full and systematic description of the construction details used in concrete structures.

While I paid particular attention to construction details in my previous books, all dealing primarily with structural design and engineering, I was naturally unable to address the issue in depth in any of them.

I have decided to do so today, acknowledging the importance of detailing and convinced that it is one of the areas of expertise that professionals must quickly learn to master. Construction details have a substantial impact not only on the quality of both the design and the building processes, but on concrete structure maintenance and durability as well.

Forty-five to fifty per cent of the problems arising around concrete structures are widely known to be attributable to the design stage. That half of those problems are due to errors in, or the lack of, construction details is a fact much less generally recognised.

Detailing is always the outcome of a synthesis of four areas of knowledge:

- a command of the theory underlying structural concrete engineering
- on-site professional practice
- experimental information obtained from laboratory trials
- the experience obtained in forensic engineering studies.

The extraordinary complexity resulting from such diversity is deftly reflected in the expression 'the art of detailing', which alludes to the mix of technical skill and creativity entailed in good detailing.

Someone inevitably decides how details are to be built: otherwise construction could not proceed. But the task is actually incumbent upon the designer. The further 'downstream' the detailing is done, the greater is the risk of malfunction.

This book begins with an introductory chapter that summarises specifications on concrete cover, reinforcing bar placement and spacing, hook bending radii, anchorages and bar welding. It also briefly discusses questions that have been scantily addressed in most countries' codes, such as how bars should be tied or spacers and chairs placed.

The second chapter is a description of the 213 construction details that comprise the book. Divided into 15 groups, they embrace what I believe to be a sufficient range of issues arising in reinforced concrete construction.

In this chapter each page on the left shows a drawing of a construction detail. The Notes set out on the page opposite on the right contains further information, as specified below.

(a)   A series of Recommendations that supplement and help to interpret the drawing, in some cases with concise reference to specific engineering questions.

(b)   Reference to Statutory Legislation in the European Union.

(c)   Reference to Recommended Alternative Codes to enable the reader to fill in the gaps where no statutory legislation is in place in the European Union, or in a number of specific cases, to resort to variations of interest.

(d)   Finally, a list of Specific References that deal explicitly and directly with the detail in question.

Version 2005 AutoCAD software is furnished with the book to enable designers to adapt each detail to the reinforcing bars used in their designs and print the results on a printer or plotter.

In closing, I owe a word of thanks to the people who collaborated in the preparation of this book. My gratitude goes to Antonio Machado for coordinating the draughting, Maribel González and Mercedes Julve for the typing; and Antonio Machado, Fernando Marcos and Julio César López for draughting the details from my sketches, which were not always as carefully drawn as would have been desired.

Many thanks as well to Margaret Clark and Mc LEHM Language Services for the translation of my original Spanish manuscript into English.

I am also indebted to Jorge Ley for his assistance in many respects.

Lastly, I wish to express my very special gratitude to Taylor & Francis for the support received in connection with the publication of this book, and particularly to Tony Moore and Siobhán Poole for their assistance.

José Calavera
Madrid, March 2011

The author

José Calavera graduated in civil engineering in 1960 and earned his doctorate in the field in 1967, both from the School of Civil Engineering, Polytechnic University of Madrid. From 1960 to 1967 he headed the Engineering Department at Tetracero, a Spanish producer of ribbed bars for reinforced concrete.

In 1967 he founded the Technical Institute of Materials and Construction (INTEMAC – Instituto Técnico de Materiales y Construcciones), an independent quality control organisation that covers design, materials and workmanship in both building and civil engineering. He is presently the Institute's Honorary President.

In 1982 he was appointed Professor of the Building and Precasting Department at the School of Civil Engineering, Polytechnic University of Madrid, where he is now Emeritus Professor.

He is a Fellow of the American Concrete Institute (ACI), the American Society of Civil Engineers (ASCE) and the International Association for Bridge and Structural Engineering (IABSE). He holds the International Federation for Structural Concrete's (FIB) Medal of Honour, and has been awarded the Italian Association of the Prefabrication Prize for Outstanding Achievement in Engineering and the Eduardo Torroja Medal.

He has written 15 books in Spanish, one in Italian and two in English on structural concrete-related subjects. His most prominent designs include the Fuente Dé Aerial Cableway, the roof over the Real Madrid Sports Centre and the space frame roofs over the National Livestock Market at Torrelavega. He is also a renowned specialist in forensic engineering.

Author's curriculum vitae

José Calavera

- PhD in Civil Engineering.
- BSc in Civil Engineering.
- Emeritus Professor of Building and Prefabrication at the School of Civil Engineering, Polytechnic University of Madrid.
- Honorary President of the Technical Institute of Materials and Construction (INTEMAC – Instituto Técnico de Materiales y Construcciones).
- Member of the Commission on Prefabrication of the International Federation for Structural Concrete (Fédération Internationale du Béton – FIB).
- Member of the Working Group 2.2 on Design by Testing of the Commission 2 on Safety and Performance Concepts of the International Federation for Structural Concrete (FIB).
- Editor for Europe of the international Council on Tall Buildings.
  Previously, he was:
  - Chairman of Commission VII on Reinforcement: Technology and Quality Control of the Euro-International Concrete Committee (Comité Euro-International du Béton – CEB).
  - Chairman of the Joint Committee on Tolerances (CEB – FIB).
  - Chairman of the Working Group on Precast Beam-Block Floor Systems of the International Federation for Structural Concrete (FIB).
  - Member of the Administrative Council of CEB.
  - Member of the Model Code CEB–FIB 1990 Drafting Committee.
  - Chairman of the Eurocode Drafting Committee for the Design of Concrete Foundations.
  - Chairman of the Working Group on Precast Prestressed Bridges of the International Federation for Structural Concrete (FIB).
  - Chairman of the Working Group on Treatment of Imperfections in Precast Concrete of the International Federation for Structural Concrete (FIB).
  - Chairman of Scientific-Technical Association of Structural Concrete (ACHE)
- Medal of the Spanish Technical Association for Prestressing (ATEP) (1978).
- Honorary Professor of the Civil Construction Faculty, Pontifical Catholic University of Chile (1980).
- Member of Honour of the Engineering Faculty, Pontifical Catholic University of Chile (1980).
- Elected Fellow of the American Concrete Institute (ACI) (1982).
- Medal of Honour of the Civil Engineering College (1987).
- Eduardo Torroja Medal (1990).
- Medal of the Spanish Road Association (1991).
- Honorary Doctorate of the Polytechnic University of Valencia (1992).

- Institutional Medal of the Lisandro Alvarado' Central Western University, Venezuela (1993).
- Medal of the International Federation for Structural Concrete (FIB) (1999).
- Medal of Honour of the Fundación García-Cabrerizo (1999).
- Award of the Spanish Group of IABSE (2000).
- Great Figures of Engineering Award of the Italian Association of Prefabrication (CTE) (2000).
- Award of the Spanish National Association of Reinforced Bars Manufacturers (ANIFER) (2001).
- Member of Honour of the Spanish Association of Structural Consultants (ACE) (2001).
- Honorary Member of the Academy of Sciences and Engineering of Lanzarote (2003).
- Member of Honour of the Argentine Structural Engineering Association (2004).
- Camino de Santiago Award of Civil Engineering (2004).
- Elected Fellow of IABSE (International Association for Bridge and Structural Engineering) (2006).
- Member of the Board of Trustees of the Fundación Juanelo Turriano (2006).
- Member of Honour of the Association of BSc Civil Engineers (2008).
- Best Professional Profile in Forensic Construction Engineering Award of the Latino American Association of Quality Control and Forensics Engineering (ALCONPAT) (2009).
- Elected Fellow of ASCE (American Society of Civil Engineers) (2009).
- Among his most important projects are the Fuente Dé Aerial Cableway (Cantabria), the space frame roofs of the Real Madrid Sports Centre and the Mahou Beer Factory (Madrid), the space frame roofs of the National Livestock Market of Torrelavega (Santander) and numerous industrial buildings, especially for paper manufacturers and the prefabrication of concrete and steel industry.
- He is author of 15 books in Spanish, two in English and one in Italian, three monographs and 176 publications on matters concerning structural design, reinforced and prestressed concrete, structural safety, prefabrication, quality control and pathology of structures. He has been thesis director for 27 doctoral theses.

Acknowledgements

The publishers wish to thank the Instituto Técnico de Materiales y Construcciones (INTEMAC) for granting permission to reproduce parts of the following books authored by J. Calavera.

Manual de detalles constructivos en obras de hormigón armado, [Manual for detailing reinforced concrete structures], Madrid, 1993.

Cálculo de estructuras de cimentación [Foundation concrete design], 4th edn, Madrid, 2000.

Muros de contención y muros de sótano [Retaining walls and basement walls], 3rd edn, Madrid, 2000.

Cálculo, construcción, patología y rehabilitación de forjados de edificación [Design, construction, pathology and strengthening of slabs in buildings], 5th edn, Madrid, 2002.

Proyecto y cálculo de estructuras de hormigón [Structural concrete design], 2nd edn, Madrid, 2008.

Citations

- Symbols and abbreviations. The conventions adopted in Eurocode 2 (EC2) have been used as a rule, except in Group 15 (Special construction details for earthquake zones), where the Eurocode 8 (EC8) conventions were followed.

- For greater clarity and brevity, references are shown as a number in brackets, which matches the number under which the publication is listed in the References at the end of the book.

  For instance: (3) refers to the third reference, namely EN ISO 3766:2003, *Construction Drawings. Simplified Representation of Concrete Reinforcement.*

- References to sections of the book itself are cited directly.

  For instance: 1.2 refers to section 1.2, Tying bars, in Chapter 1, General rules for bending, placing, anchoring and welding reinforcing bars.

- References to other construction details cite the designation shown at the top of each page.

  For instance: see CD – 01.02 refers to detail CD – 01.02, Wall footing supporting a brick wall.

- References to recommendations sometimes specify another CD. For instance: R-3 in 01.03 refers to Recommendation 3 in CD – 01.03. On occasion, the word 'recommendation' is written out in full, rather than as the abbreviation 'R'.

- References to formulas are placed in square brackets.

  For instance: [1.1] is the first formula in Chapter 1, item 1.1.1.

- The figures in Chapter 1 are designated as Figures 1-1 to 1-45.

- When figures are (very occasionally) shown in the Notes, they are designated by letters: (a), (b) and so on.

- The book is logically subject to EC2 specifications in particular and European Committee for Standardization (CEN) standards in general. When a given subject is not included in the CEN system of standards, explicit mention is made of that fact and an alternative standard is suggested.

- Inevitably, as in any code, the author's opinion occasionally differs from the criteria set out in CEN standards. Such recommendations are clearly labelled **AR** (author's recommendation). In these cases readers are invited to use their own judgement.

General notes

1. Chapter 1 summarises the specifications in Eurocode 2 on concrete cover, bar spacing, bending radii, spacer placement and welding, or alternative codes when no CEN standard is in place (for spacers and tying bars, for instance).

2. Many details assume a 2.5 cm or 1 ϕ cover (abbreviated throughout this book as a lower case r), which is the value for the most usual case, i.e. exposure classes XC2/XC3 in structural class S4. For other conditions, the cover can be changed as described in 1.1.3.

   The cover values in the drawings are the C_{min} values. *A further 10 mm must be added to accommodate the spacers.* (Members cast against the ground are an exception: in such cases the 7.5 cm specified includes the 10-mm margin.) The minimum cover value was not simply enlarged by 10 mm, because while this is the EC2 *recommended* value, countries are free to set their own value in their National Annexes.

3. Details on spacers and tying are indicative only. Their number and specific position are given in Chapter 1. The symbols for spacers and chairs are shown in Figure 1-21.

4. In some cases, more than one page was required to describe a detail. This is clearly specified in the heading ('1 of 2', for instance). In all such cases, the same Notes apply to both drawings and are repeated on the page opposite on the right for the reader's convenience.

5. In keeping with standard terminology in many English-speaking countries, in this book the word 'stirrups' has been used to designate transverse reinforcement in beams and 'ties' to signify transverse reinforcement in columns. In Eurocode EC2, the word 'links' is applied in both situations.

   In most structures these two types of reinforcement serve very different purposes, and perhaps for that reason, in (US) English, French and Spanish, different terms are used for each.

The three golden rules for pouring concrete on site

Construction details have a heavy impact on the actual quality of the concrete in a structure.

RULE No. 1

CONCRETE MUST BE CAST INTO ITS FINAL POSITION ACROSS AN ESSENTIALLY VERTICAL PATH

Horizontal paths must be avoided. This must be taken into consideration in the design drawings for reinforcing bar arrangements.

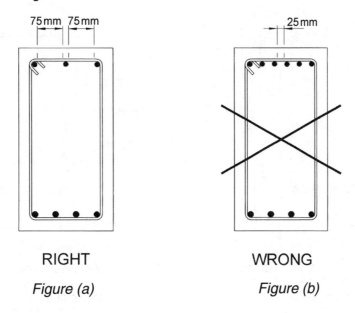

RIGHT	WRONG
Figure (a)	*Figure (b)*

Figure (a) depicts the right way to reinforce a beam to ensure that the forms are filled speedily and satisfactorily. If the bars are arranged as depicted in Figure (b), the coarse aggregate will not pass readily between them. As a result, the concrete will have to flow horizontally, inducing segregation and lengthening the time needed to fill the formwork. Moreover, such arrangements leave insufficient space for the vibrator.

Concrete should not be dumped in a pile for subsequent spreading with vibrators. Rather, it should be poured in each and every spot where it is needed.

Figure (c) shows the right way to reinforce a beam. With 65-mm spacing (somewhat smaller on site due to the height of the ribs), a standard 50-mm vibrator will be able to reach the bottom reinforcement. The solution depicted in Figure (d) is wrong, for it leaves insufficient room for the vibrator.

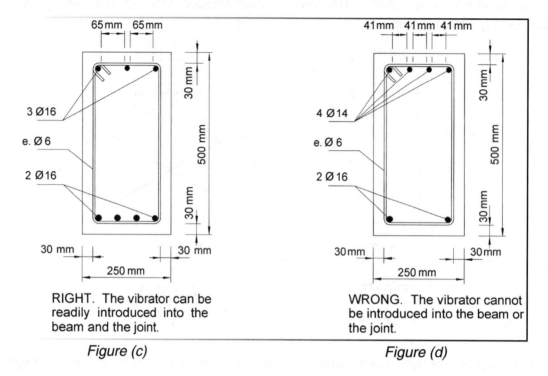

RIGHT. The vibrator can be readily introduced into the beam and the joint.

Figure (c)

WRONG. The vibrator cannot be introduced into the beam or the joint.

Figure (d)

Figures (e) and (f) show two further cases in which the vibrator is able, or unable, to reach the bottom reinforcement.

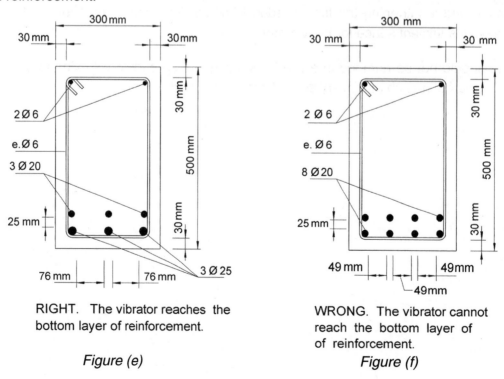

RIGHT. The vibrator reaches the bottom layer of reinforcement.

Figure (e)

WRONG. The vibrator cannot reach the bottom layer of of reinforcement.

Figure (f)

RULE No. 3

CONCRETE CONSISTENCY MUST BE IN KEEPING WITH THE REINFORCEMENT
ARRANGEMENT. AS A GENERAL RULE THE CONCRETE SLUMP SHOULD BE NO
SMALLER THAN 60 mm

Unless the reinforcement is arranged very openly and spaciously or powerful vibration methods are used, overly dry concrete is characterised by the following.

- The required strength can be reached in test specimens (but not on site) with a lower proportion of cement. The real strength of concrete can be lower.

- Since the control specimens can be compacted with no difficulty, the laboratory trials will furnish good information.

- *In situ* placement and a good surrounding of the reinforcement will be difficult to achieve and the loose consistency will lower actual on-site strength.

1 General rules for bending, placing, anchoring and welding reinforcing bars

1 INTRODUCTION

The present summary, based largely on Eurocode 2 (EC2), covers details of a general nature whose inclusion in all the relevant chapters of this book would be unnecessarily repetitious.

Nonetheless, certain characteristics specific to each type of structural member are addressed in the respective chapters.

Eurocode 2 (EC2) is supplemented by a number of other standards, the following in particular:

EN 10080	*Steel for the reinforcement of concrete* (1)
EN ISO 17760	*Permitted welding process for reinforcement* (2)
EN ISO 3766:2003	*Construction drawings. Simplified representation of concrete rein-forcement* (3)
Eurocode 8	*Design of structures for earthquake resistance* (4).

Further to EC2 (5), for buildings located in seismic areas, the construction details in this and the following chapter may be modified as described in Chapter 2, Group 15 below.

1.1 SUMMARY OF CODES AND STANDARDS ON CONSTRUCTION DETAILS

1.1.1 PERMISSIBLE MANDREL DIAMETERS FOR BENT BARS (see EC2, 8.3)

EC2 (5) stipulates that:

- the minimum diameter to which a bar may be bent shall be defined as the smallest diameter at which no bending cracks appear in the bar and which ensures the integrity of the concrete inside the bend of the bar;
- in order to avoid damage to the reinforcement, the diameter to which the bar is bent (mandrel diameter) should not be less than $\phi_{m,min}$.

TABLE T-1.1
MINIMUM MANDREL DIAMETER TO PREVENT DAMAGE
TO REINFORCEMENT (EC2)

(a)   for bars and wire

Bar diameter	Minimum mandrel diameter for bends, hooks and loops
$\phi \leq 16$ mm	4ϕ
$\phi > 16$ mm	7ϕ

(b)   for bent welded reinforcement and wire mesh bent after welding

Minimum mandrel diameter	
or	or
5ϕ	$d \geq 3\phi$: 5ϕ $d < 3\phi$ or welding within the curved zone: 20ϕ

Note: The mandrel size for welding within the curved zone may be reduced to 5ϕ where welding is performed as specified in EN ISO 17660 Annex B (2).

The mandrel diameter need not be checked to avoid concrete failure if the following conditions exist:

- the length of the bar anchorage beyond the end of the bend is not over 5ϕ;
- the bar is not in an end position (plane of bend close to concrete face) and a cross bar with a diameter $\geq \phi$ is duly anchored inside the bend;
- the mandrel diameter is at least equal to the recommended values given in Table T-1.1.

Otherwise, the mandrel diameter, $\phi_{m,min}$, must be increased as per Expression [1.1]

$$\phi_{m,min} \geq F_{bt}((1/a_b) + 1/(2\phi))/f_{cd} \qquad\qquad [1.1]$$

where:

F_{bt} is the tensile force of the ultimate loads in a bar or bundle of bars at the start of a bend

a_b for a given bar (or group of bars in contact) is half of the centre-to-centre distance between bars (or groups of bars) perpendicular to the plane of the bend. For a bar or bundle of bars adjacent to the face of the member, a_b should be taken as the cover plus $\phi/2$.

The value of f_{cd} must not be taken to be greater than the value for class C55/67 concrete.

TABLE T-1.2
MANDREL DIAMETERS FOR REINFORCING BARS
IN ACCORDANCE WITH TABLE T-1.1 (B 400 or B 500 steel) (in mm) (EC2)

case (a)   case (b)

ϕ mm	MANDREL (mm)	ϕ mm	MANDREL (mm)	MANDREL (mm) $d \geq 3\varnothing$	$d < 3\varnothing$ or welding within the curved zone
6	24	6	30	30	120
8	32	8	40	40	160
10	40	10	50	50	200
12	48	12	60	60	240
14	56	14	70	70	280
16	64	16	80	80	320
20	140	20	100	100	400
25	175	25	125	125	500
28	196	28	140	140	560
32	224	32	160	160	640
40	280	40	200	200	800
50	350	50	250	250	1000

The same rules are applicable to ties and stirrups.

AR. In cases routinely found in practice, such as depicted in Figure 1-1, the use of 12-mm ϕ or larger stirrups leaves the corner unprotected. Consequently, joining two smaller diameter stirrups is preferable to using 14-mm and especially 16-mm ϕ elements.

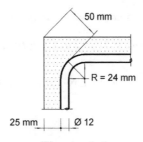

Figure 1-1

1.1.2   STANDARD BENDS, HOOKS AND LOOPS

EC2 specifies the bends, hooks and loops depicted in Figure 1-2 for rebar in general. Ties and stirrups call for special shapes, as specified in EC2, 8.4 and 8.5, and shown in Figure 1-3.

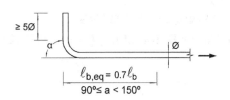

(a) Equivalent anchorage length for standard bend

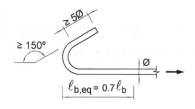

(b) Equivalent anchorage length for standard hook

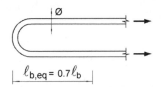

(c) Equivalent anchorage length for standard loop

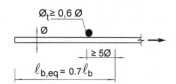

(d) Equivalent anchorage length for welded transverse bar

Figure 1-2. Anchorage methods other than straight bars

(See EC2, Table 8.21 for side cover specifications)

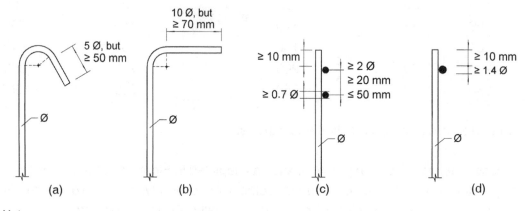

Notes:

AR. For (a) usually the hook is at 45°

For (c) and (d) the cover should not be less than either 3 Ø or 50 mm. See 1.6 for more detail.

Figure 1-3. Link anchorage

Welding must be performed as specified in EN ISO 17660 (2) and the welding capacity must be as stipulated in 1.6.

1.1.3 COVER

The EC2 specifications are summarised below.

(a) General

The concrete cover is the distance between the surface of the reinforcement closest to the nearest concrete surface (including ties and stirrups and surface reinforcement where relevant) and the nearest concrete surface. For these intents and purposes, groups of reinforcing bars cannot be replaced by the equivalent circle: rather, the cover refers to the actual bars in the group.

4

The nominal cover shall be specified on the drawings [*]. Such cover is defined as the minimum cover, c_{min}, plus a design allowance for deviation, Δc_{dev}.

$$c_{nom} = c_{min} + \Delta c_{dev} \qquad [1.2]$$

(b) Minimum cover, c_{min}

Minimum concrete cover, c_{min}, shall be provided in order to ensure:

- the safe transmission of bond forces
- the protection of the steel against corrosion
- adequate fire resistance.

The greatest value of c_{min} that meets the requirements for both bond and environmental conditions shall be used.

$$c_{min} = \max \{c_{min,b}\, ;\, c_{min,dur}\, ;\, 10 \text{ mm}\} \qquad [1.3]$$

where:

$\quad c_{min,b}$ = minimum cover stipulated to meet the bond requirement
$\quad c_{min,dur}$ = minimum cover stipulated for the environmental conditions.

(For the use of stainless steel or additional protection, see EC2 4.4.1.2)

The minimum cover, $c_{min,b}$, needed to transmit bond forces is given in Table T-1.3

TABLE T-1.3
MINIMUM COVER, $c_{min,b}$, BOND-RELATED REQUIREMENTS (EC2)

Bond requirement	
Arrangement of bars	Minimum cover $c_{min,b}$ [**]
Discrete	Bar diameter
Bundled	Equivalent diameter (ϕ_n)

(**) If the nominal maximum aggregate size is greater than 32 mm, $c_{min,b}$ must be increased by 5 mm.

The minimum cover for reinforcement in normal weight concrete for each exposure and structural class is given by $c_{min,dur}$.

Note: The recommended structural class (50-year design service life) is S4 for the indicative concrete strengths given in EC2, Annex E, while the recommended modifications to the structural class are given in Table T-1.4. The recommended minimum structural class is S1.

(*) Given that the EC2 'recommends' $\Delta c_{dev} = 10$ mm, but allows the choice to each Member State of the European Union to determine another value in its National Annex, in this manual are indicated the values of c_{min}.

The recommended values of $c_{min,dur}$ are given in Table T-1.5 (reinforcing steel).

TABLE T-1.4
RECOMMENDED STRUCTURAL CLASSIFICATION (EC2)

Criterion	Structural class						
	Exposure class (see EC2, Table 4.1 and notes)						
	X0	XC1	XC2/XC3	XC4	XD1	XD2/XS1	XD3/XS2/XS3
Design working life of 100 years	increase class by 2	increase class by 2	increase class by 2	increase class by 2	increase class by 2	increase class by 2	increase class by 2
Strength class	≥ C30/37 reduce class by 1	≥ C30/37 reduce class by 1	≥ C35/45 reduce class by 1	≥ C40/50 reduce class by 1	≥ C40/50 reduce class by 1	≥ C40/50 reduce class by 1	≥ C45/55 reduce class by 1
Member with slab geometry (position of reinforcement not affected by construction process)	reduce class by 1	reduce class by 1	reduce class by 1	reduce class by 1	reduce class by 1	reduce class by 1	reduce class by 1
Special quality control of the concrete production ensured	reduce class by 1	reduce class by 1	reduce class by 1	reduce class by 1	reduce class by 1	reduce class by 1	reduce class by 1

TABLE T-1.5
VALUES OF MINIMUM COVER, $c_{min,dur}$, REQUIREMENTS WITH REGARD TO DURABILITY FOR REINFORCEMENT STEEL IN ACCORDANCE WITH EN 10080 (1) (EC2)*

Structural class	Environmental requirement for $c_{min,dur}$ (mm)						
	Exposure class (see EC2, Table 4.1)						
	X0	XC1	XC2/XC3	XC4	XD1/XS1	XD2/XS2	XD3/XS3
S1	10	10	10	15	20	25	30
S2	10	10	15	20	25	30	35
S3	10	10	20	25	30	35	40
S4	10	15	25	30	35	40	45
S5	15	20	30	35	40	45	50
S6	20	25	35	40	45	50	55

(*) Fire safety may call for higher values.

AR. Except in special cases, covers of under 15 mm should not be used.

Where *in situ* concrete is placed on other (precast or *in situ*) concrete elements, the minimum concrete cover from the reinforcement to the interface may be reduced to a value meeting the bond requirement only, providing that:

* the concrete strength class is at least C25/30;
* the concrete surface is exposed to an outdoor environment for only a short time (< 28 days);
* the interface is roughened.

For uneven surfaces (e.g. exposed aggregate) the minimum cover should be increased by at least 5 mm.

AR. If the surface is roughened mechanically, this value should be 20 mm, for mechanical treatment generates microcracks in the concrete surface.

(c) Allowance in design for deviation

When calculating the design cover, c_{nom}, the minimum cover must be increased to allow for deviations (Δc_{dev}) by the absolute value of the tolerance for negative deviation. The recommended allowance is 10 mm.

For concrete poured onto uneven surfaces, the minimum cover should generally be increased by allowing larger deviations in design. The increase should compensate for the difference deriving from the unevenness, maintaining a minimum cover of 40 mm for concrete poured onto prepared ground (including blinding) and 75 mm for concrete poured directly onto the soil.[*]

The value

$$c_{nom} = c_{min} + c_{dev} \qquad [1.4]$$

reflects the spacer size.

AR. The 40-mm cover for blinding would appear to be excessive. Where the blinding is reasonably flat, 25 mm or 1 ϕ would appear to suffice.

1.1.4 BAR SPACING

Bars must be spaced in such a way that the concrete can be poured and compacted for satisfactory bonding and strength development. See the 'three golden rules of concrete pouring' that precede Chapter 1.

The clear (horizontal and vertical) distance between individual parallel bars or horizontal layers of parallel bars should not be less than the larger of (d_g + 5 mm), where d_g is the maximum aggregate size, and 20 mm (Figure 1-4).

Where bars are positioned in separate horizontal layers, the bars in each successive layer should be vertically aligned with the bars in the layer below. Sufficient space must be left between the resulting columns of bars for vibrator access and good concrete compaction.

Lapped bars may be allowed to touch one another within the lap length.

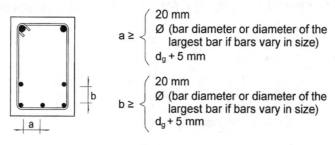

Figure 1-4

(*) Under certain circumstances, European standard EN 1536 (24) allows smaller values for bored piles.

AR. The 20-mm limit for *a* and *b* is too narrow to ensure satisfactory concrete casting. For single layers, a 25-mm space is suggested, and 35 mm for two or more: *a* should be 2.5 times the diameter of the vibrator needle for bars in any other than the bottom layer in the beam. Note that the longitudinal ribs on bars usually constitute 0.07 to 0.10 of the diameter and that bar placement inevitably entails deviations.

1.1.5 BUNDLED BARS

(a) Bundled bars versus large diameter bars

The standard series of large diameter ($\phi \geq 32$ mm) reinforcing bars includes two diameters in Europe, 40 mm and 50 mm, and three in the United States, 11 (ϕ 35 mm), 14 (ϕ 44 mm) and 18 (ϕ 57 mm).[*] While using these diameters provides for more compact reinforcement, which is a clear advantage, it also entails two drawbacks. On the one hand, the substantial load transfers generated call for carefully designed anchorage. On the other, since such large diameter bars cannot be lap spliced, construction is more complex and costly. Indeed, even lap splicing, if it were allowed, would be extremely expensive because of the extra steel needed for the long overlap lengths that would be required.

The alternative solution is to use bundled bars, which afford the advantages of compact distribution without the aforementioned drawbacks.

(b) Possibly usable bundles

The EC2 specifications are summarised below.

- As a rule, no more than three bars can be bundled (and their axes must not be in the same plane).
- In overlap areas and when using compressed bars in vertically cast members in which no splicing is needed, four bars are required.
- The equivalent diameter (for the ideal bar whose area is the same as the area of the bundle) must not be over 55 mm.

(c) Equivalent diameters, areas and mechanical strength

The specifications laid down in Tables T-1.6 and T-1.7 are applicable to bars with an equivalent diameter $\phi_n = \phi \sqrt{n_b}$, where n_b is the number of bars and ϕ is the diameter of each individual bar (Table T-1.8).

TABLE T-1.6.
MAXIMUM BUNDLES GENERAL SPECIFICATIONS

	ϕ in mm			
n	25	28	32	40
2	YES	YES	YES	NO
3	YES	YES	YES	NO
4	Only in overlaps	NO	NO	NO

(*) In the United States and Canada, bars are designated by their diameter expressed in eighths of an inch.

TABLE T-1.7
BUNDLES COMPRESSED BARS IN VERTICALLY CAST MEMBERS AND OVERLAP AREAS IN GENERAL

	ϕ in mm			
n	25	28	32	40
2	YES	YES	YES	NO
3	YES	YES	YES	NO
4	YES	NO	NO	NO

TABLE T-1.8
BUNDLES EQUIVALENT DIAMETERS in mm

	ϕ in mm			
n	25	28	32	40
2	35	40	45	57
3	43	48	55	69
4	50	56	64	80

(d)   Cross-sectional arrangement of bundles

- *Distances between bundles or bundles and bars.* The provisions of 1.1.4 apply. The minimum distance must be equal to the equivalent diameter, ϕ_n, whose values are given in Table T-1.8. Note that the minimum spacing between bundles is *the physical space between two points on the perimeter of the bar closest to the nearest bar in another bundle.* The space between two bundles should always be large enough to accommodate a vibrator during concrete casting.
- Cover. The cover must be at least equal to the equivalent diameter, ϕ_n, measured as the distance to the closest bar.
- For *anchorage and overlaps* in bundled bars, see item 8.9.3 of EC2.

1.1.6   SURFACE REINFORCEMENT

For bars with a diameter of over 32 mm, the following rules supplement the specifications in EC2, 8.4 and 8.7.

When such large diameter bars are used, cracking may be controlled either with surface reinforcement or by calculating crack widths.

As a general rule, large diameter bars should not be lapped. Exceptions include sections whose smallest dimension is 1.0 m or where the stress is no greater than 80 per cent of the design's ultimate strength. In any event, such bars should be lapped with mechanical devices.

In addition to shear reinforcement, transverse reinforcement should be placed in anchorage zones with no transverse compression.

(a) Additional reinforcement

For straight anchorage lengths (see Figure 1-5), such additional reinforcement should be at least as described below.

(i)   In the direction parallel to the stressed surface:

$$A_{sh} = 0.25\, A_s n_1.$$ [1.5]

(ii) In the direction perpendicular to the stressed surface:

$$A_{sv} = 0.25\, A_s n_2,$$ [1.6]

where:

A_s is the cross-sectional area of the anchored bar
n_1 is the number of layers with bars anchored at the same point in the member
n_2 is the number of bars anchored in each layer.

The additional transverse reinforcement should be uniformly distributed in the anchorage area and bars should not be spaced at more than five times the diameter of the longitudinal reinforcement.

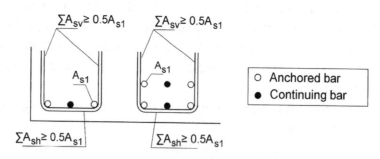

Example: On the left, $n_1 = 1$, and $n_2 = 2$ and on the right,
$n_1 = 2$, $n_2 = 2$

ADDITIONAL REINFORCEMENT IN AN ANCHORAGE FOR LARGE
DIAMETER BARS WHERE THERE IS NO TRANSVERSE COMPRESSION

Figure 1-5

For surface reinforcement, (i) and (ii) apply, but the area of the surface reinforcement should not be less than $0.01\, A_{ct,ext}$ in the direction perpendicular, and $0.02\, A_{ct,ext}$ in the direction parallel, to the large bars. $A_{ct,ext}$ is the area of the tensile concrete external to the stirrups (see Figure 1-6).

(b) Additional surface reinforcement

Surface reinforcement may be needed either to control cracking or to ensure adequate cover resistance to spalling. EC2 addresses this question in Informative Annex J, summarised below.

Surface reinforcement to resist spalling should be used where the main reinforcement comprises:

- bars with diameters of over 32 mm;
- bundled bars with an equivalent diameter of over 32 mm.

The surface reinforcement, which should consist of welded-wire mesh or narrow bars, should be placed outside the stirrups, as shown in Figure 1-6.

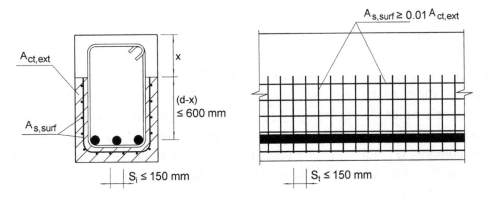

x is the depth of the neutral axis at ULS

Figure 1-6. Example of surface reinforcement

The area of the surface reinforcement $A_{s,surf}$ should not be less than 0.01 $A_{ct,ext}$ in the directions parallel and perpendicular to the tension reinforcement in the beam.

Where the reinforcement cover is over 70 mm, similar surface reinforcement should be used, with an area of 0.005 $A_{ct,ext}$ in each direction for enhanced durability.

The longitudinal bars in surface reinforcement may be regarded as constituting reinforcement to resist any other action effects whatsoever.

AR. If such additional reinforcement is included, concrete with a suitable slump should be used and poured and compacted with utmost care.

1.2 TYING BARS

This issue is not addressed in the EC2 or in any CEN standard. Three excellent References ((6), (7) and (8)) are available on the subject:

- The Concrete Society, *Spacers for reinforced concrete* Report CS 101, Concrete Society, 1989.
- Comité Euro-International du Béton (CEB), *Spacers, chairs and tying of steel reinforcement*, Lausanne, 1990.
- BS 7973:2001, *Spacers and chairs for steel reinforcement and their specifications.*

1.2.1 TYING METHOD

Reinforcing bars are always tied with tempered steel wire, generally 1.6 mm in diameter.

Figure 1-7   *Figure 1-8*

Standard practice is to use wires fitted with hooks (Figure 1-7) that are marketed in three or four lengths to adjust to the standard bar diameters. They are tied with the tool depicted in Figure 1-8, consisting of a worm spindle with which the steel fixer first hooks the two loops in the wire together (Figure 1-7) and then pulls outward on the tool, joining the bars in just two or three movements (Figure 1-9). The use of tongs leads to bonds such as that depicted in Figure 1-10, which often work loose. A valid alternative is mechanical joiners that tie the bars securely (Figure 1-11).

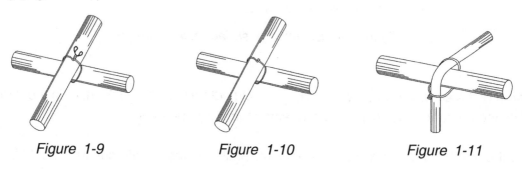

Figure 1-9   *Figure 1-10*   *Figure 1-11*

1.2.2 TIE POINTS

The following recommendations are made.

- *Slabs and plates.* All the intersections between bars around the perimeter of the reinforcement panel should be tied.

  In the rest of the panel, where the bar diameter is 20 mm or less, every second intersection should be tied. Where the bars are 25 mm or larger, the distance between tied intersections should not exceed 50 diameters (Figure 1-12) of the thinnest tied bar.

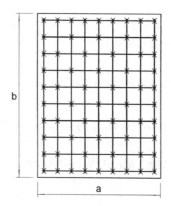

Figure 1-12

- *Beams.* All the corners of the stirrups must be tied to the main reinforcing bars. If welded-wire fabric reinforcement is used to form the stirrups, the main reinforcing bars at the corners should be tied at intervals no larger than 50 times the diameter of the main reinforcing bars.

All the bars not located in the corner of the stirrup should be tied at intervals no larger than 50 times the bar diameter.

Multiple stirrups should be tied together (Figure 1-13).

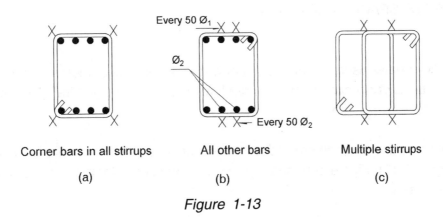

Corner bars in all stirrups	All other bars	Multiple stirrups
(a)	(b)	(c)

Figure 1-13

- *Columns.* All the ties should be tied to the main reinforcement at the intersections. When welded-wire fabric cages are used, the vertical wires should be tied to the main reinforcement at intervals measuring 50 times the bar diameter.
- *Walls.* The bars are tied at every second intersection.

For the intents and purposes of tying reinforcing bars, precast walls manufactured with the mid-plane in a horizontal position are regarded to be slabs.

The rules for slabs and plates are applicable to walls cast *in situ* (Figure 1-14).

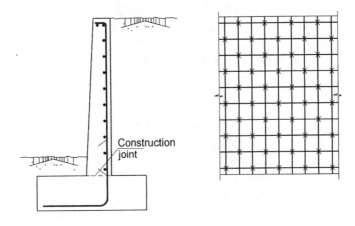

Figure 1-14

- *Footings.* The horizontal part of the starter bars should be secured at each right-angle intersection between starter bar and foundation reinforcement. All the ties in footings should be secured to the vertical part of the starter bars.

AR. The footing assembly should have at least two tie bars.

13

1.3 SPACERS AND CHAIRS

This subject is not addressed in the EC2 or in any CEN standard. Here also, references (6), (7) and (8) are excellent sources of information.

Spacers and chairs are elements made of sundry materials used to ensure a suitable concrete cover or to hold bars in position. The specific definition for each is given below (see (1.3.1a) and (1.3.1b)).

1.3.1 TYPES OF SPACER AND CHAIR

(a) Spacers

The are plastic or galvanised or stainless steel wire or steel plate or mortar elements designed to ensure a satisfactory concrete cover for reinforcing bars. Three general types of spacer can be distinguished.

- *Wheel* or *clip* spacers, which are tied with a wire or clipped to bars or act as unsecured supports (Figure 1-15). Figure 1-16 shows a manual procedure for making this type of spacer where necessary. (*)

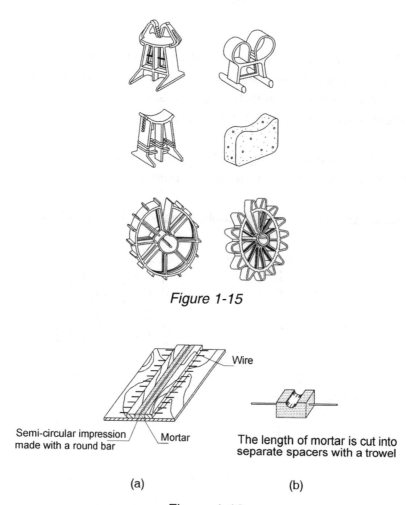

Figure 1-15

Semi-circular impression made with a round bar
Wire
Mortar

The length of mortar is cut into separate spacers with a trowel

(a)   (b)

Figure 1-16

(*) 'Home-made' spacers are not usually permitted.

- *Linear* spacers designed to support the bottom beam and slab reinforcement where the bars must be prevented from toppling (Figure 1-17).

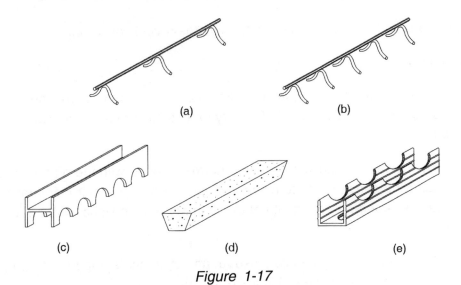

(a)   (b)

(c)   (d)   (e)

Figure 1-17

- *End* spacers (Figure 1-18), placed at the ends of bars to ensure they remain at a sufficient distance from the formwork.

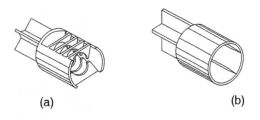

(a)   (b)

Figure 1-18

(b)   Chairs

Chairs are usually made of galvanised or stainless steel wire. They may be *discrete*, to support bars in a specific position, or *continuous*, for continuous support (Figure 1-19). Type (c), which may be made of welded-wire fabric, is especially suitable for slabs and flooring. Plastic tubes are required where the slab soffit is tobe exposed.

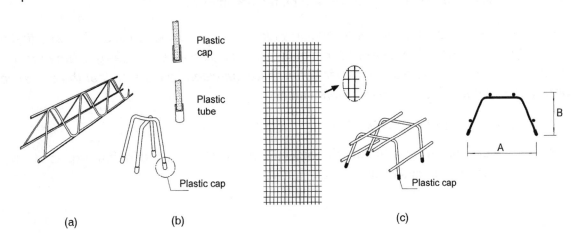

(a)   (b)   (c)

Figure 1-19

15

In all cases their purpose is to separate the top reinforcement in slabs from the formwork or the soil, or the various layers of reinforcement in walls or similar.

Linear chairs should in turn rest on suitable spacers to prevent corrosion.

(c)   Special chairs

The standard sizes in which chairs are manufactured are limited for reasons of economy. Nonetheless, similar elements are needed in very deep slabs or very wide walls to secure the reinforcing bars.

These elements, known as 'special chairs', are normally made with the off cuts metal generated when assembling concrete reinforcement (Figure 1-20 (a)). Their diameter and dimensions should be in keeping with the depth of the member and, in slabs, with the weight of the top layer of reinforcement.

If they rest on the formwork, they should be supported by spacers (Figure 1-20 (b)). They are tied to the bottom layer, when present (Figure 1-20 (c)).

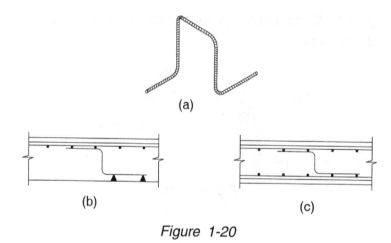

Figure 1-20

(d)   Price

The price of spacers, chairs and special chairs is usually included in the kilo of reinforcing steel assembled at the worksite. While the financial impact of these elements is small, their cost should be estimated where large slabs and plates are involved.

Important note: Spacers, chairs and special chairs must be positioned in such a way that they ensure that all cover distances are 10 mm greater than called for in the design. The reason is to provide the necessary tolerance for bar deformation between spacers so that the actual cover or distance is not less than the minimum values.

1.3.2   GRAPHIC REPRESENTATION

The conventional signs listed in the box below (Figure 1-21) are used in drawings to symbolise spacers, chairs and special chairs.

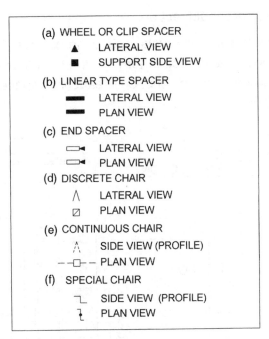

(a) WHEEL OR CLIP SPACER
   ▲   LATERAL VIEW
   ■   SUPPORT SIDE VIEW

(b) LINEAR TYPE SPACER
   ▬▬   LATERAL VIEW
   ▬▬   PLAN VIEW

(c) END SPACER
   ▭◄   LATERAL VIEW
   ▭◄   PLAN VIEW

(d) DISCRETE CHAIR
   ∧   LATERAL VIEW
   ▨   PLAN VIEW

(e) CONTINUOUS CHAIR
   ∧   SIDE VIEW (PROFILE)
   – –▭– –   PLAN VIEW

(f) SPECIAL CHAIR
   ⌐   SIDE VIEW (PROFILE)
   ⌐   PLAN VIEW

Figure 1-21

1.3.3 PLACEMENT RULES

(a) Slabs and footings

- *Bottom reinforcement* layers should rest on and be attached to spacers positioned at intervals measuring no more than 50 times the bar diameter and never over 1000 mm (Figure 1-22).

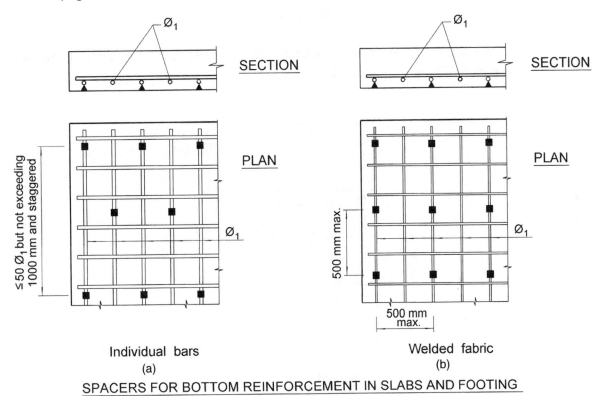

SPACERS FOR BOTTOM REINFORCEMENT IN SLABS AND FOOTING

Figure 1-22

- The *top reinforcement* should rest on:
  - continuous chairs set at intervals measuring not over 50 times the bar diameter. These continuous chairs rest, in turn, on the bottom reinforcement, if any (Figure 1-23 (a)) and where there is none, on linear chairs spaced at no more than 500 mm (Figure 1-23 (b));

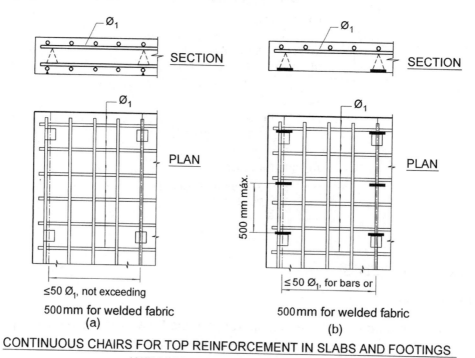

CONTINUOUS CHAIRS FOR TOP REINFORCEMENT IN SLABS AND FOOTINGS WITH BOTTOM REINFORCEMENT

Figure 1-23

  - discrete chairs spaced in both directions at no more than 50 times the diameter of the supported bar (Figure 1-24);

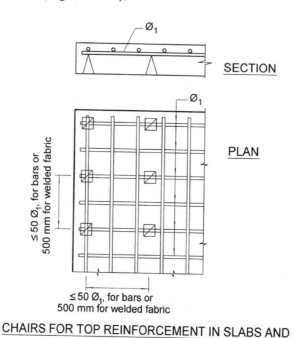

CHAIRS FOR TOP REINFORCEMENT IN SLABS AND FOOTINGS WITHOUT BOTTOM REINFORCEMENT

Figure 1-24

– special chairs spaced at the intervals shown in Figure 1-25.

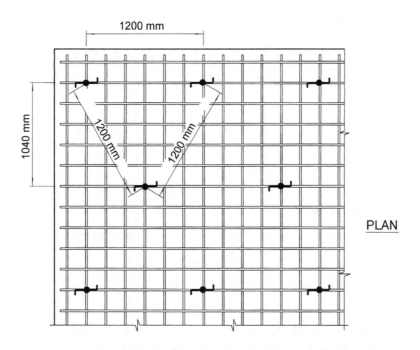

SPECIAL CHAIRS FOR TOP REINFORCEMENT IN SLABS

Figure 1-25

(b) Spacers on slab edges

- In the *absence of horizontal reinforcement* along the slab edges, spacers are placed on alternate bars, as in Figure 1-26 (a).

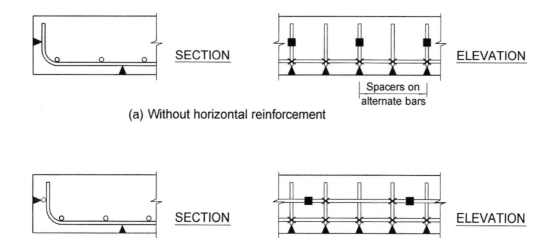

(a) Without horizontal reinforcement

(b) With horizontal reinforcement

SPACERS FOR SLAB EDGES

Figure 1-26

- When the slab *has horizontal reinforcement* on the edges, the spacers should be placed as in Figure 1-26 (b).

(c)   Spacers for ribbed slabs

Spacing is as shown in Figure 1-27.

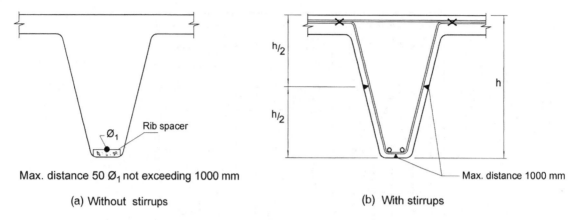

Max. distance 50 Ø₁ not exceeding 1000 mm

(a) Without stirrups

(b) With stirrups

Figure 1-27

(d)   Beams

Spacers should be set in the stirrups at intervals of no more than 1000 mm longitudinally (Figure 1-28), with a minimum of three planes of spacers per span.

Cross-wise placement must be as shown in Figure 1-28.

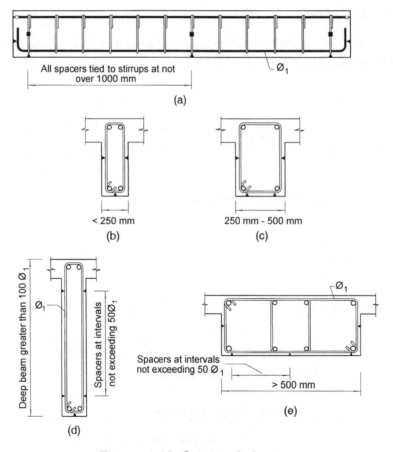

Figure 1-28 Spacers in beams

End or ordinary spacers are needed at the ends of beams in contact with the formwork, depending on the case (Figure 1-29).

Figure 1-29

(e) Columns

Lengthwise along the member, spacers should be placed in the stirrups at no more than 100 times the minimum diameter of the main reinforcement, with at least three planes of spacers per member or section (Figure 1-30).

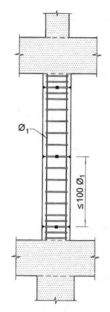

SPACER ARRANGEMENT IN COLUMNS

Figure 1-30

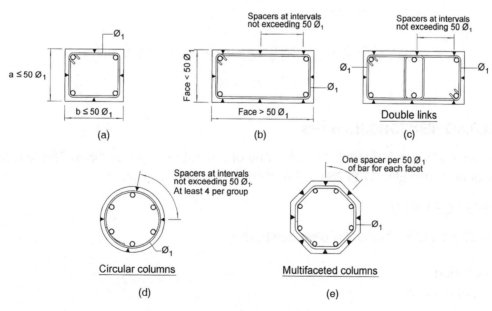

MULTIFACETED COLUMNS

Figure 1-31

Crosswise placement should be as shown in Figure 1-31.

21

(f) Walls

Spacers should be staggered in each layer of reinforcement at intervals measuring the larger of 500 mm or 50 times the diameter of the reinforcing bars. Spacers for reinforcement on opposite sides of the member must be placed at the same height. Spacing between the reinforcement on the two sides must be ensured with continuous chairs set at intervals of not over 1000 mm. In large retaining walls, special chairs must be used instead of chairs (Figure 1-32).

In precast concrete walls whose mid plane is in a horizontal position, the spacers must be set as described for slabs.

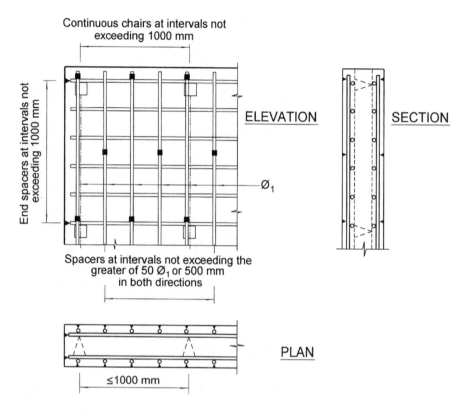

SPACERS AND CONTINUOUS CHAIRS FOR WALLS

Figure 1-32

1.4 WELDING REINFORCING BARS

Bars are commonly welded together, using any of a number of procedures. The following is a summary of standard ENISO 17660-1, *Welding of Reinforcing steel* (11).

1.4.1. TYPES OF WELD

Two types of welded joints can be distinguished:

(a) load-bearing
(b) non-load-bearing.

In addition, such joints can be classified by their position or the welding procedure used.

The welding procedures allowed for reinforcing bars are set out in Table T-1.9 (EC2, Table 3.4) and weldability requirements are established in EN 10080 (1).

TABLE T-1.9
WELDING PROCESSES PERMITTED
AND EXAMPLES OF APPLICATION

Loading conditions	Welding method	Tension bars[1]	Compression bars[1]
Predominantly static	flash welding	butt joints	
	manual metal arc welding and metal arc welding with filling electrode	butt joints with $\phi \geq 20$ mm, splice, lap and cross joints[3] and joints with other steel members	
	metal arc active welding[2]	splice, lap and cross[3] joints and joints with other steel members	
		-	butt joints with $\phi \geq 20$ mm
	friction welding	butt joints, joints with other steel members	
	resistance spot welding	lap joints[4] cross joints[2, 4]	
Not predominantly static	flash welding	butt joints	
	manual metal arc welding	-	butt joints with $\phi \geq 14$ mm
	metal arc active welding[2]	-	butt joints with $\phi \geq 14$ mm
	resistance spot welding	lap joints[4] cross joints[2, 4]	

Notes:
1. Only bars with approximately the same nominal diameter may be welded.
2. Allowable ratio for mixed diameter bars ≥ 0.57
3. For load-bearing joints with $\phi \leq 16$ mm
4. For load-bearing joints with $\phi \leq 28$ mm

All reinforcing bars shall be welded as specified in EN ISO 17760 (2).

The welds joining wires along the anchorage length of welded-wire fabric shall be strong enough to resist the design loads.

The strength of the welds in welded-wire fabric may be assumed to be adequate if each welded joint can withstand a shearing force equal to at least 25 per cent of the characteristic yield stress times the nominal cross-sectional area of the thicker of the two wires, if they differ in size.

AR. In cross joints with hardened concrete around the bars, the strength of the joint is 25 per cent higher than in exposed joints.

1.4.2. WELDED JOINT DETAILS

(a) Butt joints

In all such joints, the edges are prepared as shown in Figure 1-33.

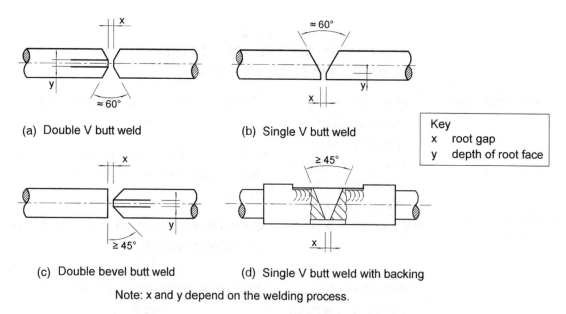

(a) Double V butt weld

(b) Single V butt weld

Key
x   root gap
y   depth of root face

(c) Double bevel butt weld

(d) Single V butt weld with backing

Note: x and y depend on the welding process.

EXAMPLES OF BUTT JOINT PREPARATION

Figure 1-33

(b) Lap joints

These joints call for no preparation. The procedure is as shown in Figures 1-34 and 1-35.

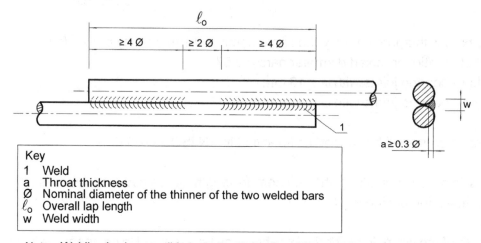

Key
1   Weld
a   Throat thickness
Ø   Nominal diameter of the thinner of the two welded bars
ℓ_o Overall lap length
w   Weld width

Note: Welding is also possible on both sides with minimum weld length of 2.5 Ø.
       A conservative estimate of the effective throat thickness can be taken as
       a ≈ 0.5 w.

LAP JOINT

Figure 1-34

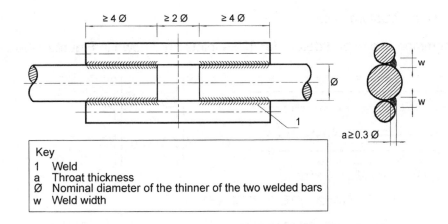

Key
1 Weld
a Throat thickness
Ø Nominal diameter of the thinner of the two welded bars
w Weld width

Note: Welding is also possible on both sides with minimum weld length of 2.5 Ø.
A conservative estimate of the effective throat thickness can be taken as
a ≈ 0.4 w.

STRAP JOINT

Figure 1-35

(c) Cross joints

Cross joints can be resistance-welded (with machines with widely varying characteristics) or metal-arc-welded (Figures 1-36 and 1-37).

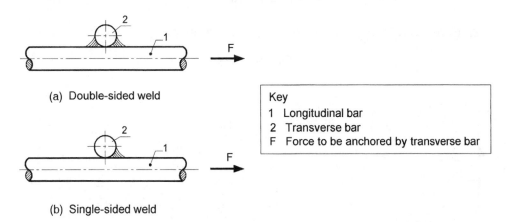

(a) Double-sided weld

Key
1 Longitudinal bar
2 Transverse bar
F Force to be anchored by transverse bar

(b) Single-sided weld

CROSS JOINTS WELDED BY RESISTANCE MANUAL ARC WELDING

Figure 1-36

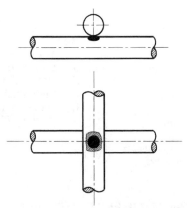

CROSS JOINTS WELDED BY RESISTANCE
SPOT WELDING

Figure 1-37

(d) Bar-to-steel structure joints

The most common use of and details for these joints are shown in Figures 1-38 and 1-39.

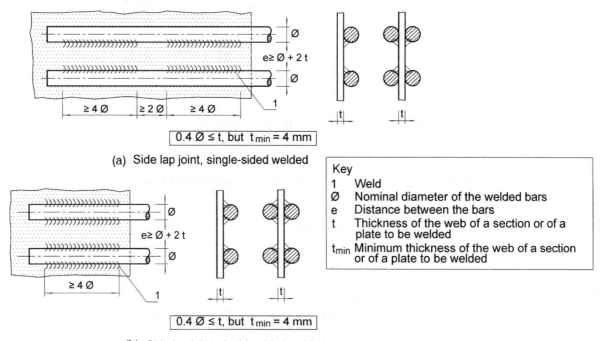

(a) Side lap joint, single-sided welded

(b) Side lap joint, double-sided welded

Key
1   Weld
∅   Nominal diameter of the welded bars
e   Distance between the bars
t   Thickness of the web of a section or of a
     plate to be welded
t_{min} Minimum thickness of the web of a section
     or of a plate to be welded

SIDE LAP JOINT ON STRAIGHT REINFORCING STEEL BARS

Figure 1-38

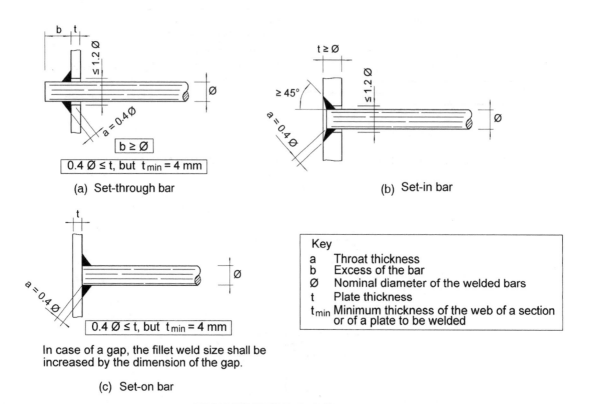

(a) Set-through bar

(b) Set-in bar

(c) Set-on bar

In case of a gap, the fillet weld size shall be
increased by the dimension of the gap.

Key
a   Throat thickness
b   Excess of the bar
∅   Nominal diameter of the welded bars
t   Plate thickness
t_{min} Minimum thickness of the web of a section
     or of a plate to be welded

TRANVERSE END PLATE JOINT

Figure 1-39

1.5 VERIFICATION OF THE ANCHORAGE LIMIT STATE

The method used is applicable to both flexible ($v \geq 2h$) and stiff footings with a ratio of $\frac{v}{h} > 1$.

The following is a more detailed description of the method than that found in EC2, 9.8.2.2.

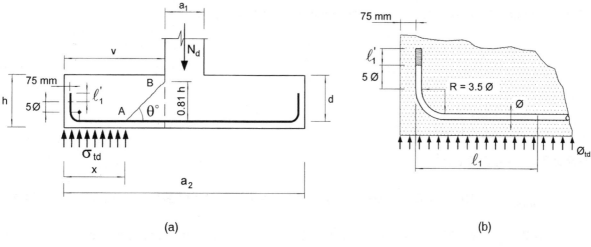

Figure 1-40

A $\theta°$ shear crack is assumed to be possible, as depicted in Figure 1-40 (a). Given the low bending reinforcement ratios, d may be accepted to be $0.9h$ and the height of the point where cracking begins, $0.9d$ or $0.81h$.

For the sake of simplification, the moments used in the calculations below are the moments on the column face. A 75-mm side cover is also assumed, as in footings poured onto soil.

The reinforcement anchor length should be as required to anchor the bar force from the point where the reinforcement intersects with the crack onward, i.e. point A. (The reinforcement is assumed to be constant across the entire width.)

Taking the moments at B

$$F_{sx} \cdot 0.81h = x\sigma_{td} \cdot \left(v - \frac{x}{2}\right)$$

where

$$x = v - 0.81h \cot\theta$$

and operating, yields

$$F_{sx} \cdot 0.81h = \frac{\sigma_{td}}{2}\left(v^2 - 0.66h^2 \cot^2\theta\right)$$

$$F_{sx} = \frac{\sigma_{td}\left(v^2 - 0.66h^2 \cot^2\theta\right)}{1.62h} \qquad [1.7]$$

In addition to the equation above and in keeping with the bending moment applied, the following must hold:

$$A_s f_{yd} \cdot 0.81h = \sigma_{td} \cdot \frac{v^2}{2} \rightarrow A_s f_{yd} = \frac{\sigma_{td} v^2}{1.62h}$$

from which it follows that:

$$\ell_{b,net} = \frac{F_{sx}}{A_s f_{yd}} \cdot \ell_b \cdot \frac{A_{s,req}}{A_{s,prov}} = \frac{\left(v^2 - 0.66h^2 \cot^2 \theta\right)}{v^2} \cdot \ell_b \cdot \frac{A_{s,req}}{A_{s,prov}}$$ [1.8]

(ℓ_b reflects bonding position I, given the position of the bars.)

$$\ell_{b,net} = \left[1 - 0.66\left(\frac{h}{v}\right)^2 \cot^2 \theta\right] \ell_b \cdot \frac{A_{s,req}}{A_{s,prov}} \ (*)$$ [1.9]

1.5.1 BOND ANCHORAGE

Given $x = v - 0.81h \cot \theta$:

If $\ell_{b,net} \leq x - 75 = v - 0.81h \cot \theta - 70$ $\rightarrow$ use straight anchorage [1.10]

If $0.7\,\ell_{b,net} \leq x - 75 = v - 0.81h \cot \theta - 70$ $\rightarrow$ use hook anchorage [1.11]

If $0.7\,\ell_{b,net} > x - 75 = v - 0.81h \cot \theta - 70$ $\rightarrow$ lengthen ℓ'_1 [1.12]
(See Figure 1-40 (b))

(Dimensions in mm)

Since length ℓ'_1 is in bonding position I:

$$\ell_{b,net} = \frac{x - 75}{0.7} + \ell'_1 = \frac{v - 0.81h \cot \theta - 75}{0.7} + \ell'_1$$

and consequently:

$$\ell'_1 = \ell_{b,net} - \frac{v - 0.81h \cot \theta - 75}{0.7}$$ [1.13]

where $\ell_{b,net}$ is calculated as per [1.9].

Since:

$$x = v - 0.81h \cot \theta:$$

The minimum value of x is found with the maximum value of $\cot \theta$ which, according to EC2, is equal to $\cot \theta = 2.5$, yielding:

$$x = v - 2.03h = h\left(\frac{v}{h} - 2.03\right)$$

or

$$\frac{x}{h} = \frac{v}{h} - 2.03$$ [1.14]

(*) [1.9] is a more general expression than the formulas set out in EC2, in which critical value x is simplified to be equal to $0.5h$. See the discussion below.

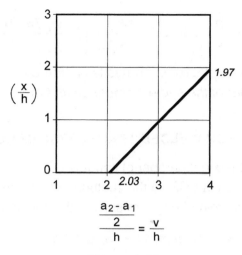

Figure 1-41

The graph in Figure 1-41 shows the distance x in terms of h for different values of $\frac{v}{h}$.

If $\frac{v}{h} \leq 2.5$, a conservative value for x is $0.5h$, as adopted in Eurocode EC2. Direct calculation is preferable.

1.5.2 WELDED TRANSVERSE BAR ANCHORAGE

In this case, the force to be anchored at the end of the bar, assuming a 75-mm cover, is defined as shown in Figure 1-42.

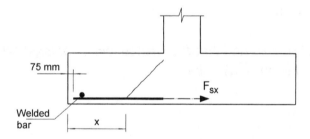

Figure 1-42

$$F_o = F_{sx} - \frac{x-75}{\ell_b} \cdot F_{sx} = F_{sx} \left[1 - \frac{x-70}{\ell_b} \right] \quad \text{where } F_{sx} \text{ was deduced from [1.7]}$$

and substituting

$$F_o = \frac{\sigma_{td} \left(v^2 - 0.66h^2 \cot^2 \theta \right)}{1.62h} \left[1 - \frac{v - 75 - 0.81h \cot \theta}{\ell_b} \right]$$

and given $\sigma_{td} = \dfrac{1.62h\, A_s f_{yd}}{v^2}$

$$F_o = A_s f_{yd} \left(1 - 0.66 \left(\frac{h}{v} \right)^2 \cot^2 \theta \right) \left[1 - \frac{v - 75 - 0.81h \cot \theta}{\ell_b} \right] \qquad [1.15]$$

hence, according to the discussion in section 1.6, with welding strength $0.5\, A_s f_{yd}$, the number of welded transverse bars required, n, is calculated as:

29

$$n = \frac{1}{2}\left(1 - 0.66\left(\frac{h}{v}\right)^2 \cot^2\theta\right)\ell_b \cdot \frac{A_{s,req}}{A_{s,prov}}\left[1 - \frac{v - 75 - 0.81h\cot\theta}{\ell_b}\right] \qquad [1.16]$$

Inasmuch as expression [1.16] is always less than one, *anchorage is attained with a welded transverse bar of the same diameter as the main bars.*

1.6 ANCHORAGE RULES FOR WELDED TRANSVERSE BARS

This issue has been researched in a number of countries. The results are summarised below[(*)] in somewhat greater detail than in EC2. In this method, the strength of the welded joint can be found assuming that the cross-joined bars are surrounded by concrete.

1.6.1 BARS WHERE $14 \le \phi_T \le 32$ mm, $14 \le \phi_L \le 32$ mm

The design value of the anchorage capacity of a welded transverse joint can be found from the following formula:

$$F_{btd} = L_{td} \cdot \phi_T \, \sigma_{td} \le F_{wd} \qquad [1.17]$$

where:

F_{btd} =   design value of the anchorage capacity of the welded transverse joint

$$L_{td} \quad = \quad 1.16\,\phi_T\sqrt{\frac{f_{yd}}{\sigma_{Td}}} \le L_T \qquad [1.18]$$

L_T =   in single transverse bars, bar length, or spacing between longitudinal bars, as appropriate

ϕ_T =   transverse bar diameter.

(The assumption is that $\phi_T \le \phi_L$ and that *the welded joint is at least 50 per cent of the design mechanical strength of the longitudinal bar*).

$$\sigma_{td} = \frac{f_{td\,0.05} + \sigma_{cd}}{y} \le 3f_{cd} \qquad [1.19]$$

$f_{td\,0,05}$ =   design characteristic tensile strength of the concrete surrounding the welded joint. Its sign is positive

f_{cd} =   design compressive strength of the concrete

σ_{cd} =   design compressive stress in the concrete in the direction normal to the axes of the two bars (positive for compression)

y = $0.015 + 0.14\,e^{(-0,18\,x)}$

$$x = 2\frac{c}{\phi_T} + 1 \qquad [1.20]$$

c =   cover in the direction perpendicular to the axes of the two bars

F_{wd} =   guaranteed design strength for the welded joint (where $\gamma_s = 1.15$).

(*) The following formulas are based essentially on research conducted by Finnish construction and reinforcing steel companies. Of particular note are the studies conducted by Statens Tekniska Forskingscentral and Pekka Nykyri (9), which were included in the Annex to the final version of Eurocode 2, Part 3 (Concrete foundations) (10). (It presently forms parts of Part of EC2 (5)).

The foregoing is illustrated graphically in Figure 1-43.

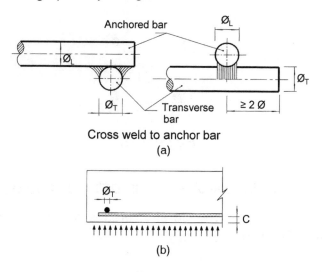

Figure 1-43

If two transverse bars are welded to opposite longitudinal sides, the anchorage strength found with the aforementioned formula is doubled (Figure 1-44 (a)). By contrast, if two transverse parallel bars are welded at the minimum spacing of $3\,\phi_T$, strength is multiplied by 1.4 (Figure 1-44 (b)).

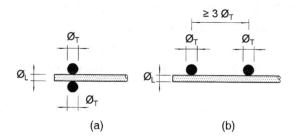

Figure 1-44

1.6.2 BARS WHERE $\phi_T \leq 12$ mm, $\phi_L \leq 12$ mm

The strength of a welded transverse joint in a concrete mass is at least 25 per cent greater than the strength of the same welded joint found with the standard uncovered joint test.

The design value of the anchorage capacity of a welded transverse joint can be found from the following formula:

$$F_{btd} = 1.25\,F_{wd} \leq 16\,A_{sL} \cdot f_{cd}\,\frac{\phi_T}{\phi_L}$$ [1.21]

where the minimum bar length or spacing between parallel longitudinal bars, as appropriate, must be $\geq 7\,\phi_L$:

F_{wd} = design guaranteed strength for welded joint

ϕ_T = transverse bar diameter, $\phi_T \leq 12$ mm

ϕ_L = anchored bar diameter, $\phi_L \leq 12$ mm

A_{sL} = area of the transverse bar cross-section.

(The assumption is that $\phi_T \leq \phi_L$ and that *the strength of the welded joint is at least 50 per cent of the design mechanical strength of the longitudinal bar.*)

31

If two transverse bars (Figure 1-45) are welded at the minimum spacing of $4\,\phi_T$, strength is multiplied by 1.4.

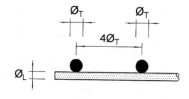

Figure 1-45

1.7   RELATIONSHIP BETWEEN CONSTRUCTION DETAILS AND DURABILITY

The insufficient durability of many concrete structures is one of the most severe problems encountered in this type of construction today.

Satisfactory construction detailing plays an essential role in concrete structure durability. Such details affect performance in any number of ways, including the aspects listed below.

(a)   They contribute to appropriate cover over reinforcement, preventing corrosion during the expected design life.
(b)   They control shrinkage and thermal contraction cracking.
(c)   They control cracking due to reinforcement elongation.
(d)   They prevent cracking in anchorage and curved areas.
(e)   They prevent cracking due to excessive local compressive stress in the concrete.

Construction detailing in the design not only has an impact on the safety and cost of built structures, but on their durability as well.

Table E.1N in Annex E of EC2 (reproduced as Table T-1.10 below) recommends the minimum concrete strength for each exposure class.

TABLE T-1.10
INDICATIVE STRENGTH CLASSES

Exposure class according to Table 4.1									
Corrosion									
	Carbonation-induced corrosion				Chloride-induced corrosion			Chloride-induced corrosion from sea water		
	XC1	XC2	XC3	XC4	XD1	XD2	XD3	XS1	XS2	XS3
Indicative strength class	C20/25	C25/30	C30/37		C30/37		C35/45	C30/37	C35/45	
Damage to concrete									
	No risk	Freeze/thaw attack			Chemical attack					
	X0	XF1	XF2	XF3	XA1	XA2	XA3			
Indicative strength class	C12/15	C30/37	C25/30	C30/37	C30/37		C35/45			

2 Constructive details

Group 01

Foundations

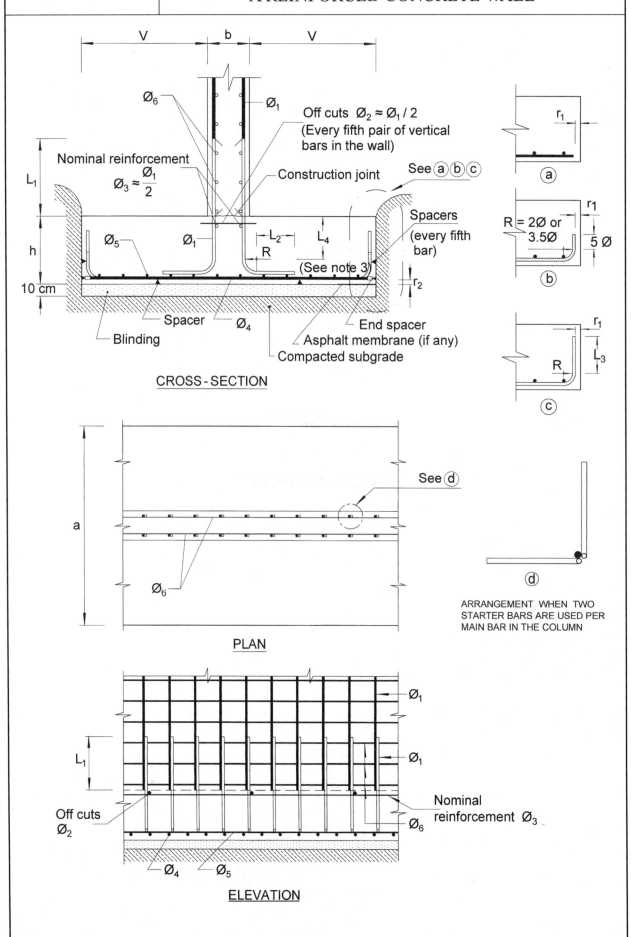

CD – 01.01

WALL FOOTING SUPPORTING A REINFORCED CONCRETE WALL

$\varnothing_6$

$\varnothing_1$

Off cuts $\varnothing_2 \approx \varnothing_1 / 2$
(Every fifth pair of vertical bars in the wall)

Nominal reinforcement
$\varnothing_3 \approx \dfrac{\varnothing_1}{2}$

Construction joint

See (a)(b)(c)

L_1

$\varnothing_5$

$\varnothing_1$

L_2

L_4

R

Spacers
(every fifth bar)

h

(See note 3)

10 cm

r_2

Spacer

$\varnothing_4$

End spacer

Asphalt membrane (if any)

Blinding

Compacted subgrade

CROSS - SECTION

r_1

(a)

r_1

$R = 2\varnothing$ or
$3.5\varnothing$

$5\varnothing$

(b)

r_1

R

L_3

(c)

a

See (d)

$\varnothing_6$

PLAN

(d)

ARRANGEMENT WHEN TWO
STARTER BARS ARE USED PER
MAIN BAR IN THE COLUMN

$\varnothing_1$

$\varnothing_1$

L_1

Nominal
reinforcement $\varnothing_3$

Off cuts
$\varnothing_2$

$\varnothing_6$

$\varnothing_4$

$\varnothing_5$

ELEVATION

1. RECOMMENDATIONS

1. The ϕ_1 bars rest directly on the top of the footing, with no hooks needed.

2. L_1 is the ϕ_1 bar lap length.

3. **AR.** The starter bars are diameter ϕ_1 bars. L_4 should be $\geq \ell_{bd}$, but length ℓ_{bd} is defined in EC2 for the least favourable case. In this case the side covers for the ϕ_1 bars are very large. See (12), p. 69. A safe value would be $\ell'_{bd} = \frac{2}{3}\ell_{bd}$. If $L_4 < \frac{2}{3}\ell_{bd}$, often a more suitable solution than increasing the depth of the footing is to use two starter bars for every ϕ_1 bar in the wall. The sum of the cross-sections of these two bars should not be less than the ϕ_1 bar cross-section, but their diameter should be such that $\frac{2}{3}\ell_{bd} < L_4$, where ℓ_{bd} is the anchorage length. The arrangement detail is as in alternative (d). This rule can be applied wherever the bar cover is $\geq 10 \phi$ but not under 10 cm.

4. $r_1 = 7.5$ cm if the concrete for the footing is poured directly against the soil.

5. $r_2 = 2.5$ cm but not less than ϕ_4.

6. See 1.5 to determine when to use anchor type a, b or c at the end of the ϕ_4 reinforcing bar.

7. Length L_2 shall suffice to tie the ϕ_1 starter bars securely to two ϕ_5 transverse bars. (It may not be less than 2s, where s is the space between ϕ_5 bars.)

8. The bars with diameters ϕ_2 and ϕ_3 are supports whose sole purpose is to keep the starter bar assembly firmly in place while the concrete is cast in the foundations. The starter bar assembly is to be tied at all cross points. Depending on the case, these ties may be supplemented or replaced by positioning reinforcement ϕ_6 in lapping length L_1 before pouring the concrete in the footing.

9. The top of the footing is smoothed with floats or a power float except in the contact area with the future wall, where a rough surface such as is generated by the vibrator is needed.

10. In soft soils, the 25 cm over the level of the future blinding should not be excavated until shortly before pouring the concrete to ensure that the soil is not softened by rainfall immediately before placement.

11. The foundation subgrade must be compacted before the blinding is poured.

12. The blinding under the footing is floated or smoothed with a power float. Its standard 10-cm thickness may be varied to absorb tolerances in foundation subgrade levelling.

13. Where the footing is to be poured on soils that constitute an aggressive medium, it should be protected by an asphalt membrane as specified.

14. Before pouring the concrete to form the wall, the joint surface should be cleaned and pressure hosed. The concrete should not be cast until the surface dries.

15. See 1.2 and 1.3 for descriptions of how to tie bars and place spacers.

WALL FOOTING SUPPORTING A BRICK WALL

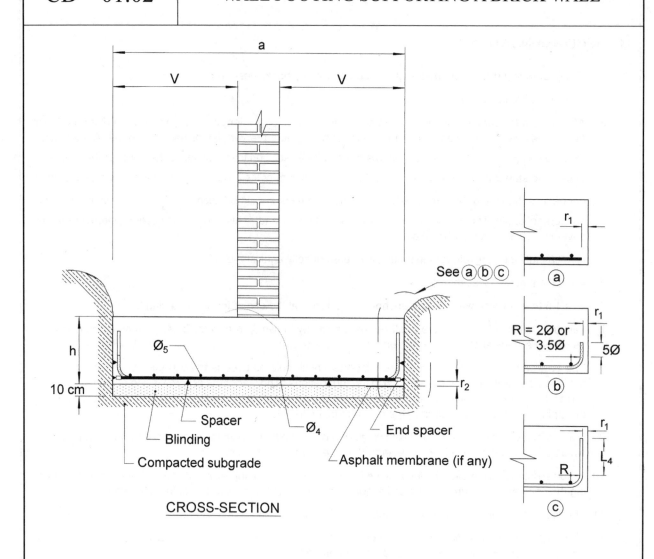

a

V   V

h

10 cm

Ø₅

Spacer

Blinding

Compacted subgrade

Ø₄

End spacer

Asphalt membrane (if any)

r₂

See ⓐ ⓑ ⓒ

r₁

ⓐ

R = 2Ø or 3.5Ø

r₁

5Ø

ⓑ

r₁

L₄

R

ⓒ

CROSS-SECTION

1. RECOMMENDATIONS

1. r_1 = 7.5 cm if the concrete for the footing is poured directly against the soil.

2. r_2 = 2.5 cm but not less than ϕ_4.

3. See 1.5 to determine when to use anchor type a, b or c at the end of the ϕ_4 reinforcing bar.

4. The top is smoothed with floats or a power float. The mortar bed under the wall that interfaces with the footing shall be 10 mm thick.

5. In soft soils, the 25 cm over the level of the future blinding should not be excavated until shortly before pouring the concrete to ensure that the soil is not softened by rainfall immediately before placement.

6. The foundation subgrade should be compacted before the blinding is poured.

7. The blinding under the footing is floated or smoothed with a power float. Its standard 10-cm thickness may be varied to absorb tolerances in foundation subgrade levelling.

8. Where the footing is to be poured on soils that constitute an aggressive medium, it should be protected by an asphalt membrane as specified.

9. Before the first bed mortar is laid under the wall, the area of the joint that will interface with it should be pressure hosed. The mortar should not be placed until the surface of the footing dries.

10. See 1.2 and 1.3 for descriptions of how to tie bars and place spacers.

2. STATUTORY LEGISLATION

This type of footing is not addressed in EC2 (5).

The articles in the Code on spread footings are partially applicable.

SPREAD FOOTING

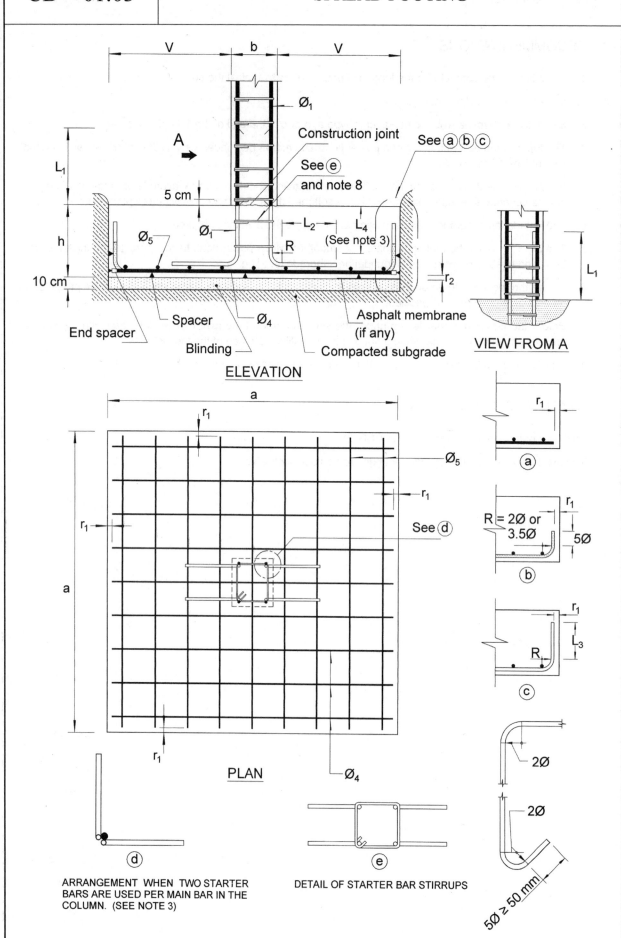

ELEVATION

VIEW FROM A

PLAN

ⓐ

R = 2Ø or 3.5Ø

ⓑ

ⓒ

ⓓ

ARRANGEMENT WHEN TWO STARTER BARS ARE USED PER MAIN BAR IN THE COLUMN. (SEE NOTE 3)

ⓔ

DETAIL OF STARTER BAR STIRRUPS

5Ø ≥ 50 mm

40

1. RECOMMENDATIONS

1. The ϕ_1 bars rest directly on the top of the footing, with no hooks needed.

2. L_1 is the ϕ_1 bar lap length.

3. **AR.** The starter bars are diameter ϕ_1 bars. L_4 should be $\geq \ell_{bd}$, but length ℓ_{bd} is defined in EC2 for the least favourable case. In this case the side covers for the ϕ_1 bars are very large. See (12), p. 69. A safe value would be $\ell'_{bd} = \frac{2}{3}\ell_{bd}$. If $L_4 < \frac{2}{3}\ell_{bd}$, often a more suitable solution than increasing the depth of the footing is to use two starter bars for every ϕ_1 bar in the wall. The sum of the cross-sections of these two bars should not be less than the ϕ_1 bar cross-section, but their diameter should be such that $\frac{2}{3}\ell_{bd} < L_4$, where ℓ_{bd} is the anchorage length. The arrangement detail is as in alternative (d). This rule can be applied wherever the bar cover is $\geq 10\phi$ but not under 10 cm.

4. $r_1 = 7.5$ cm if the concrete for the footing is poured directly against the soil.

5. $r_2 = 2.5$ cm but not less than ϕ_4.

6. See 1.5 to determine when to use anchor type a, b or c at the end of the ϕ_4 reinforcing bar.

7. Length L_2 shall suffice to tie the ϕ_1 starter bars securely to two ϕ_5 transverse bars. (It may not be less than 2s, where s is the space between the ϕ_5 bars.)

8. The starter bar ties are supports whose sole purpose is to keep the starter bar assembly firmly in place while the concrete is cast in the foundations. The starter bar assembly is to be tied at all cross points. These are not the same ties as in the column.

9. The top of the footing is smoothed with a float or power float except in the contact area with the future column, where a rough surface such as is generated by the vibrator is needed.

10. In soft soils, the 25 cm over the level of the future blinding should not be excavated until shortly before pouring the concrete to ensure that the soil is not softened by rainfall immediately before placement.

11. The foundation subgrade must be compacted before the blinding is poured.

12. The blinding under the footing is floated or smoothed with a power float. Its standard 10-cm thickness may be varied to absorb tolerances in foundation subgrade levelling.

13. Where the footing is to be poured on soils that constitute an aggressive medium, it should be protected by an asphalt membrane as specified.

14. Before pouring the column concrete, the joint surface should be cleaned and pressure hosed. The concrete should not be cast until the surface dries.

15. See 1.2 and 1.3 for descriptions of how to tie bars and place spacers.

16. Where calculations are very tight, the mid plane in pairs of lapped bars should be perpendicular to the direction of the greatest bending moment acting on the column springing.

17. Rectangular footings can be set out in exactly the same way, although the resulting grid is logically not symmetrical.

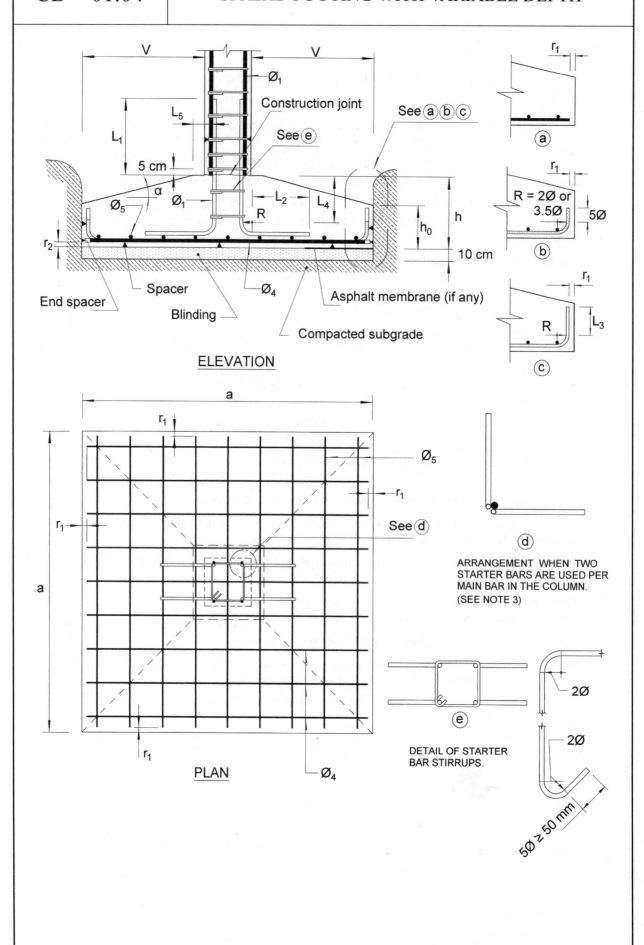

ELEVATION

PLAN

(a)

(b) R = 2Ø or 3.5Ø, 5Ø

(c)

(d)

ARRANGEMENT WHEN TWO STARTER BARS ARE USED PER MAIN BAR IN THE COLUMN. (SEE NOTE 3)

(e)

DETAIL OF STARTER BAR STIRRUPS.

$5Ø ≥ 50$ mm

Construction joint

See (e)

See (a)(b)(c)

Ø₁

5 cm

α

Ø₅

Ø₁

L_2

L_4

R

h

h_0

10 cm

r_2

End spacer

Spacer

Blinding

Ø₄

Asphalt membrane (if any)

Compacted subgrade

L_1

L_5

V

V

Ø₅

See (d)

r_1

r_1

r_1

r_1

a

a

Ø₄

2Ø

2Ø

1. RECOMMENDATIONS

1. The ϕ_1 bars rest directly on the top of the footing, with no hooks needed.

2. L_1 is the ϕ_1 bar lap length.

3. **AR.** The starter bars are diameter ϕ_1 bars. L_4 should be $\geq \ell_{bd}$, but length ℓ_{bd} is defined in EC2 for the least favourable case. In this case the side covers for the ϕ_1 bars are very large. See (12), p. 69. A safe value would be $\ell'_{bd} = \frac{2}{3}\ell_{bd}$. If $L_4 < \frac{2}{3}\ell_{bd}$, often a more suitable solution than increasing the depth of the footing is to use two starter bars for every ϕ_1 bar in the wall. The sum of the cross-sections of these two bars should not be less than the ϕ_1 bar cross-section, but their diameter should be such that $\frac{2}{3}\ell_{bd} < L_4$, where ℓ_{bd} is the anchorage length. The arrangement detail is as in alternative (d).

   Particular care should be taken with edge or corner columns, because there either $L_4 \geq L_1$, or the two starter bars per ϕ_1 bar arrangement should be adopted (detail (d)).

4. $r_1 = 7.5$ cm if the concrete for the footing is poured directly against the soil.

5. $r_2 = 2.5$ cm but not less than ϕ_4.

6. See 1.5 to determine when to use anchor type a, b or c at the end of the ϕ_4 reinforcing bar.

7. The square base with side length L_5 should be at least 15 cm² to provide for assembly of the column formwork.

8. Length L_2 shall suffice to tie the ϕ_1 starter bars securely to two ϕ_5 transverse bars. (It may not be less than 2s, where s is the space between ϕ_5 bars.)

9. The sole purpose of the starter bar ties is to keep the starter bar assembly firmly in place while the concrete is cast in the foundations. The starter bar assembly is to be tied at all cross points. These are not the same ties as in the column.

10. The top of the footing is smoothed with a float or power float except in the contact area with the future column, where a rough surface such as is generated by the vibrator is needed.

11. In soft soils, the 25 cm over the level of the future blinding should not be excavated until shortly before pouring the concrete to ensure that the soil is not softened by rainfall immediately before placement.

12. The foundation subgrade must be compacted before the blinding is poured.

13. The blinding under the footing is floated or smoothed with a power float. Its standard 10-cm thickness may be varied to absorb tolerances in foundation subgrade levelling.

14. Where the footing is to be poured on soils that constitute an aggressive medium, it should be protected by an asphalt membrane as specified.

15. Before pouring the column concrete, the joint surface should be cleaned and pressure hosed. The concrete should not be cast until the surface dries.

16. See 1.2 and 1.3 for descriptions of how to tie bars and place spacers.

17. Where calculations are very tight, the mid plane in pairs of lapped bars should be perpendicular to the direction of the greatest bending moment acting on the column springing.

18. Rectangular spread footings can be set out in exactly the same way, although the resulting grid is logically not symmetrical.

19. This type of footing is usually cost-effective for large loads. Even in that case, $\alpha \leq 25°$ to ensure that the concrete slope can be maintained during vibration.

20. In large footings, top surface reinforcement may be advisable to prevent drying shrinkage and thermal contraction.

NOTE: This footing, out of use for years, has recently made a comeback in windmill foundations. See also CD-01.06 for a variation of considerable interest.

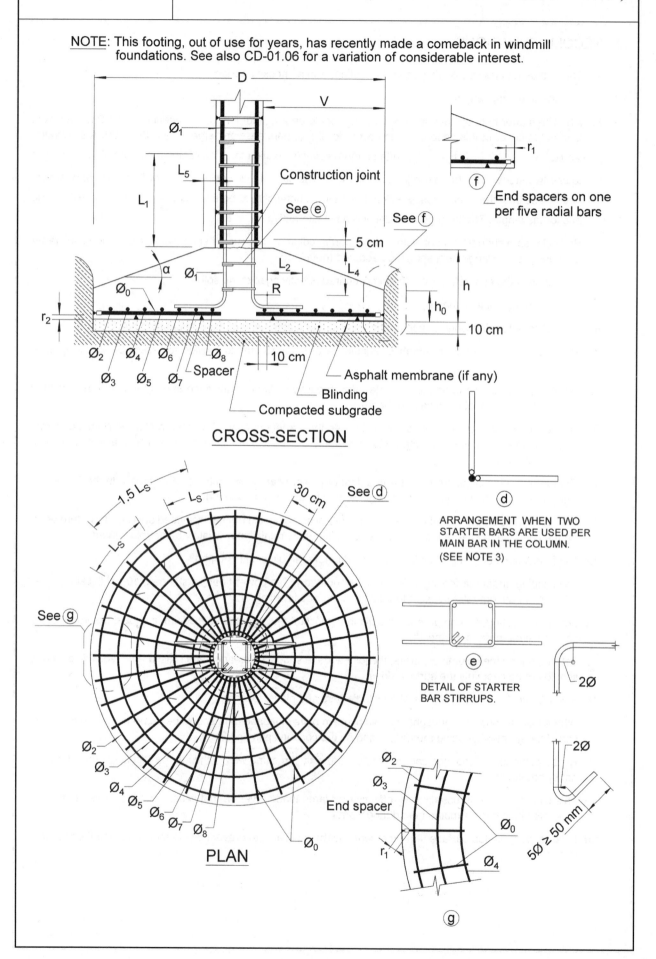

CROSS-SECTION

PLAN

End spacers on one per five radial bars

(f)

ARRANGEMENT WHEN TWO STARTER BARS ARE USED PER MAIN BAR IN THE COLUMN. (SEE NOTE 3)

(d)

DETAIL OF STARTER BAR STIRRUPS.

(e)

$2\emptyset$

$2\emptyset$

$5\emptyset \geq 50\ mm$

(g)

End spacer

1. RECOMMENDATIONS

1. The ϕ_1 bars rest directly on the top of the footing, with no hooks needed.

2. L_1 is the ϕ_1 bar lap length.

3. **AR.** The starter bars are diameter ϕ_1 bars. L_4 should be $\geq \ell_{bd}$, but length ℓ_{bd} is defined in EC2 for the least favourable case. In this case the side covers for the ϕ_1 bars are very large. See (12), p. 69. A safe value would be $\ell'_{bd} = \frac{2}{3}\ell_{bd}$. If $L_4 < \frac{2}{3}\ell_{bd}$, often a more suitable solution than increasing the depth of the footing is to use two starter bars for every ϕ_1 bar in the wall. The sum of the cross-sections of these two bars should not be less than the ϕ_1 bar cross-section, but their diameter should be such that $\frac{2}{3}\ell_{bd} < L_4$, where ℓ_{bd} is the anchorage length. The arrangement detail is as in alternative (d). This rule can be applied wherever the bar cover is $\geq 10\phi$ but not under 10 cm.

4. $r_1 = 7.5$ cm if the concrete for the footing is poured directly against the soil.

5. $r_2 = 2.5$ cm but not less than ϕ_0.

6. The L_5 wide circular base should measure at least 15 cm to provide a base for the column formwork.

7. Length L_2 shall suffice to tie the starter bars securely to the reinforcing bars. (Normally $L_2 = 2s$, where s is the spacing for circular reinforcement.)

8. The sole purpose of the starter bar ties is to keep the starter bar assembly firmly in place while the concrete is cast in the foundation. The starter bar assembly is to be tied at all cross points. These are not the same ties as in the column.

9. The top of the footing is smoothed with a float or power float except in the contact area with the future column, where a rough surface such as is generated by the vibrator is needed.

10. In soft soils, the 25 cm over the level of the future blinding should not be excavated until shortly before pouring the concrete to ensure that the soil is not softened by rainfall immediately before placement.

11. The foundation subgrade must be compacted before the blinding is poured.

12. The blinding under the footing is floated or smoothed with a power float. Its standard 10-cm thickness may be varied to absorb tolerances in foundation subgrade levelling.

13. Where the footing is to be poured on soils that constitute an aggressive medium, it should be protected by an asphalt membrane as specified.

14. Before pouring the column concrete, the joint surface should be cleaned and pressure hosed. The concrete should not be cast until the surface dries.

15. See 1.2 and 1.3 for descriptions of how to tie bars and place spacers.

16. Where calculations are very tight, the mid plane in pairs of lapped bars should be perpendicular to the direction of the greatest bending moment acting on the column springing.

17. This type of footing is usually cost-effective for large loads. Even in that case, $\alpha \leq 25°$ to ensure that the concrete slope can be maintained during vibration.

18. In large footings, top surface reinforcement may be advisable to prevent drying shrinkage and thermal contraction.

AR. According to EC2, lapping in circumferential reinforcement should be spaced at L_s. Any other length is preferable. (See (13), (14)).

2. STATUTORY LEGISLATION

This type of footing is not addressed in EC2 (5).

CIRCULAR FOOTING (REINFORCED WITH TWO WELDED PANELS)

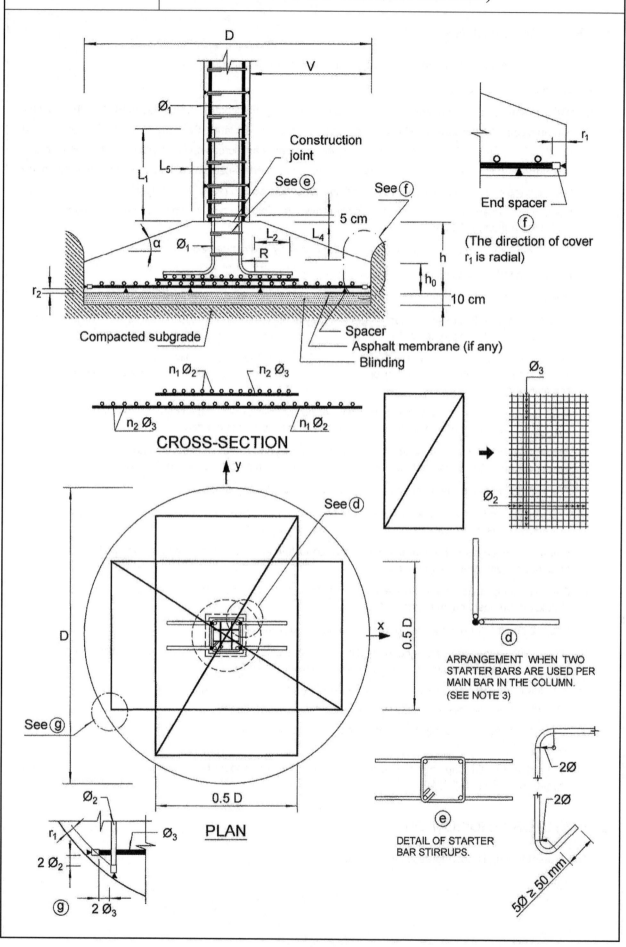

Construction joint

See (e)

See (f)

5 cm

See (f)

r_1

End spacer

(f)

(The direction of cover r_1 is radial)

α

$\emptyset_1$

L_2

R

L_4

h

h_0

10 cm

r_2

Compacted subgrade

Spacer

Asphalt membrane (if any)

Blinding

$n_1\,\emptyset_2$

$n_2\,\emptyset_3$

$n_2\,\emptyset_3$

$n_1\,\emptyset_2$

$\emptyset_3$

$\emptyset_2$

CROSS-SECTION

y

See (d)

x

0.5 D

D

See (g)

(d)

ARRANGEMENT WHEN TWO STARTER BARS ARE USED PER MAIN BAR IN THE COLUMN. (SEE NOTE 3)

0.5 D

0.5 D

PLAN

$\emptyset_2$

r_1

$\emptyset_3$

$2\,\emptyset_2$

(g)

$2\,\emptyset_3$

2Ø

2Ø

(e)

DETAIL OF STARTER BAR STIRRUPS.

$5Ø \gtrless 50\ mm$

1. RECOMMENDATIONS

1. The ϕ_1 bars rest directly on the top of the footing, with no hooks needed.

2. L_1 is the ϕ_1 bar lap length.

3. **AR.** The starter bars are diameter ϕ_1 bars. L_4 should be $\geq \ell_{bd}$, but length ℓ_{bd} is defined in EC2 for the least favourable case. In this case the side covers for the ϕ_1 bars are very large. See (12), p. 69. A safe value would be $\ell'_{bd} = \frac{2}{3}\ell_{bd}$. If $L_4 < \frac{2}{3}\ell_{bd}$, often a more suitable solution than increasing the depth of the footing is to use two starter bars for every ϕ_1 bar in the wall. The sum of the cross-sections of these two bars should not be less than the ϕ_1 bar cross-section, but their diameter should be such that $\frac{2}{3}\ell_{bd} < L_4$, where ℓ_{bd} is the anchorage length. The arrangement detail is as in alternative (d). This rule can be applied wherever the bar cover is $\geq 10 \phi$ but not under 10 cm.

4. $r_1 = 7.5$ cm if the concrete for the footing is poured directly against the soil.

5. $r_2 = 2.5$ cm but not less than ϕ_2.

6. The L_5-wide circular base should measure at least 15 cm to provide a base for the column formwork.

7. Length L_2 shall suffice to tie the starter bars securely to the reinforcing bars. (It should not be less than 2s, where s is the space between bars ϕ_2).

8. The sole purpose of the starter bar ties is to keep the starter bar assembly firmly in place while the concrete is cast in the foundations. The starter bar assembly is to be tied at all cross points. These are not the same ties as in the column.

9. The top of the footing is smoothed with a float or power float except in the contact area with the future column, where a rough surface such as is generated by the vibrator is needed.

10. In soft soils, the 25 cm over the level of the future blinding should not be excavated until shortly before pouring the concrete to ensure that the soil is not softened by rainfall immediately before placement.

11. The foundation subgrade must be compacted before the blinding is poured.

12. The blinding under the footing is floated or smoothed with a power float. Its standard 10-cm thickness may be varied to absorb tolerances in foundation subgrade levelling.

13. Where the footing is to be poured on soils that constitute an aggressive medium, it should be protected by an asphalt membrane as specified.

14. Before pouring the column concrete, the joint surface should be cleaned and pressure hosed. The concrete should not be cast until the surface dries.

15. See 1.2 and 1.3 for descriptions of how to tie bars and place spacers.

16. See 1.6 for panel welding.

17. Where calculations are very tight, the mid plane in pairs of lapped bars should be perpendicular to the direction of the greatest bending moment acting on the column springing.

18. In large footings, top surface reinforcement may be advisable to prevent drying shrinkage and thermal contraction.

2. SPECIFIC REFERENCES

This solution was first used in northern Europe, together with welded panels (see 1.6). It was included in Part 3 of EC2 in 1998 (10). Part 3 presently forms part of EC2 (see 9.8.2.1). This solution is only cost-effective in small footings where the circular trenches can be dug mechanically.

The design is given in (12). Further to those calculations, the total longitudinal reinforcement in each panel can be found as:

$$A_{s,tot,x} = A_{s,tot,y} = \frac{F_d D}{3\pi d f_{yd}}$$

where F_d is the design load on the column.

CD – 01.07 SPREAD FOOTING AND EXPANSION JOINT

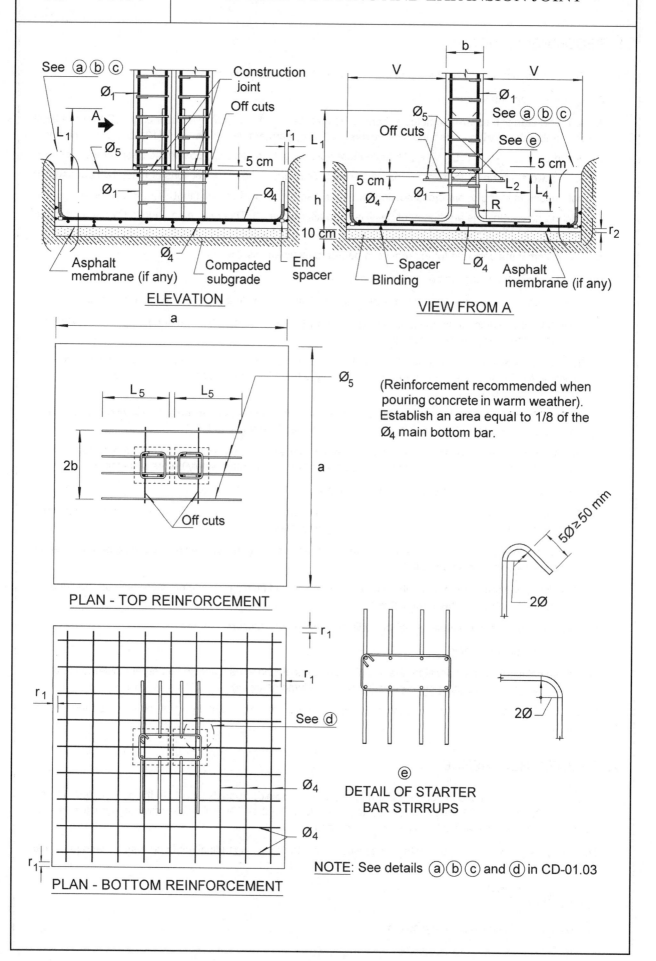

See ⓐ ⓑ ⓒ
Ø₁
Construction joint
Off cuts
A
L₁
Ø₅
Ø₁
5 cm
r₁
L₁
Ø₄
Asphalt membrane (if any)
Compacted subgrade
End spacer
Ø₄
10 cm
h

ELEVATION

b
V
V
Ø₁
Ø₅
See ⓐ ⓑ ⓒ
Off cuts
See ⓔ
5 cm
5 cm
Ø₁
L₂
L₁
L₄
Ø₄
R
Spacer
Blinding
Ø₄
Asphalt membrane (if any)
r₂

VIEW FROM A

a
L₅
L₅
Ø₅
2b
a
Off cuts

PLAN - TOP REINFORCEMENT

(Reinforcement recommended when pouring concrete in warm weather). Establish an area equal to 1/8 of the Ø₄ main bottom bar.

r₁
r₁
r₁
See ⓓ
Ø₄
Ø₄
r₁

PLAN - BOTTOM REINFORCEMENT

5Ø≥50 mm
2Ø
2Ø

ⓔ
DETAIL OF STARTER BAR STIRRUPS

<u>NOTE</u>: See details ⓐⓑⓒ and ⓓ in CD-01.03

1. RECOMMENDATIONS

1. The ϕ_1 bars rest directly on the top of the footing, with no hooks needed.

2. L_1 is the ϕ_1 bar lap length.

3. **AR.** The starter bars are diameter ϕ_1 bars. L_4 should be $\geq \ell_{bd}$, but length ℓ_{bd} is defined in EC2 for the least favourable case. In this case the side covers for the ϕ_1 bars are very large. See (12), p. 69. A safe value would be $\ell'_{bd} = \frac{2}{3}\ell_{bd}$. If $L_4 < \frac{2}{3}\ell_{bd}$, often a more suitable solution than increasing the depth of the footing is to use two starter bars for every ϕ_1 bar in the wall. The sum of the cross-sections of these two bars should not be less than the ϕ_1 bar cross-section, but their diameter should be such that $\frac{2}{3}\ell_{bd} < L_4$, where ℓ_{bd} is the anchorage length. The arrangement detail is as in alternative (d). This rule can be applied wherever the bar cover is $\geq 10\,\phi$ but not under 10 cm.

4. $r_1 = 7.5$ cm if the concrete for the footing is poured directly against the soil.

5. $r_2 = 2.5$ cm but not less than ϕ_4.

6. See 1.5 to determine when to use anchor type a, b or c at the end of the ϕ_4 reinforcing bar.

7. Length L_2 shall suffice to tie the ϕ_1 starter bars securely to two ϕ_4 transverse bars. (It may not be less than 2s, where s is the space between ϕ_4 bars.)

8. The sole purpose of the starter bar ties is to keep the starter bar assembly firmly in place while the concrete is cast in the foundations. The starter bar assembly is to be tied at all cross points.

9. The top of the footing is smoothed with a float or power float except in the contact area with the future column, where a rough surface such as is generated by the vibrator is needed.

10. In soft soils, the 25 cm over the level of the future blinding should not be excavated until shortly before pouring the concrete to ensure that the soil is not softened by rainfall immediately before placement.

11. The foundation subgrade must be compacted before the blinding is poured.

12. The blinding under the footing is floated or smoothed with a power float. Its standard 10-cm thickness may be varied to absorb tolerances in foundation subgrade levelling.

13. Where the footing is to be poured on soils that constitute an aggressive medium, it should be protected by an asphalt membrane as specified.

14. Before pouring the column concrete, the joint surface should be cleaned and pressure hosed. The concrete should not be cast until the surface dries.

15. See 1.2 and 1.3 for descriptions of how to tie bars and place spacers.

16. Where calculations are very tight, the mid plane in pairs of lapped bars should be perpendicular to the direction of the greatest bending moment acting on the column springing.

17. Rectangular footings can be set out in exactly the same way, although the resulting grid is logically not symmetrical.

18. Unless calculated in the design, the width of the expansion joint should be 20–30 mm.

19. Reinforcement ϕ_s is placed to control possible cracking on the top surface between two columns as the temperature drops in the two frames, which pull in opposite directions.

STRAP FOOTING

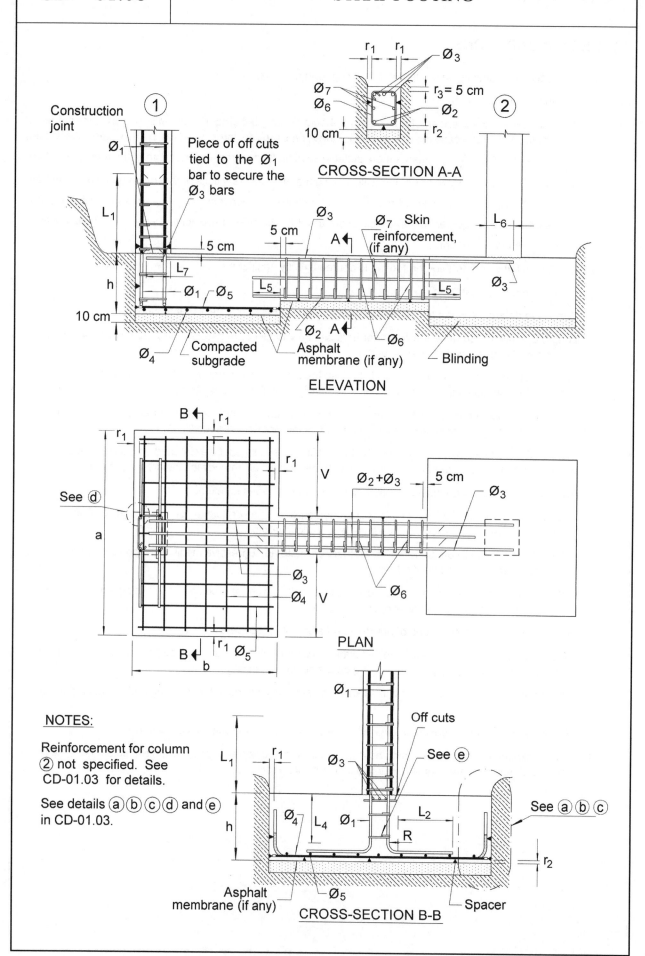

Construction joint

Piece of off cuts tied to the $\emptyset_1$ bar to secure the $\emptyset_3$ bars

$\emptyset_1$

L_1

5 cm

L_7

h

$\emptyset_1$ $\emptyset_5$

10 cm

$\emptyset_4$

Compacted subgrade

Asphalt membrane (if any)

5 cm

$\emptyset_3$

L_5

$\emptyset_2$ A

$\emptyset_7$ Skin reinforcement, (if any)

$\emptyset_6$

Blinding

L_5

$\emptyset_3$

L_6

ELEVATION

r_1 r_1 $\emptyset_3$

$\emptyset_7$
$\emptyset_6$

$r_3 = 5$ cm

$\emptyset_2$

10 cm

r_2

CROSS-SECTION A-A

B r_1

r_1

r_1

V

$\emptyset_2 + \emptyset_3$

5 cm

$\emptyset_3$

See ⓓ

a

$\emptyset_3$

$\emptyset_4$ V

$\emptyset_6$

B r_1 $\emptyset_5$

b

PLAN

$\emptyset_1$

Off cuts

$\emptyset_3$

See ⓔ

L_1 r_1

$\emptyset_3$

h

$\emptyset_4$ L_4 $\emptyset_1$

L_2

R

See ⓐⓑⓒ

r_2

Asphalt membrane (if any)

$\emptyset_5$

Spacer

CROSS-SECTION B-B

NOTES:

Reinforcement for column ② not specified. See CD-01.03 for details.

See details ⓐ ⓑ ⓒ ⓓ and ⓔ in CD-01.03.

1. RECOMMENDATIONS

1. The ϕ_1 bars rest directly on the top of the footing, with no hooks needed.

2. L_1 is the ϕ_1 lap bar length.

3. The starter bars are diameter ϕ_1 bars, except where $L_4 < \ell_{bd}$. If $L_4 < \ell_{bd}$, often a more suitable solution than increasing the depth of the footing is to use two starter bars for every ϕ_1 bar in the column. The sum of the cross-sections of these two bars should not be less than the ϕ_1 bar cross-section, but their diameter should be such that ℓ_{bd}, their anchorage length, should be $< L_4$.

4. $r_1 = 7.5$ cm if the concrete for the footing is poured directly against the soil.

5. $r_2 = 2.5$ cm but not less than ϕ_4.

6. See 1.5 to determine when to use anchor type a, b or c at the end of the ϕ_4 reinforcing bar.

7. Length L_2 must suffice to tie the ϕ_1 starter bars securely to two ϕ_5 transverse bars. (It may not be less than 2s, where s is the space between ϕ_5 bars.)

8. Length L_5 is the ϕ_2 bar anchorage length.

9. If the tops of the beam and footing are flush, the beam should be at least 5 cm less deep to ensure that its reinforcement does not interfere with the reinforcing bars in the footing.

10. In this case the column starter bars need ties of the same diameter and maximum spacing as in the column. They differ in size.

11. The top of the footing is smoothed with a float or power float except in the contact area with the future column, where a rough surface such as is generated by the vibrator is needed.

12. In soft soils, the 25 cm over the level of the future blinding should not be excavated until shortly before pouring the concrete to ensure that the soil is not softened by rainfall immediately before placement.

13. The subgrade is compacted before pouring the blinding for the footing and beam.

14. The blinding is floated or smoothed with a power float. Its standard 10-cm thickness may be varied to absorb tolerances in foundation subgrade levelling.

15. Where the footing is to be poured on soils that constitute an aggressive medium, it should be protected by an asphalt membrane as specified.

16. Before pouring the column concrete, the joint surface should be cleaned and pressure hosed. The concrete should not be cast until the surface dries.

17. See 1.2 and 1.3 for descriptions of how to tie bars and place spacers.

2. STATUTORY LEGISLATION

EC2 (5).

SELF-CENTRED EDGE FOOTING

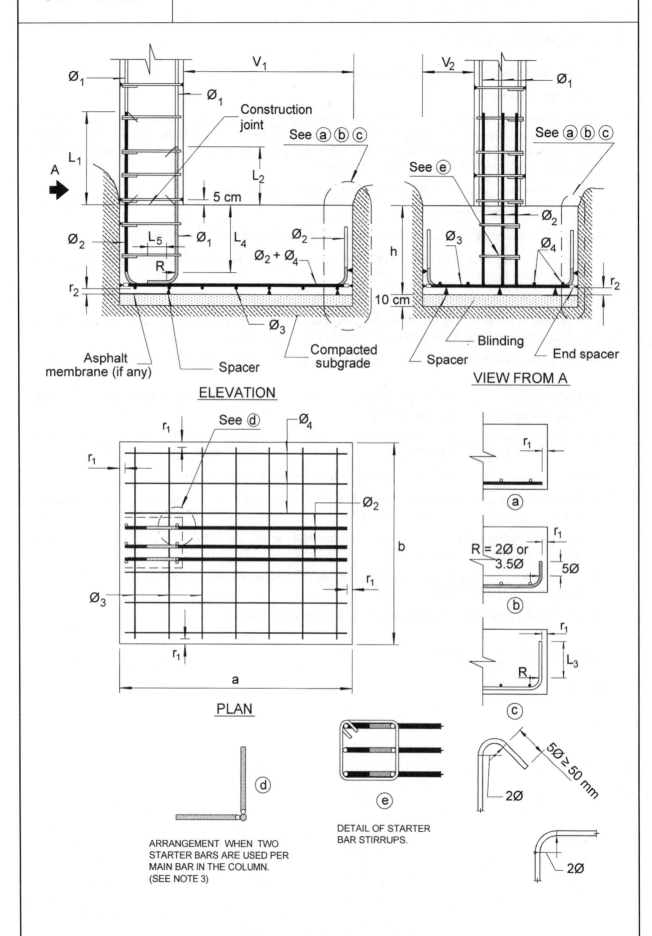

ELEVATION

VIEW FROM A

PLAN

ARRANGEMENT WHEN TWO
STARTER BARS ARE USED PER
MAIN BAR IN THE COLUMN.
(SEE NOTE 3)

DETAIL OF STARTER
BAR STIRRUPS.

1. RECOMMENDATIONS

1. The ϕ_1 bars rest directly on the top of the footing, with no hooks needed.

2. L_1 is the lap length in the larger of the reinforcing bars, ϕ_1 or ϕ_2. L_2 is the ϕ_1 bar lap length.

3. The starter bars are diameter ϕ_1 bars, except where $L_4 < L_1$. If $L_4 < L_1$, often a more suitable solution than increasing the depth of the footing is to use two starter bars for every ϕ_2 bar in the column. The sum of the cross-sections of these two bars should not be less than the ϕ_2 bar cross-section, but their diameter should be such that ℓ_{bd}, their anchorage length, should be $< L_4$.

4. $r_1 = 7.5$ cm if the concrete for the footing is poured directly against the soil.

5. $r_2 = 2.5$ cm but not less than ϕ_3.

6. See 1.5 to determine when to use anchor type a, b or c at the end of the ϕ_2 and ϕ_3 reinforcing bars.

7. The ϕ_4 distribution steel has a straight extension at the end.

8. Length L_5 should suffice to tie ϕ_1 starter bars to ϕ_2 bars; 20 cm is the suggested length.

9. In this case the column starter bars need ties of the same diameter and maximum spacing as in the column. They differ in size.

10. The top of the footing is smoothed with a float or power float except in the contact area with the future column, where a rough surface such as is generated by the vibrator is needed.

11. In soft soils, the 25 cm over the level of the future blinding should not be excavated until shortly before pouring the concrete to ensure that the soil is not softened by rainfall immediately before placement.

12. The subgrade is compacted before pouring the blinding for the footing and beam.

13. The blinding is floated or smoothed with a power float. Its standard 10-cm thickness may be varied to absorb tolerances in foundation subgrade levelling.

14. Where the footing is to be poured on soils that constitute an aggressive medium, it should be protected by an asphalt membrane as specified.

15. Before pouring the column concrete, the joint surface should be cleaned and pressure hosed. The concrete should not be cast until the surface dries.

16. See 1.2 and 1.3 for descriptions of how to tie bars and place spacers.

2. STATUTORY LEGISLATION

EC2 (5).

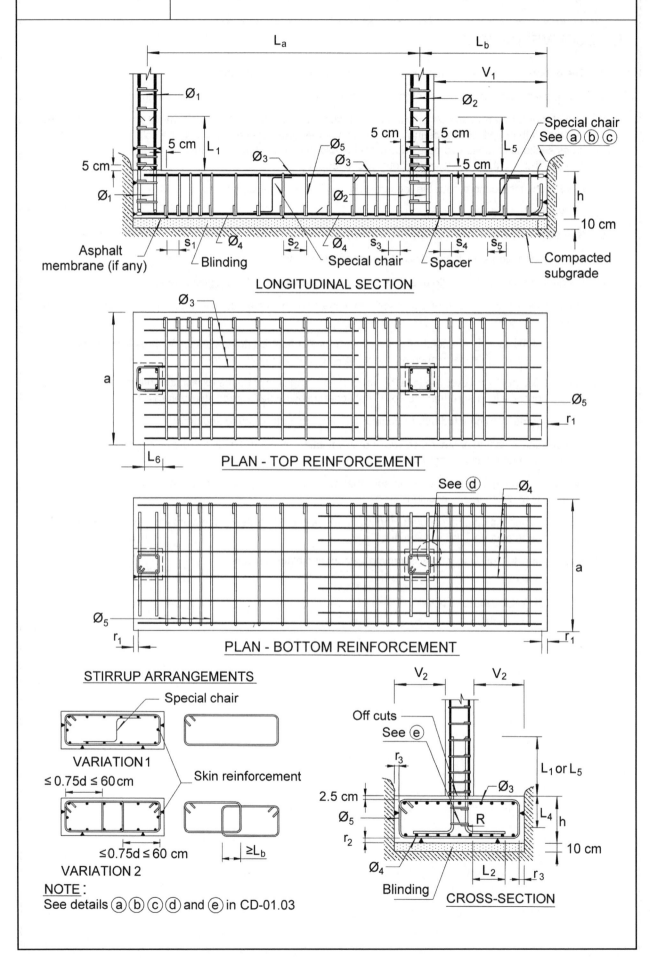

LONGITUDINAL SECTION

PLAN - TOP REINFORCEMENT

PLAN - BOTTOM REINFORCEMENT

STIRRUP ARRANGEMENTS

VARIATION 1

≤ 0.75d ≤ 60 cm

VARIATION 2

≤ 0.75d ≤ 60 cm

NOTE :
See details ⓐ ⓑ ⓒ ⓓ and ⓔ in CD-01.03

CROSS-SECTION

1. RECOMMENDATIONS

1. The ϕ_1 and ϕ_2 bars rest directly on the top of the footing, with no hooks needed.

2. L_1 is the lap length in the ϕ_1 bar and L_5 is the lap length in the ϕ_2 bar.

3. **AR.** The starter bars in the inner column are diameter ϕ_2 bars. L_4 should be $\geq \ell_{bd}$, but length ℓ_{bd} is defined in EC2 for the least favourable case. In this case the side covers for the ϕ_2 bars are very large. See (12), p. 69. A safe value would be $\ell'_{bd} = \frac{2}{3} \ell_{bd}$. If $L_4 < \frac{2}{3} \ell_{bd}$, often a more suitable solution than increasing the depth of the footing is to use two starter bars for every ϕ_2 bar in the column. The sum of the cross-sections of these two bars should not be less than the ϕ_2 bar cross-section, but their diameter should be such that $\frac{2}{3} \ell_{bd} < L_4$, where ℓ_{bd} is the anchorage length. The arrangement detail is as in alternative (d).

   Particular care should be taken with edge and corner columns, because there either $L_4 \geq L_1$, or the two starter bars per ϕ_2 bar arrangement should be adopted (detail (d)).

4. r_1 and r_3 are equal to 7.5 cm if the concrete for the footing is poured directly against the soil.

5. $r_2 = 2.5$ cm but not less than ϕ_5.

6. See 1.5 to determine when to use anchor type a, b or c at the end of the ϕ_4 reinforcing bar.

7. Length L_2 shall suffice to tie ϕ_1 or ϕ_2 starter bars to two ϕ_4 transverse bars. (It may not be less than 2s, where s is the space between the ϕ_4 bars.)

8. In this case the edge ϕ_1 column starter bars need ties of the same diameter and maximum spacing as in the column. They differ in size. See CD – 01.03 for inner column ties.

9. Sole stirrups may only be used where a ≤ 65 cm. If a > 65 cm, multiple stirrups are needed, except where the concrete absorbs the shear stress entirely. The vertical legs must be spaced at no more than 0.75d ≤ 60 cm.

10. If multiple stirrups are used, the horizontal legs must overlap along a distance of at least L_b, where L_b is the ϕ_5 bar lap length. (This is required for the horizontal leg of the stirrup to act as transverse bending reinforcement.)

11. The top of the footing is smoothed with a float or power float except in the contact area with the future column, where a rough surface such as is generated by the vibrator is needed.

12. In soft soils, the 25 cm over the level of the future blinding should not be excavated until shortly before pouring the concrete to ensure that the soil is not softened by rainfall immediately before placement.

13. The subgrade is compacted before pouring the blinding for the footing and beam.

14. The blinding is floated or smoothed with a power float. Its standard 10-cm thickness may be varied to absorb tolerances in foundation subgrade levelling.

15. Where the footing is to be poured on soils that constitute an aggressive medium, it should be protected by an asphalt membrane as specified.

16. Before pouring the column concrete, the joint surface should be cleaned and pressure hosed. The concrete should not be cast until the surface dries.

17. See 1.2 and 1.3 for descriptions of how to tie bars and place spacers.

2. STATUTORY LEGISLATION

This type of footing is not addressed in EC2.

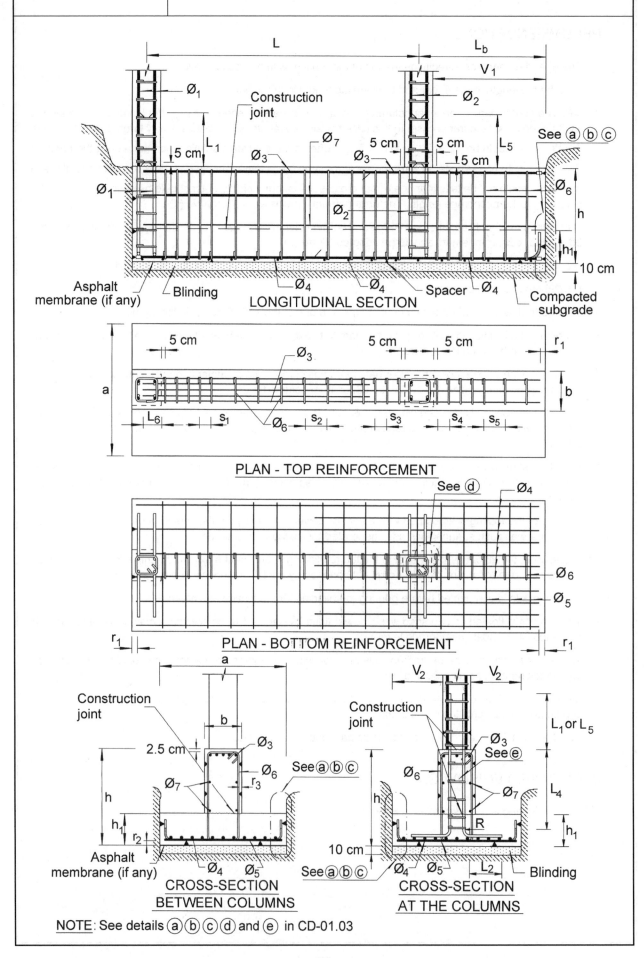

CD – 01.11 COMBINED EDGE FOOTING (VARIATION 2)

LONGITUDINAL SECTION

PLAN - TOP REINFORCEMENT

PLAN - BOTTOM REINFORCEMENT

CROSS-SECTION BETWEEN COLUMNS

CROSS-SECTION AT THE COLUMNS

NOTE: See details ⓐ ⓑ ⓒ ⓓ and ⓔ in CD-01.03

1. RECOMMENDATIONS

1. The ϕ_1 and ϕ_2 bars rest directly on the top of the footing, with no hooks needed.

2. L_1 is the lap length in the ϕ_1 bar and L_5 is the lap length in the ϕ_2 bar.

3. **AR.** The starter bars in the inner column are diameter ϕ_2 bars. L_4 should be $\geq \ell_{bd}$, but length ℓ_{bd} is defined in EC2 for the least favourable case. In this case the side cover on the ϕ_2 bars is usually very large. See (12), p. 69. A safe value would be $\ell'_{bd} = \frac{2}{3}\ell_{bd}$. If $L_4 < \frac{2}{3}\ell_{bd}$, often a more suitable solution than increasing the depth of the footing is to use two starter bars for every ϕ_2 bar in the wall. The sum of the cross-sections of these two bars should not be less than the ϕ_2 bar cross-section, but their diameter should be such that $\frac{2}{3}\ell_{bd} < L_4$, where ℓ_{bd} is the anchorage length. The arrangement detail is as in alternative (d). (If the side cover on the web in the ϕ_1 or ϕ_2 bars is less than ten times their diameter, or less than 10 cm, substitute $\frac{2}{3}\ell_{bd}$ for ℓ_{bd} in the above formulas.

    Particular care should be taken with edge and corner columns, because there either $L_4 \geq L_1$, or the two starter bars per ϕ_2 bar arrangement should be adopted (alternative (d)).

4. $r_1 = 7.5$ cm if the concrete for the footing is poured directly against the soil.

5. $r_2 = 2.5$ cm $\geq \phi_5$; $r_3 = 2.5$ cm $\geq \phi_6$.

6. See 1.5 to determine when to use anchor type a, b or c at the end of the ϕ_4 and ϕ_5 reinforcing bars.

7. Length L_2 shall suffice to tie ϕ_1 or ϕ_2 starter bars to two ϕ_4 transverse bars. (It may not be less than 2s, where s is the space between ϕ_4 bars.)

8. In this case the edge column starter bars need ties of the same diameter and maximum spacing as in the column. The starter bars on the inner columns need ties of the same diameter and maximum spacing as in the column if the side cover over the T-section web is less than ten times their diameter, or less than 10 cm. Otherwise two ties suffice (see CD – 01.03) across depth h_1 and two along the height of the web.

9. The top of the footing is smoothed with a float or power float except in the contact area with the future column, where a rough surface such as is generated by the vibrator is needed. The flange-to-web joint is treated similarly.

10. In soft soils, the 25 cm over the level of the future blinding should not be excavated until shortly before pouring the concrete to ensure that the soil is not softened by rainfall immediately before placement.

11. The subgrade is compacted before pouring the blinding for the footing and beam.

12. The blinding is floated or smoothed with a power float. Its standard 10-cm thickness may be varied to absorb tolerances in foundation subgrade levelling.

13. Where the footing is to be poured on soils that constitute an aggressive medium, it should be protected by an asphalt membrane as specified.

14. Before pouring the column concrete, the joint surface should be cleaned and pressure hosed. The concrete should not be cast until the surface dries. The flange-to-web joint should be treated similarly.

15. See 1.2 and 1.3 for descriptions of how to tie bars and place spacers.

2. STATUTORY LEGISLATION

This type of footing is not addressed in EC2 (5).

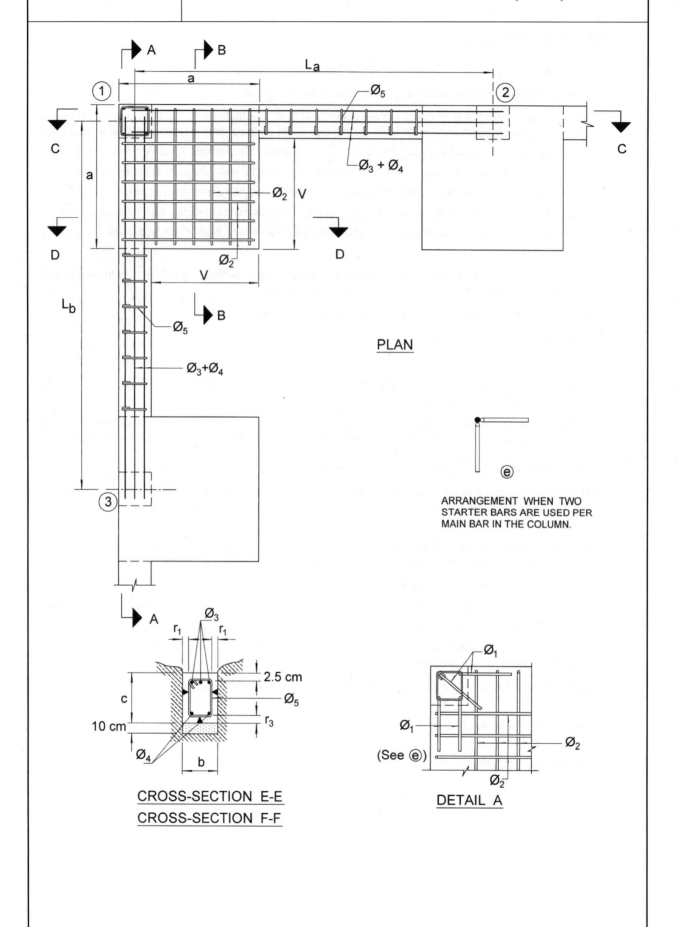

PLAN

ARRANGEMENT WHEN TWO
STARTER BARS ARE USED PER
MAIN BAR IN THE COLUMN.

CROSS-SECTION E-E
CROSS-SECTION F-F

DETAIL A

1. RECOMMENDATIONS

1. The ϕ_1 bars rest directly on the top of the footing, with no hooks needed.

2. L_1 is the ϕ_1 bar lap length.

3. The starter bars are diameter ϕ_1 bars, except where $L_4 < L_1$. If $L_4 < L_1$, often a more suitable solution than increasing the depth of the footing is to use two starter bars for every ϕ_1 bar in the column. The sum of the cross-sections of these two bars should not be less than the ϕ_1 bar cross-section, but their diameter should be such that $L_b < L_4$, where L_b is the anchorage length. The arrangement detail is as in alternative (e).

4. $r_1 = 7.5$ cm if the concrete for the footing is poured directly against the soil.

5. $r_2 = 2.5$ cm $\geq \phi_4$.

6. See 1.5 to determine when to use anchor type a, b or c at the end of the ϕ_2 reinforcing bars.

7. Length L_2 must suffice to tie the ϕ_1 starter bars securely to two ϕ_2 transverse bars. (It may not be less than 2s, where s is the space between ϕ_2 bars.)

8. L_6 is the ϕ_4 bar anchorage length.

9. In this case the column starter bars need ties of the same diameter and maximum spacing as in the column. They differ in size.

10. The top of the footing is smoothed with a float or power float except in the contact area with the future column, where a rough surface such as is generated by the vibrator is needed.

11. In soft soils, the 25 cm over the level of the future blinding should not be excavated until shortly before pouring the concrete to ensure that the soil is not softened by rainfall immediately before placement.

12. The subgrade is compacted before pouring the blinding for the footing and beam.

13. The blinding is floated or smoothed with a power float. Its standard 10-cm thickness may be varied to absorb tolerances in foundation subgrade levelling.

14. Where the footing is to be poured on soils that constitute an aggressive medium, it should be protected by an asphalt membrane as specified.

15. Before pouring the column concrete, the joint surface should be cleaned and pressure hosed. The concrete should not be cast until the surface dries.

16. See 1.2 and 1.3 for descriptions of how to tie bars and place spacers.

2. STATUTORY LEGISLATION

This type of footing is not addressed in EC2.

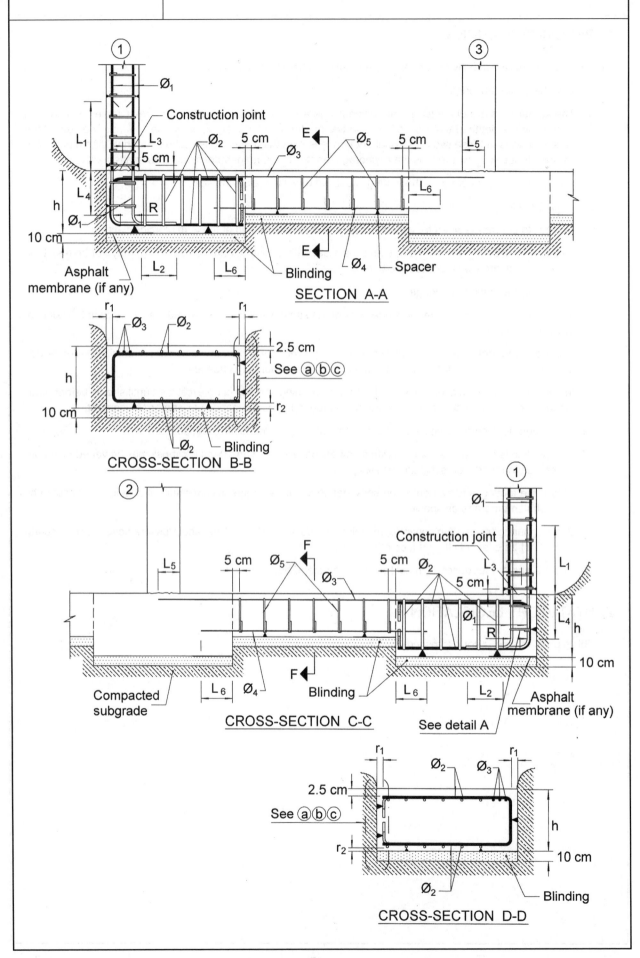

SECTION A-A

CROSS-SECTION B-B

CROSS-SECTION C-C

CROSS-SECTION D-D

1. RECOMMENDATIONS

1. The ϕ_1 bars rest directly on the top of the footing, with no hooks needed.

2. L_1 is the ϕ_1 bar lap length.

3. The starter bars are diameter ϕ_1 bars, except where $L_4 < L_1$. If $L_4 < L_1$, often a more suitable solution than increasing the depth of the footing is to use two starter bars for every ϕ_1 bar in the column. The sum of the cross-sections of these two bars should not be less than the ϕ_1 bar cross-section, but their diameter should be such that $L_b < L_4$, where L_b is the anchorage length. The arrangement detail is as in alternative (e).

4. $r_1 = 7.5$ cm if the concrete for the footing is poured directly against the soil.

5. $r_2 = 2.5$ cm $\geq \phi_4$.

6. See 1.5 to determine when to use anchor type a, b or c at the end of the ϕ_2 reinforcing bars.

7. Length L_2 must suffice to tie the ϕ_1 starter bars securely to two ϕ_2 transverse bars. (It may not be less than 2s, where s is the space between ϕ_2 bars.)

8. L_6 is the ϕ_4 bar anchorage length.

9. In this case the column starter bars need ties of the same diameter and maximum spacing as in the column. They differ in size.

10. The top of the footing is smoothed with a float or power float except in the contact area with the future column, where a rough surface such as is generated by the vibrator is needed.

11. In soft soils, the 25 cm over the level of the future blinding should not be excavated until shortly before pouring the concrete to ensure that the soil is not softened by rainfall immediately before placement.

12. The subgrade is compacted before pouring the blinding for the footing and beam.

13. The blinding is floated or smoothed with a power float. Its standard 10-cm thickness may be varied to absorb tolerances in foundation subgrade levelling.

14. Where the footing is to be poured on soils that constitute an aggressive medium, it should be protected by an asphalt membrane as specified.

15. Before pouring the column concrete, the joint surface should be cleaned and pressure hosed. The concrete should not be cast until the surface dries.

16. See 1.2 and 1.3 for descriptions of how to tie bars and place spacers.

2. STATUTORY LEGISLATION

This type of footing is not addressed in EC2 (5).

SELF-CENTRED CORNER FOOTING

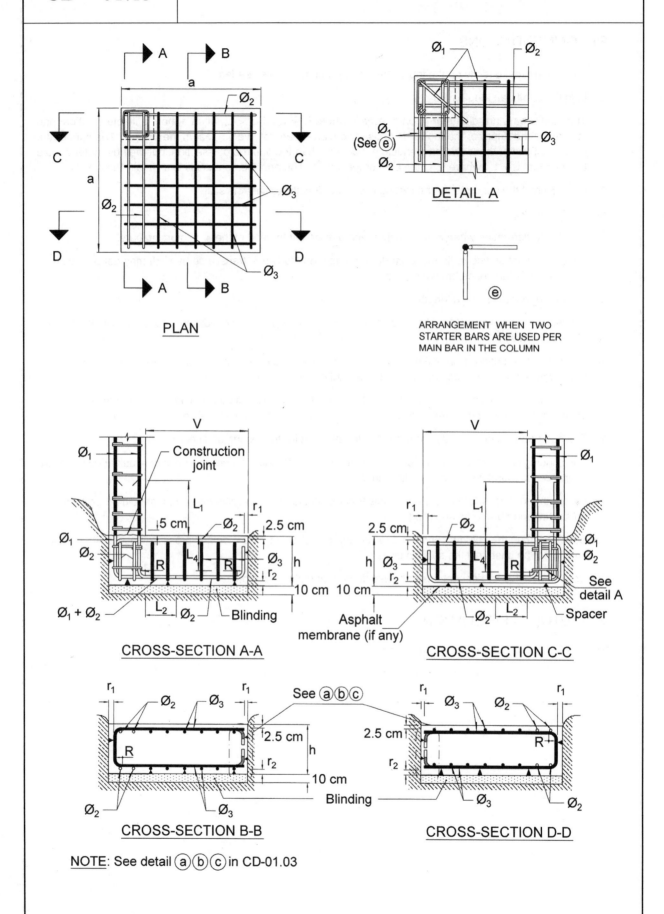

PLAN

DETAIL A

ARRANGEMENT WHEN TWO
STARTER BARS ARE USED PER
MAIN BAR IN THE COLUMN

CROSS-SECTION A-A

CROSS-SECTION C-C

CROSS-SECTION B-B

CROSS-SECTION D-D

NOTE: See detail ⓐⓑⓒ in CD-01.03

1. RECOMMENDATIONS

1. The ϕ_1 bars rest directly on the top of the footing, with no hooks needed.

2. L_1 is the ϕ_1 bar lap length.

3. The starter bars are diameter ϕ_1 bars, except where $L_4 < L_1$. If $L_4 < L_1$, often a more suitable solution than increasing the depth of the footing is to use two starter bars for every ϕ_1 bar in the column. The sum of the cross-sections of these two bars should not be less than the ϕ_1 (bar cross-section, but their diameter should be such that $L_b < L_4$, where L_b is the anchorage length. The arrangement detail is as in alternative (e).

4. $r_1 = 7.5$ cm if the concrete for the footing is poured directly against the soil.

5. $r_2 = 2.5$ cm but not less than ϕ_2.

6. See 1.5 to determine when to use anchor type a, b or c at the end of the ϕ_2 and ϕ_3 reinforcing bars.

7. Length L_2 shall suffice to tie the ϕ_1 starter bars securely to two ϕ_3 transverse bars. (It may not be less than 2s, where s is the space between ϕ_3 bars.)

8. In this case the column starter bars need ties of the same diameter and maximum spacing as in the column. They differ in size.

9. The top of the footing is smoothed with a float or power float except in the contact area with the future column, where a rough surface such as is generated by the vibrator is needed.

10. In soft soils, the 25 cm over the level of the future blinding should not be excavated until shortly before pouring the concrete to ensure that the soil is not softened by rainfall immediately before placement.

11. The subgrade is compacted before pouring the blinding for the footing and beam.

12. The blinding is floated or smoothed with a float or power float. Its standard 10-cm thickness may be varied to absorb tolerances in foundation subgrade levelling.

13. Where the footing is to be poured on soils that constitute an aggressive medium, it should be protected by an asphalt membrane as specified.

14. Before pouring the column concrete, the joint surface should be cleaned and pressure hosed. The concrete should not be cast until the surface dries.

15. See 1.2 and 1.3 for descriptions of how to tie bars and place spacers.

2. STATUTORY LEGISLATION

This type of footing is not addressed in EC2 (5).

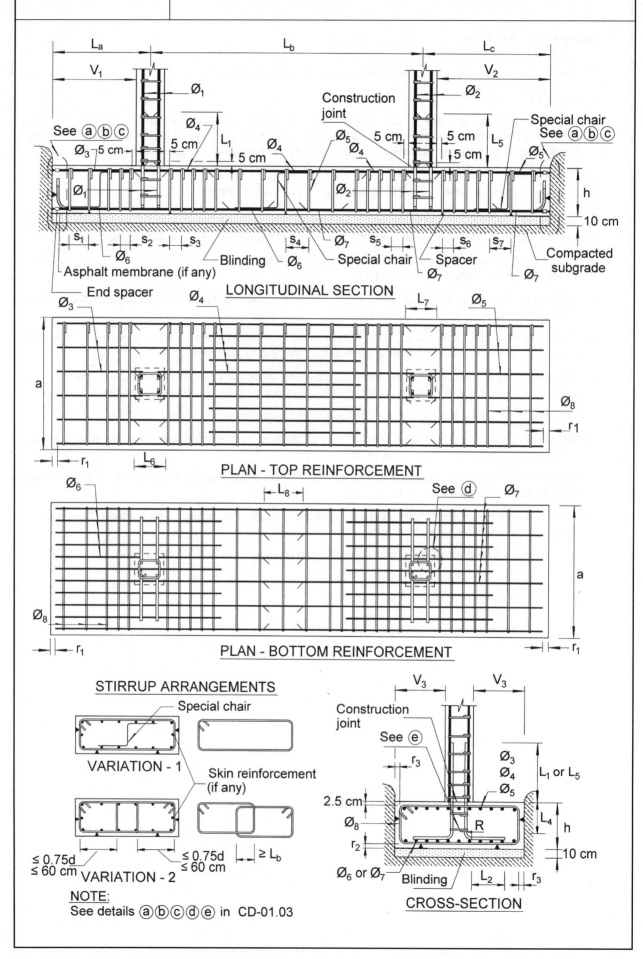

CD – 01.14

COMBINED FOOTING (VARIATION 1)

L_a L_b L_c

V_1 V_2

$\varnothing_1$ $\varnothing_2$

Construction joint

See ⓐⓑⓒ

$\varnothing_3$ 5 cm $\varnothing_4$ $\varnothing_5$ 5 cm 5 cm Special chair See ⓐⓑⓒ

5 cm L_1 $\varnothing_4$ $\varnothing_4$ 5 cm L_5 $\varnothing_5$

5 cm 5 cm

$\varnothing_1$ $\varnothing_2$ h

$\varnothing_1$ 10 cm

s_1 s_2 s_3 s_4 s_5 s_6 s_7 Compacted subgrade

$\varnothing_6$ Blinding $\varnothing_6$ Special chair Spacer

Asphalt membrane (if any) $\varnothing_7$ $\varnothing_7$

End spacer $\varnothing_7$

LONGITUDINAL SECTION

$\varnothing_3$ $\varnothing_4$ L_7 $\varnothing_5$

a $\varnothing_8$

r_1

r_1 L_6

PLAN - TOP REINFORCEMENT

$\varnothing_6$ L_8 See ⓓ $\varnothing_7$

a

$\varnothing_8$

r_1 r_1

PLAN - BOTTOM REINFORCEMENT

STIRRUP ARRANGEMENTS

Special chair

VARIATION - 1

Skin reinforcement (if any)

VARIATION - 2

≤ 0.75d ≤ 0.75d ≥ L_b
≤ 60 cm ≤ 60 cm

NOTE:
See details ⓐⓑⓒⓓⓔ in CD-01.03

V_3 V_3

Construction joint

See ⓔ

r_3 $\varnothing_3$
$\varnothing_4$ L_1 or L_5
$\varnothing_5$

2.5 cm L_4

$\varnothing_8$ R h

r_2 10 cm

$\varnothing_6$ or $\varnothing_7$ Blinding L_2 r_3

CROSS-SECTION

64

1. RECOMMENDATIONS

1. The ϕ_1 and ϕ_2 bars rest directly on the top of the footing, with no hooks needed.

2. L_1 is the lap length in the ϕ_1 bar and L_5 is the lap length in the ϕ_2 bar.

3. **AR.** The starter bars are diameter ϕ_1 bars. L_4 should be $\geq \ell_{bd}$, but length ℓ_{bd} is defined in EC2 for the least favourable case. In this case the side covers for the ϕ_1 bars are very large. See (12), p. 69. A safe value would be $\ell'_{bd} = \frac{2}{3} \ell_{bd}$. If $L_4 < \frac{2}{3} \ell_{bd}$, often a more suitable solution than increasing the depth of the footing is to use two starter bars for every ϕ_1 bar in the column. The sum of the cross-sections of these two bars should not be less than the ϕ_1 bar cross-section, but their diameter should be such that $\frac{2}{3} \ell_{bd} < L_4$, where ℓ_{bd} is the anchorage length. The arrangement detail is as in alternative (d). This rule can be applied wherever the bar cover is $\geq 10 \phi$ but not under 10 cm.

4. r_1 and r_3 are equal to 7.5 cm if the concrete for the footing is poured directly against the soil.

5. $r_2 = 2.5$ cm.

6. See 1.5 to determine when to use anchor type a, b or c at the end of the ϕ_6 and ϕ_7 reinforcing bars.

7. Length L_2 shall suffice to tie ϕ_1 or ϕ_2 starter bars to two ϕ_6 or ϕ_7 transverse bars. (It may not be less than 2s, where s is the space between ϕ_6 or ϕ_7 bars.)

8. L_6 is the lap length in the larger of the ϕ_3 or ϕ_4 bars, L_7 is the lap length in the larger of the ϕ_4 or ϕ_5 bars and L_8 is the lap length in the larger of the ϕ_6 or ϕ_7 bars.

9. Sole stirrups may only be used where a $\leq$ 65 cm. If a > 65 cm, multiple stirrups are needed, except where the concrete absorbs the shear stress entirely.

10. If multiple stirrups are used, the horizontal legs should overlap along a length of at least L_b, where L_b is the anchorage length of the ϕ_8 bars. (This is required for the horizontal leg of the stirrup to act as transverse bending reinforcement.)

11. The sole purpose of the starter bar ties is to keep the starter bar assembly firmly in place while the concrete is cast in the foundations. The starter bar assembly is to be tied at all cross points. These are not the same ties as in the column.

12. The top of the footing is smoothed with a float or power float except in the contact area with the future column, where a rough surface such as is generated by the vibrator is needed.

13. In soft soils, the 25 cm over the level of the future blinding should not be excavated until shortly before pouring the concrete to ensure that the soil is not softened by rainfall immediately before placement.

14. The subgrade is compacted before pouring the blinding.

15. The blinding is floated or smoothed with a float or power float. Its standard 10-cm thickness may be varied to absorb tolerances in foundation subgrade levelling.

16. Where the footing is to be poured on soils that constitute an aggressive medium, it should be protected by an asphalt membrane as specified.

17. Before pouring the column concrete, the joint surface should be cleaned and pressure hosed. The concrete should not be cast until the surface dries.

18. See 1.2 and 1.3 for descriptions of how to tie bars and place spacers.

2. STATUTORY LEGISLATION

This type of footing is not addressed in EC2 (5).

COMBINED FOOTING (VARIATION 2)

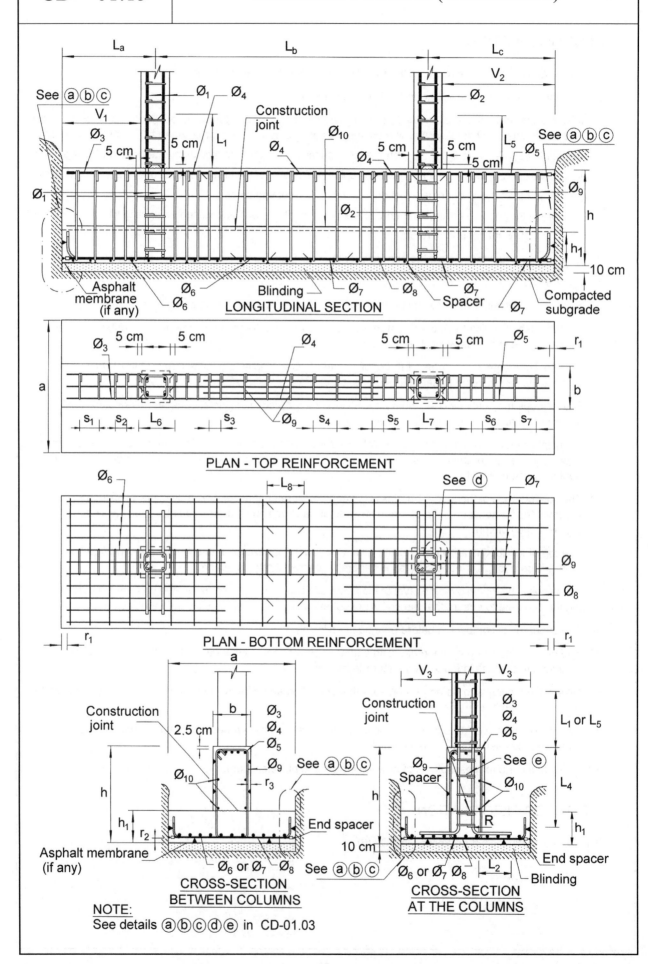

LONGITUDINAL SECTION

PLAN - TOP REINFORCEMENT

PLAN - BOTTOM REINFORCEMENT

CROSS-SECTION BETWEEN COLUMNS

CROSS-SECTION AT THE COLUMNS

NOTE:
See details ⓐⓑⓒⓓⓔ in CD-01.03

1. RECOMMENDATIONS

1. The ϕ_1 and ϕ_2 bars rest directly on the top of the footing, with no hooks needed.

2. L_1 is the lap length in the ϕ_1 bar and L_5 is the lap length in the ϕ_2 bar.

3. The starter bars are diameter ϕ_1 bars, except where $L_4 < \frac{2}{3} L_1$. If $L_4 < \frac{2}{3} L_1$, often a more suitable solution than increasing the depth of the footing is to use two starter bars for every ϕ_1 bar in the column. The sum of the cross-sections of these two bars should not be less than the ϕ_1 bar cross-section, but their diameter should be such that $\frac{2}{3} L_b < L_4$, where L_b is the anchorage length. The arrangement detail is as in alternative (d) The foregoing is equally applicable to ϕ_2 bars. (If the side cover on the web in the ϕ_1 or ϕ_2 bars is less than ten times their diameter, or less than 10 cm, substitute $\frac{2}{3} L_1$ for L_1 in the above formulas).

4. $r_1 = 7.5$ cm if the concrete for the footing is poured directly against the soil.

5. $r_2 = 2.5$ cm $\geq \phi_8$; $r_3 = 2.5$ cm $\geq \phi_9$.

6. See 1.5 to determine when to use anchor type a, b or c at the end of the ϕ_6, ϕ_7 and ϕ_8 reinforcing bars.

7. Length L_2 shall suffice to tie ϕ_1 or ϕ_2 starter bars to two ϕ_6 or ϕ_7 transverse bars. (It may not be less than 2s, where s is the space between ϕ_6 or ϕ_7 bars.)

8. L_6 is the lap length in the larger of the ϕ_3 or ϕ_4 bars, L_7 is the lap length in the larger of the ϕ_4 or ϕ_5 bars and L_8 is the lap length in the larger of the ϕ_6 or ϕ_7 bars. (Laps need not be used if the commercial length of the bars suffices.)

9. The starter bars on the columns need ties of the same diameter and maximum spacing as in the column if the side cover over the T-section web is less than ten times their diameter, or less than 10 cm. Otherwise two ties suffice (see CD – 01.03) across depth h_1 and two along the height of the web.

10. The top of the footing is smoothed with a float or power float except in the contact area with the future column, where a rough surface such as is generated by the vibrator is needed. The flange-to-web joint is treated similarly.

11. In soft soils, the 25 cm over the level of the future blinding should not be excavated until shortly before pouring the concrete to ensure that the soil is not softened by rainfall immediately before placement.

12. The subgrade is compacted before pouring the blinding for the footing and beam.

13. The blinding is floated or smoothed with a float or power float. Its standard 10-cm thickness may be varied to absorb tolerances in foundation subgrade levelling.

14. Where the footing is to be poured on soils that constitute an aggressive medium, it should be protected by an asphalt membrane as specified.

15. Before pouring the column concrete, the joint surface should be cleaned and pressure hosed. The concrete should not be cast until the surface dries. The flange-to-web joint is treated similarly.

16. See 1.2 and 1.3 for descriptions of how to tie bars and place spacers.

2. STATUTORY LEGISLATION

This type of footing is not addressed in EC2 (5).

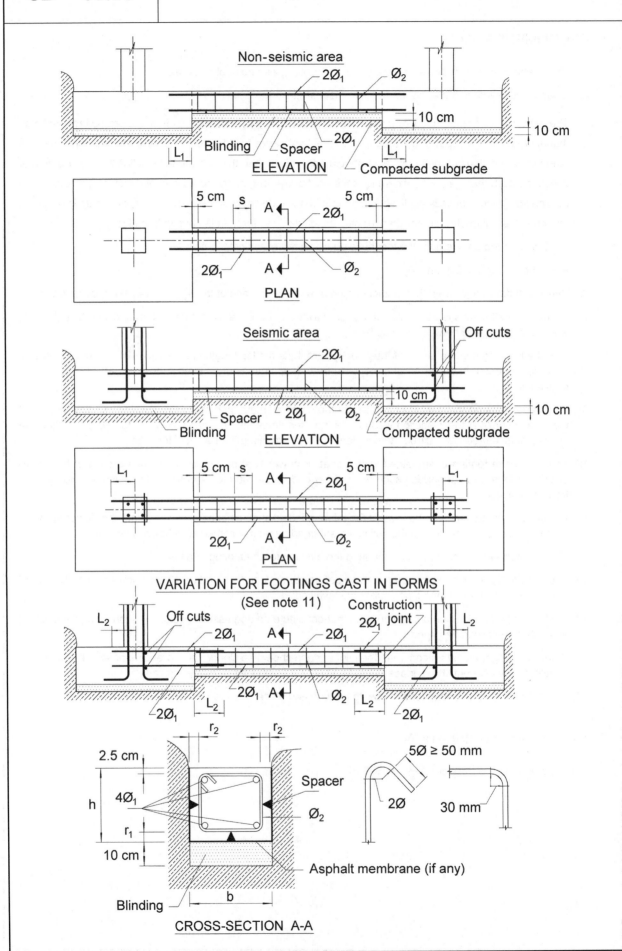

Non-seismic area

$2\emptyset_1$ $\emptyset_2$

10 cm

10 cm

Blinding   Spacer $2\emptyset_1$

ELEVATION   Compacted subgrade

L_1 L_1

5 cm   s   A $2\emptyset_1$ 5 cm

$2\emptyset_1$ A $\emptyset_2$

PLAN

Seismic area

$2\emptyset_1$ Off cuts

10 cm

10 cm

Spacer $2\emptyset_1$ $\emptyset_2$

Blinding   ELEVATION   Compacted subgrade

L_1 5 cm   s   A $2\emptyset_1$ 5 cm L_1

$2\emptyset_1$ A $\emptyset_2$

PLAN

VARIATION FOR FOOTINGS CAST IN FORMS
(See note 11)

L_2 Off cuts $2\emptyset_1$ A $2\emptyset_1$ Construction joint $2\emptyset_1$ L_2

$2\emptyset_1$ A $\emptyset_2$ L_2 $2\emptyset_1$

$2\emptyset_1$ L_2

r_2 r_2

$5\emptyset \geq 50$ mm

2.5 cm

$4\emptyset_1$ Spacer

h $\emptyset_2$

$2\emptyset$ 30 mm

r_1

10 cm   Asphalt membrane (if any)

Blinding   b

CROSS-SECTION A-A

68

1. RECOMMENDATIONS

1. $r_1 = 2.5$ cm $\geq \phi_1 - \phi_2$.

2. $r_2 = 7.5$ cm if the concrete for the footing is poured directly against the soil.

3. In beams cast in an excavation width b should be at least 40 cm, for mechanised excavation.

4. L_1 is the ϕ_1 bar anchorage length.

5. The top of the beam is floated or smoothed with a power float.

6. The bottom of the excavation is compacted before pouring the blinding.

7. The blinding under the footing is floated or smoothed with a power float. Its standard 10-cm thickness may be varied to absorb tolerances in foundation subgrade levelling.

8. Where the footing is to be poured on soils that constitute an aggressive medium, it should be protected by an asphalt membrane as specified.

9. See 1.2 and 1.3 for descriptions of how to tie bars and place spacers.

10. Care should be taken to avoid beam breakage when compacting fill over tie beams.

11. If the footing is cast in forms, footing and tie beams may be cast separately as shown. L_2 is the ϕ_1 bar anchorage length.

2. STATUTORY LEGISLATION

See EC2, 9.8.3 (5).

FOUNDATION BEAM (VARIATION 1)

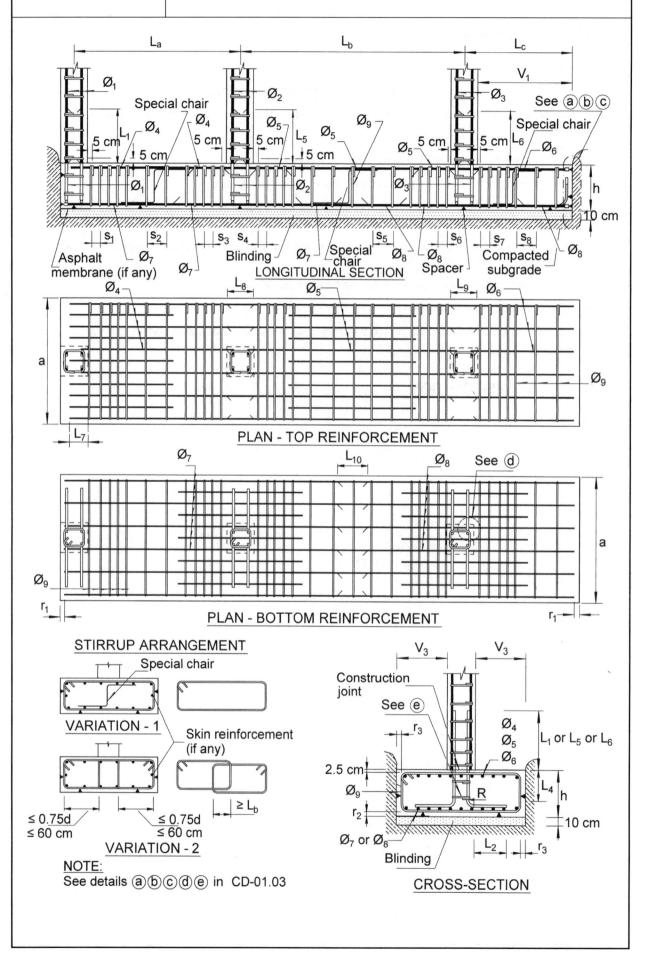

LONGITUDINAL SECTION

PLAN - TOP REINFORCEMENT

PLAN - BOTTOM REINFORCEMENT

STIRRUP ARRANGEMENT

VARIATION - 1

VARIATION - 2

NOTE:
See details ⓐⓑⓒⓓⓔ in CD-01.03

CROSS-SECTION

1. RECOMMENDATIONS

1. The ϕ_1, ϕ_2 and ϕ_3 bars rest directly on the top of the footing, with no hooks needed.

2. L_1 is the lap length in the ϕ_1 bar, L_5 is the lap length in the ϕ_2 bar and L_6 is the lap length in the ϕ_3 bar.

3. The starter bars are diameter ϕ_1 bars, except where $L_4 < L_1$. If $L_4 < L_1$, often a more suitable solution than increasing the depth of the footing is to use two starter bars for every ϕ_1 bar in the column. The sum of the cross-sections of these two bars should not be less than the ϕ_1 bar cross-section, but their diameter should be such that $L_b < L_4$, where L_b is the anchorage length. The arrangement detail is as in alternative (d). For the inner columns, see CD – 01.03.

4. L_8 is the lap length in the larger of the ϕ_4 or ϕ_5 bars, L_9 in the larger of the ϕ_5 or ϕ_6 bars and L_{10} in the larger of the ϕ_7 or ϕ_8 bars.

5. r_1 and r_3 are equal to 7.5 cm if the concrete for the footing is poured directly against the soil.

6. $r_2 = 2.5 \text{ cm} \geq \phi_9$.

7. See 1.5 to determine when to use anchor type a, b or c as end anchors.

8. Length L_2 shall suffice to tie ϕ_1, ϕ_2 or ϕ_3 starter bars to two ϕ_7 or ϕ_8 transverse bars. (It may not be less than 2s, where s is the space between ϕ_7 or ϕ_8 bars.)

9. The edge column starter bars need ties of the same diameter and maximum spacing as in the column. For the inner column starter bars, see CD – 01.03.

10. The top of the footing is smoothed with a float or power float except in the contact area with the future column, where a rough surface such as is generated by the vibrator is needed.

11. In soft soils, the 25 cm over the level of the future blinding should not be excavated until shortly before pouring the concrete to ensure that the soil is not softened by rainfall immediately before placement.

12. The subgrade is compacted before pouring the blinding for the footing and beam.

13. The blinding is floated or smoothed with a power float. Its standard 10-cm thickness may be varied to absorb tolerances in foundation subgrade levelling.

14. Where the footing is to be poured on soils that constitute an aggressive medium, it should be protected by an asphalt membrane as specified.

15. Before pouring the column concrete, the joint surface should be cleaned and pressure hosed. The concrete should not be cast until the surface dries.

16. See 1.2 and 1.3 for descriptions of how to tie bars and place spacers.

2. STATUTORY LEGISLATION

This type of footing is not addressed in EC2 (5).

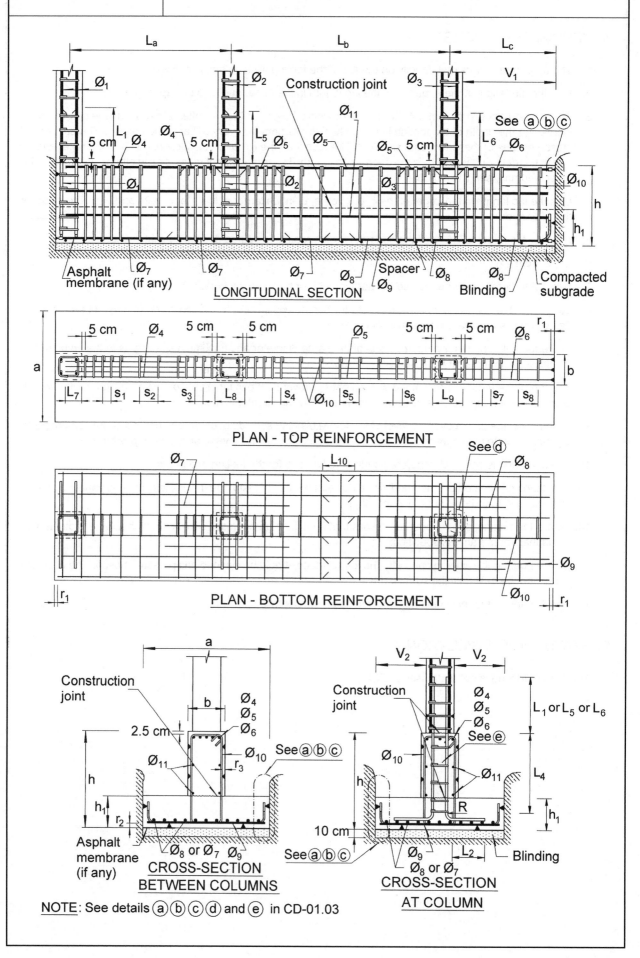

LONGITUDINAL SECTION

PLAN - TOP REINFORCEMENT

PLAN - BOTTOM REINFORCEMENT

CROSS-SECTION BETWEEN COLUMNS

CROSS-SECTION AT COLUMN

NOTE: See details ⓐⓑⓒⓓ and ⓔ in CD-01.03

1. RECOMMENDATIONS

1. The ϕ_1, ϕ_2 and ϕ_3 bars rest directly on the top of the footing, with no hooks needed.

2. L_1 is the lap length in the ϕ_1 bar, L_5 is the lap length in the ϕ_2 bar and L_6 is the lap length in the ϕ_3 bar.

3. The starter bars are diameter ϕ_1 bars, except where $L_4 < L1$. If $L_4 < L_1$, often a more suitable solution than increasing the depth of the footing is to use two starter bars for every ϕ_1 bar in the column. The sum of the cross-sections of these two bars should not be less than the ϕ_1 bar cross-section, but their diameter should be such that $L_b < L_4$, where L_b is the anchorage length. The arrangement detail is as in alternative (d). For the inner columns, see CD − 01.03. (If the side cover on the web in the ϕ_2 bars is greater than or equal to ten times their diameter, or more than 10 cm, the condition is $L_4 > \dfrac{2}{3} L_5$. If it is less, $L_4 > L_5$.) The same applies to ϕ_3.

4. L_8 is the lap length in the larger of the ϕ_4 or ϕ_5 bars, L_9 is the lap length in the larger of the ϕ_5 or ϕ_6 bars and L_{10} is the lap length in the larger of the ϕ_7 or ϕ_8 bars. (Anchorages need not be used if the commercial length of the bars suffices.)

5. $r_1 = 7.5$ cm if the concrete for the footing is poured directly against the soil.

6. $r_2 = 2.5$ cm but not less than ϕ_9; $r_3 = 2.5$ cm but not less than ϕ_{10}.

7. See 1.5 to determine when to use anchor type a, b or c as end anchors.

8. Length L_2 shall suffice to tie ϕ_1, ϕ_2 or ϕ_3 starter bars to two ϕ_7 or ϕ_8 transverse bars. (It may not be less than 2s, where s is the space between ϕ_7 or ϕ_8 bars.)

9. The edge column starter bars need ties of the same diameter and maximum spacing as in the column. The starter bars on the inner columns need ties of the same diameter and maximum spacing as in the column if the side cover over the T-section web is less than ten times their diameter, or less than 10 cm. Otherwise two ties suffice (see CD − 01.03) across depth h_1 and two along the height of the web.

10. The top of the footing is smoothed with a float or power float except in the contact area with the future column, where a rough surface such as is generated by the vibrator is needed. The flange-to-web joint is treated similarly.

11. In soft soils, the 25 cm over the level of the future blinding should not be excavated until shortly before pouring the concrete to ensure that the soil is not softened by rainfall immediately before placement.

12. The subgrade is compacted before pouring the blinding.

13. The blinding is floated or smoothed with a power float. Its standard 10-cm thickness may be varied to absorb tolerances in foundation subgrade levelling.

14. Where the footing is to be poured on soils that constitute an aggressive medium, it should be protected by an asphalt membrane as specified.

15. Before pouring the column concrete, the joint surface should be cleaned and pressure hosed. The concrete should not be cast until the surface dries. The flange-to-web joint should be treated similarly.

16. See 1.2 and 1.3 for descriptions of how to tie bars and place spacers.

2. STATUTORY LEGISLATION

This type of footing is not addressed in EC2 (5).

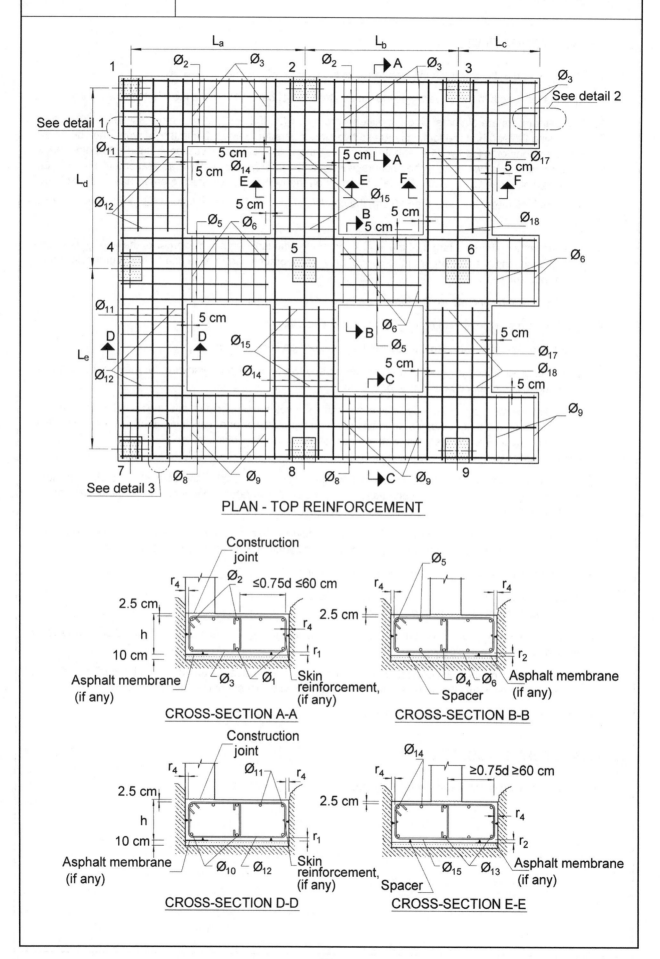

PLAN - TOP REINFORCEMENT

CROSS-SECTION A-A

CROSS-SECTION B-B

CROSS-SECTION D-D

CROSS-SECTION E-E

1. RECOMMENDATIONS

1. $r_1 = 2.5$ cm $\geq \phi_1 - \phi_3$.

2. $r_2 = 2.5$ cm.

3. r_4 and r_5 are equal to 7.5 cm if the concrete for the footing is poured directly against the soil.

4. The ϕ_{10}, ϕ_{13} and ϕ_{16} bars lie on the ϕ_1, ϕ_4 and ϕ_7 bars, respectively.

5. The ϕ_{11}, ϕ_{14} and ϕ_{17} bars lie on the ϕ_2, ϕ_5 and ϕ_8 bars, respectively.

6. If the shear stress cannot be entirely absorbed by the concrete, the stirrups must be arranged in such a way that their vertical legs are not separated by more than 60 cm or 0.75d. Cross-ties, as shown, or multiple stirrups (see CD – 01.17) may be used.

7. If multiple stirrups are used, the horizontal legs must be lapped along a distance of at least L_b, where L_b is the stirrup anchorage length. (This is required for the horizontal arm of the stirrup to act as transverse bending reinforcement; see CD – 01.17.)

8. The corner column starter bars need ties of the same diameter and maximum spacing as in the column. Two stirrups suffice for the inner column starter bars (see CD – 01.03).

9. The top of the footing is smoothed with a float or power float except in the contact area with the future column, where a rough surface such as is generated by the vibrator is needed.

10. In soft soils, the 25 cm over the level of the future blinding should not be excavated until shortly before pouring the concrete to ensure that the soil is not softened by rainfall immediately before placement.

11. The subgrade is compacted before pouring the blinding.

12. The blinding is floated or smoothed with a power float. Its standard 10-cm thickness may be varied to absorb tolerances in foundation subgrade levelling.

13. Where the footing is to be poured on soils that constitute an aggressive medium, it should be protected by an asphalt membrane as specified.

14. Before pouring the column concrete, the joint surface should be cleaned and pressure hosed. The concrete should not be cast until the surface dries.

15. See 1.2 and 1.3 for descriptions of how to tie bars and place spacers.

2. STATUTORY LEGISLATION

This type of footing is not addressed in EC2 (5).

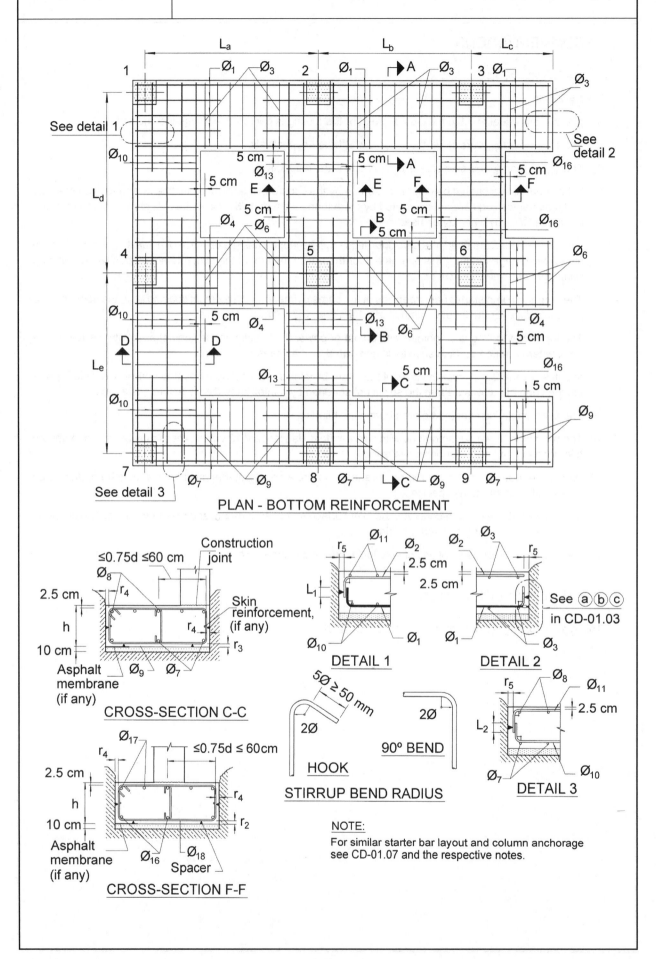

PLAN - BOTTOM REINFORCEMENT

CROSS-SECTION C-C

DETAIL 1

DETAIL 2

DETAIL 3

CROSS-SECTION F-F

HOOK

90° BEND

STIRRUP BEND RADIUS

NOTE:

For similar starter bar layout and column anchorage see CD-01.07 and the respective notes.

1. RECOMMENDATIONS

1. $r_1 = 2.5$ cm $\geq \phi_1 - \phi_3$.

2. $r_2 = 2.5$ cm.

3. r_4 and r_5 are equal to 7.5 cm if the concrete for the footing is poured directly against the soil.

4. The ϕ_{10}, ϕ_{13} and ϕ_{16} bars lie on the ϕ_1, ϕ_4 and ϕ_7 bars, respectively.

5. The ϕ_{11}, ϕ_{14} and ϕ_{17} bars lie on the ϕ_2, ϕ_5 and ϕ_8 bars, respectively.

6. If the shear stress cannot be entirely absorbed by the concrete, the stirrups must be arranged in such a way that their vertical legs are not separated by more than 60 cm or 0.75d. Cross-ties, as shown, or multiple stirrups (see CD – 01.17) may be used.

7. If multiple stirrups are used, the horizontal legs must be lapped along a distance of at least L_b, where L_b is the stirrup anchorage length. (This is required for the horizontal arm of the stirrup to act as transverse bending reinforcement; see CD – 01.17.)

8. The corner column starter bars need ties of the same diameter and maximum spacing as in the column. Two stirrups suffice for the inner column starter bars (see CD – 01.03).

9. The top of the footing is smoothed with a float or power float except in the contact area with the future column, where a rough surface such as is generated by the vibrator is needed.

10. In soft soils, the 25 cm over the level of the future blinding should not be excavated until shortly before pouring the concrete to ensure that the soil is not softened by rainfall immediately before placement.

11. The subgrade is compacted before pouring the blinding.

12. The blinding is floated or smoothed with a power float. Its standard 10-cm thickness may be varied to absorb tolerances in foundation subgrade levelling.

13. Where the footing is to be poured on soils that constitute an aggressive medium, it should be protected by an asphalt membrane as specified.

14. Before pouring the column concrete, the joint surface should be cleaned and pressure hosed. The concrete should not be cast until the surface dries.

15. See 1.2 and 1.3 for descriptions of how to tie bars and place spacers.

2. STATUTORY LEGISLATION

This type of footing is not addressed in EC2 (5).

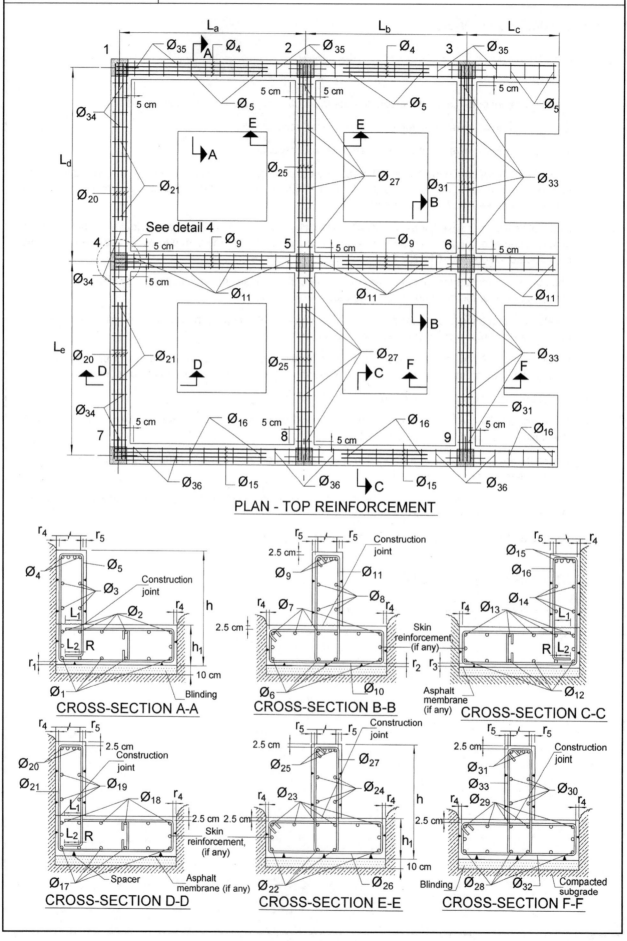

PLAN - TOP REINFORCEMENT

CROSS-SECTION A-A

CROSS-SECTION B-B

CROSS-SECTION C-C

CROSS-SECTION D-D

CROSS-SECTION E-E

CROSS-SECTION F-F

1. RECOMMENDATIONS

1. $r_1 = 2.5$ cm.

2. $r_2 = 2.5$ cm.

3. $r_3 = 2.5$ cm.

4. r_4 and r_6 are equal to 7.5 cm if the concrete for the footing is poured directly against the soil.

5. $r_5 = 2.5$ cm.

6. The ϕ_{17}, ϕ_{22} and ϕ_{28} bars lie on the ϕ_1, ϕ_6 and ϕ_{12} bars, respectively.

7. The ϕ_{18}, ϕ_{23} and ϕ_{29} bars lie on the ϕ_2, ϕ_7 and ϕ_{13} bars, respectively.

8. The ϕ_{20}, ϕ_{25} and ϕ_{31} bars lie on the ϕ_4, ϕ_9 and ϕ_{15} bars, respectively.

9. $L_2 = 20$ cm.

10. The corner and edge column starter bars need ties of the same diameter and maximum spacing as in the column. The starter bars on the inner columns need ties of the same diameter and maximum spacing as in the column if the side cover over the T-section web is less than ten times their diameter, or less than 10 cm. Otherwise two ties suffice (see CD – 01.03) across depth h_1 and two along the height of the web.

11. The top of the footing is smoothed with a float or power float except in the contact area with the future column, where a rough surface such as is generated by the vibrator is needed.

12. In soft soils, the 25 cm over the level of the future blinding should not be excavated until shortly before pouring the concrete to ensure that the soil is not softened by rainfall immediately before placement.

13. The subgrade is compacted before pouring the blinding.

14. The blinding is floated or smoothed with a power float. Its standard 10-cm thickness may be varied to absorb tolerances in foundation subgrade levelling.

15. Where the footing is to be poured on soils that constitute an aggressive medium, it should be protected by an asphalt membrane as specified.

16. Before pouring the column concrete, the joint surface should be cleaned and pressure hosed. The concrete should not be cast until the surface dries. The flange-to-web joint should be treated similarly

17. See 1.2 and 1.3 for descriptions of how to tie bars and place spacers.

2. STATUTORY LEGISLATION

This type of footing is not addressed in EC2 (5).

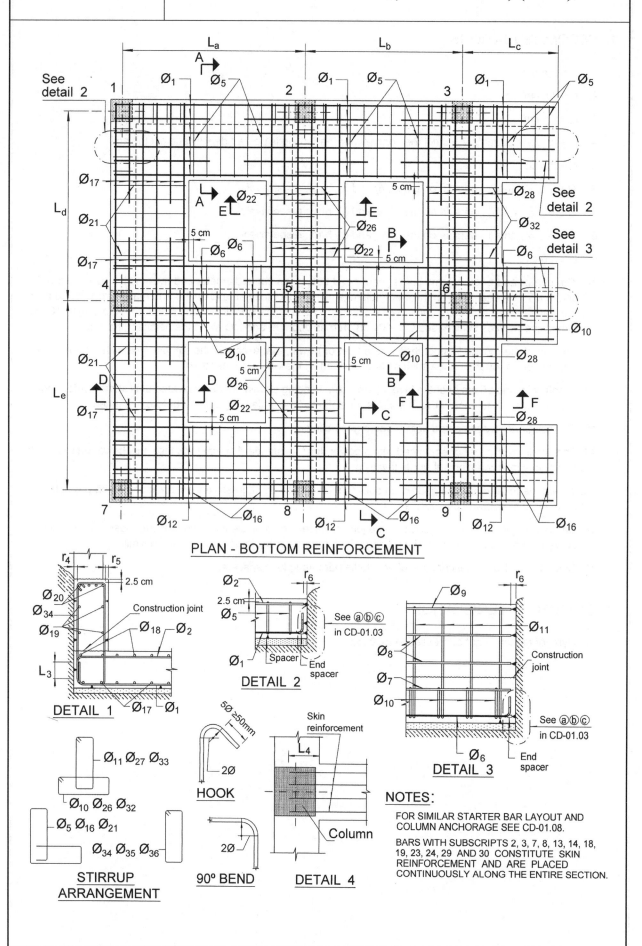

PLAN - BOTTOM REINFORCEMENT

DETAIL 1

DETAIL 2

DETAIL 3

STIRRUP ARRANGEMENT

HOOK

90° BEND

DETAIL 4

NOTES:

FOR SIMILAR STARTER BAR LAYOUT AND COLUMN ANCHORAGE SEE CD-01.08.

BARS WITH SUBSCRIPTS 2, 3, 7, 8, 13, 14, 18, 19, 23, 24, 29 AND 30 CONSTITUTE SKIN REINFORCEMENT AND ARE PLACED CONTINUOUSLY ALONG THE ENTIRE SECTION.

1. RECOMMENDATIONS

1. $r_1 = 2.5$ cm.

2. $r_2 = 2.5$ cm.

3. $r_3 = 2.5$ cm.

4. r_4 and r_6 are equal to 7.5 cm if the concrete for the footing is poured directly against the soil.

5. $r_5 = 2.5$ cm.

6. The ϕ_{17}, ϕ_{22} and ϕ_{28} bars lie on the ϕ_1, ϕ_6 and ϕ_{12} bars, respectively.

7. The ϕ_{18}, ϕ_{23} and ϕ_{29} bars lie on the ϕ_2, ϕ_7 and ϕ_{13} bars, respectively.

8. The ϕ_{20}, ϕ_{25} and ϕ_{31} bars lie on the ϕ_4, ϕ_9 and ϕ_{15} bars, respectively.

9. $L_2 = 20$ cm.

10. The corner and edge column starter bars need ties of the same diameter and maximum spacing as in the column. The starter bars on the inner columns need ties of the same diameter and maximum spacing as in the column if the side cover over the T-section web is less than ten times their diameter, or less than 10 cm. Otherwise two ties suffice (see CD – 01.03) across depth h_1 and two along the height of the web.

11. The top of the footing is smoothed with a float or power float except in the contact area with the future column, where a rough surface such as is generated by the vibrator is needed.

12. In soft soils, the 25 cm over the level of the future blinding should not be excavated until shortly before pouring the concrete to ensure that the soil is not softened by rainfall immediately before placement.

13. The subgrade is compacted before pouring the blinding.

14. The blinding is floated or smoothed with a power float. Its standard 10-cm thickness may be varied to absorb tolerances in foundation subgrade levelling.

15. Where the footing is to be poured on soils that constitute an aggressive medium, it should be protected by an asphalt membrane as specified.

16. Before pouring the column concrete, the joint surface should be cleaned and pressure hosed. The concrete should not be cast until the surface dries. The flange-to-web joint should be treated similarly.

17. See 1.2 and 1.3 for descriptions of how to tie bars and place spacers.

2. STATUTORY LEGISLATION

This type of footing is not addressed in EC2 (5).

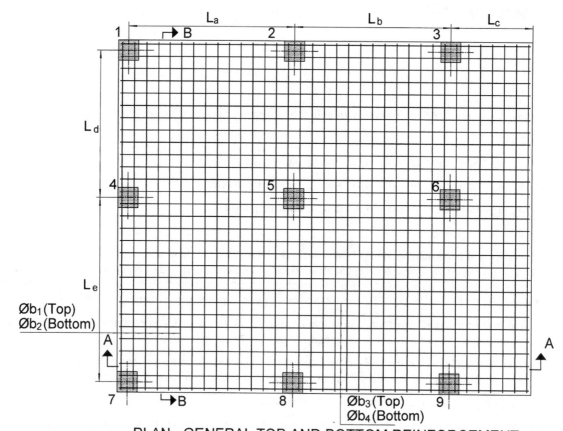

PLAN - GENERAL TOP AND BOTTOM REINFORCEMENT

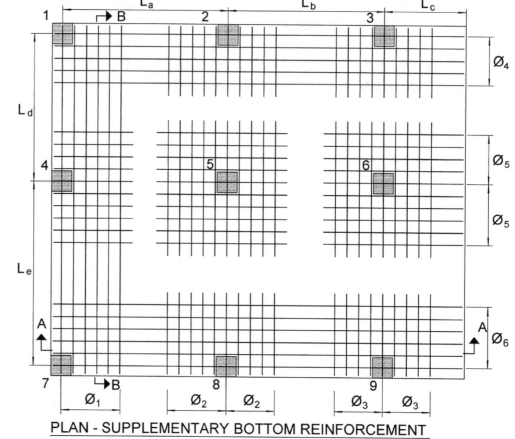

PLAN - SUPPLEMENTARY BOTTOM REINFORCEMENT

1. RECOMMENDATIONS

1. r_1 = 7.5 cm if the concrete for the footing is poured directly against the soil.

2. r_2 = 2.5 cm. (The cover over the main reinforcement should not be smaller than its diameter.)

3. L_1 is the lap length in the larger of the ϕ_{b3} or ϕ_{b4} bars.

4. L_2 is the lap length in the larger of the ϕ_{b1} or ϕ_{b2} bars.

5. The corner and edge column starter bars need ties of the same diameter and maximum spacing as in the column. Two stirrups suffice for the inner column starter bars (see CD – 01.03).

6. The top of the footing is smoothed with a float or power float except in the contact area with the future column, where a rough surface such as is generated by the vibrator is needed.

7. In soft soils, the 25 cm over the level of the future blinding should not be excavated until shortly before pouring the concrete to ensure that the soil is not softened by rainfall immediately before placement.

8. The subgrade is compacted before pouring the blinding.

9. The blinding is floated or smoothed with a power float. Its standard 10-cm thickness may be varied to absorb tolerances in foundation subgrade levelling.

10. Where the footing is to be poured on soils that constitute an aggressive medium, it should be protected by an asphalt membrane as specified.

11. Before pouring the column concrete, the joint surface should be cleaned and pressure hosed. The concrete should not be cast until the surface dries.

12. See 1.2 and 1.3 for descriptions of how to tie bars and place spacers.

13. Placement of a polyethylene membrane over the blinding may be advisable to prevent thermal contraction and shrinkage-induced tensile stress generated by the bond between the slab and the blinding concrete.

2. STATUTORY LEGISLATION

This type of footing is not addressed in EC2 (5).

3. SPECIFIC REFERENCES

See (18), (20) and (23).

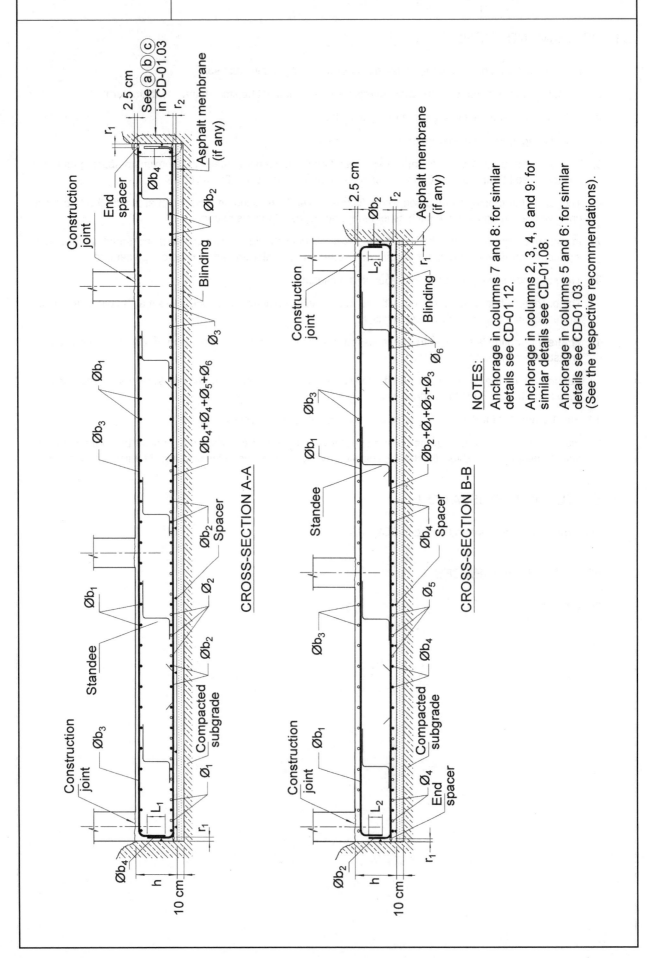

CROSS-SECTION A-A

CROSS-SECTION B-B

NOTES:

Anchorage in columns 7 and 8: for similar details see CD-01.12.

Anchorage in columns 2, 3, 4, 8 and 9: for similar details see CD-01.08.

Anchorage in columns 5 and 6: for similar details see CD-01.03.
(See the respective recommendations).

1. RECOMMENDATIONS

1. $r_1 = 7.5$ cm if the concrete for the footing is poured directly against the soil.

2. $r_2 = 2.5$ cm. (The cover over the main reinforcement should not be smaller than its diameter.)

3. L_1 is the lap length in the larger of the ϕ_{b3} or ϕ_{b4} bars.

4. L_2 is the lap length in the larger of the ϕ_{b1} or ϕ_{b2} bars.

5. The corner and edge column starter bars need ties of the same diameter and maximum spacing as in the column. Two stirrups suffice for the inner column starter bars (see CD – 01.03).

6. The top of the footing is smoothed with a float or power float except in the contact area with the future column, where a rough surface such as is generated by the vibrator is needed.

7. In soft soils, the 25 cm over the level of the future blinding should not be excavated until shortly before pouring the concrete to ensure that the soil is not softened by rainfall immediately before placement.

8. The subgrade is compacted before pouring the blinding.

9. The blinding is floated or smoothed with a power float. Its standard 10-cm thickness may be varied to absorb tolerances in foundation subgrade levelling.

10. Where the footing is to be poured on soils that constitute an aggressive medium, it should be protected by an asphalt membrane as specified.

11. Before pouring the column concrete, the joint surface should be cleaned and pressure hosed. The concrete should not be cast until the surface dries.

12. See 1.2 and 1.3 for descriptions of how to tie bars and place spacers.

13. Placement of a polyethylene membrane over the blinding may be advisable to prevent thermal contraction and shrinkage-induced tensile stress generated by the bond between the slab and the blinding concrete.

2. STATUTORY LEGISLATION

This type of footing is not addressed in EC2 (5).

3. SPECIFIC REFERENCES

See (18), (20) and (23).

CAISSON

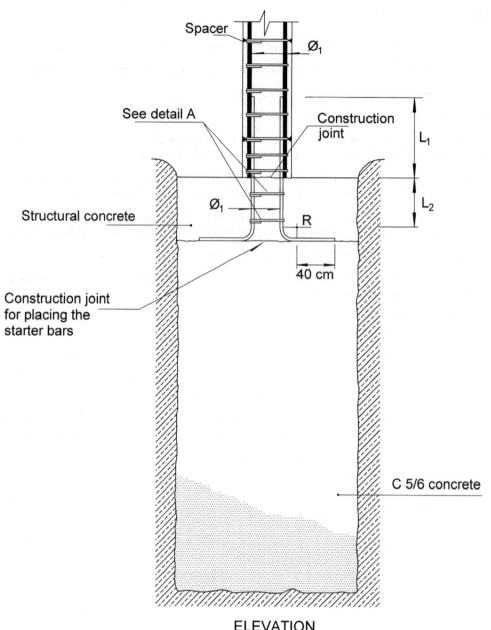

ELEVATION

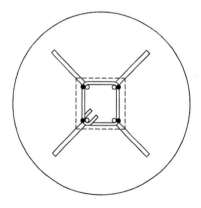

DETAIL A

1. RECOMMENDATIONS

1. The ϕ_1 bars in the column rest directly on the top of the foundation shaft.

2. L_1 is the ϕ_1 bar lap length.

3. L_2 is the ϕ_1 bar anchorage length.

4. The starter bar ties are supports whose sole purpose is to keep the starter bar assembly firmly in place while the concrete is cast in the foundations. These are not the same ties as in the column.

5. The ϕ_1 diameter starter bars should be positioned as shown in Detail A. Stability should be monitored during concrete pouring.

6. In soft soils, the bottom 35 cm should not be excavated until shortly before placing the concrete to ensure that the soil is not softened by rainfall immediately before pouring the blinding.

7. The foundation shaft subgrade must be compacted before the blinding is poured.

8. The top surface of the C 5/6 concrete should not be smoothed. (The last layer should be vibrated.)

9. The top surface of the foundation is smoothed with a float or power float except in the contact area with the future column, where a rough surface such as is generated by the vibrator is needed.

10. Before pouring the column concrete, the joint surface should be cleaned and pressure hosed. The concrete should not be cast until it dries.

11. See 1.2 and 1.3 for descriptions of how to tie bars and place spacers.

12. Where calculations are very tight, the mid plane in pairs of lapped bars should be perpendicular to the direction of the greatest bending moment acting on the column springing.

2. STATUTORY LEGISLATION

None, to the author's knowledge.

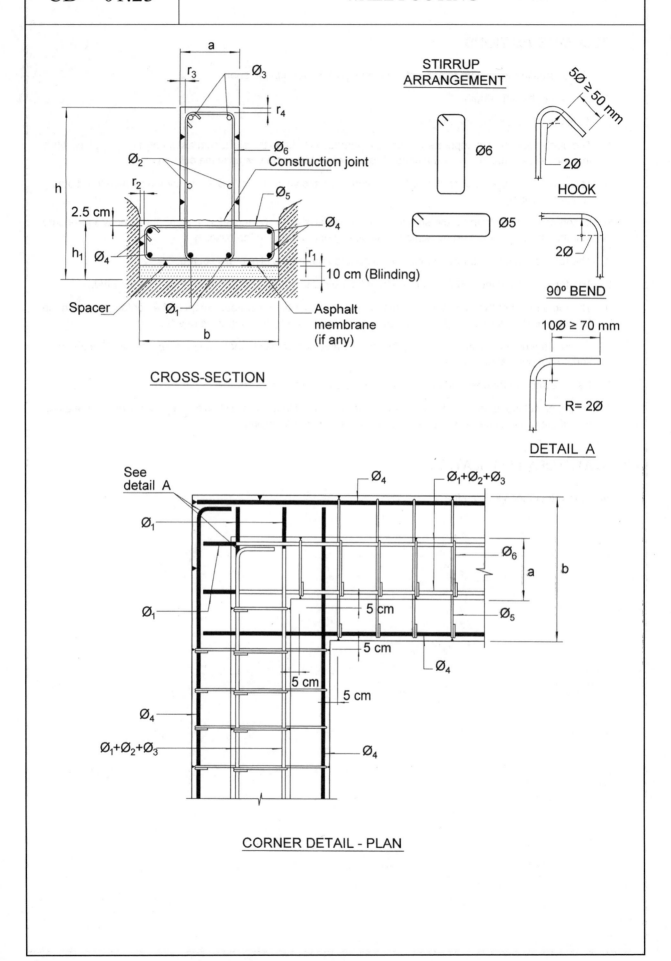

CD – 01.23

WALL FOOTING

CROSS-SECTION

a

r_3

$\varnothing_3$

r_4

$\varnothing_6$

Construction joint

$\varnothing_2$

$\varnothing_5$

r_2

$\varnothing_4$

h

2.5 cm

h_1

$\varnothing_4$

r_1

10 cm (Blinding)

Spacer

$\varnothing_1$

Asphalt membrane (if any)

b

STIRRUP ARRANGEMENT

$\varnothing_6$

$\varnothing_5$

$5\varnothing \geq 50$ mm

$2\varnothing$

HOOK

$2\varnothing$

90º BEND

$10\varnothing \geq 70$ mm

R= $2\varnothing$

DETAIL A

CORNER DETAIL - PLAN

See detail A

$\varnothing_1$

$\varnothing_1$

$\varnothing_4$

$\varnothing_1+\varnothing_2+\varnothing_3$

$\varnothing_6$

a

b

5 cm

$\varnothing_5$

5 cm

$\varnothing_4$

5 cm

5 cm

$\varnothing_4$

$\varnothing_1+\varnothing_2+\varnothing_3$

$\varnothing_4$

1. RECOMMENDATIONS

1. $r_1 = 2.5$ cm.

2. $r_2 = 7.5$ cm if the concrete for the footing is poured directly against the soil.

3. $r_3 = 2.5$ cm.

4. $r_4 = 2.5$ cm.

5. The top of the stem and base are smoothed with a float or power float except at the joint, where a rough surface such as is generated by the vibrator is needed.

6. In soft soils, the 25 cm over the level of the future blinding should not be excavated until shortly before pouring the concrete to ensure that the soil is not softened by rainfall immediately before placement.

7. The foundation subgrade must be compacted before the blinding is poured.

8. The blinding is floated or smoothed with a power float. Its standard 10-cm thickness may be varied to absorb tolerances in foundation subgrade levelling.

9. Where the footing is to be poured on soils that constitute an aggressive medium, it should be protected by an asphalt membrane as specified.

10. Before casting the concrete, the joint surface should be cleaned and pressure hosed. The concrete should not be cast until the surface dries.

11. See 1.2 and 1.3 for descriptions of how to tie bars and place spacers.

2. STATUTORY LEGISLATION

None in place.

CD – 01.24	BORED PILE

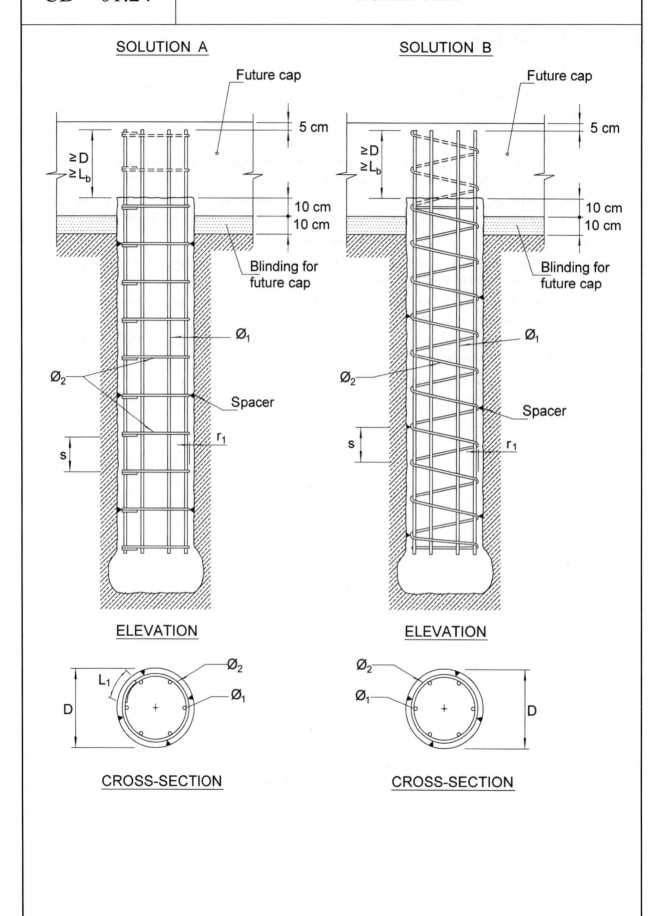

SOLUTION A

Future cap

≥D
≥L_b

5 cm

10 cm
10 cm

Blinding for future cap

Ø₁

Ø₂

Spacer

r₁

s

ELEVATION

Ø₂

L₁

D

Ø₁

+

CROSS-SECTION

SOLUTION B

Future cap

≥D
≥L_b

5 cm

10 cm
10 cm

Blinding for future cap

Ø₁

Ø₂

Spacer

r₁

s

ELEVATION

Ø₂

Ø₁

+

D

CROSS-SECTION

1. RECOMMENDATIONS

1. Cover r_1 is taken to be 7.5 cm. Under certain circumstances, European standard EN 1536 (24) allows smaller values for bored piles.

2. Transverse reinforcement shall be tied or welded at all intersections with the main reinforcing bars.

3. The reinforcement should protrude upward from the top of the pile after removal of the concrete above the cutoff level.

4. At least six ϕ_1 longitudinal bars should be used. (Five may exceptionally suffice in small diameter piles.)

5. $\phi_2 \geq \dfrac{1}{4} \phi_1$ ($\phi_1 \geq 16$ mm). (See EN 1536 ((24)), which allows smaller diameters.)

6. $s \leq 20 \phi_1$.

7. L_1 is the ϕ_2 bar lap length.

8. L_b is the ϕ_1 bar anchorage length.

9. The standard 15-cm blinding depth may be changed to absorb levelling tolerances at the bottom of the cap.

10. Wheel spacers should be used and attached to the transverse reinforcement with sufficient clearance to turn.

2. STATUTORY LEGISLATION

EC2 (5), EN 1536 (24).

PILE CAP

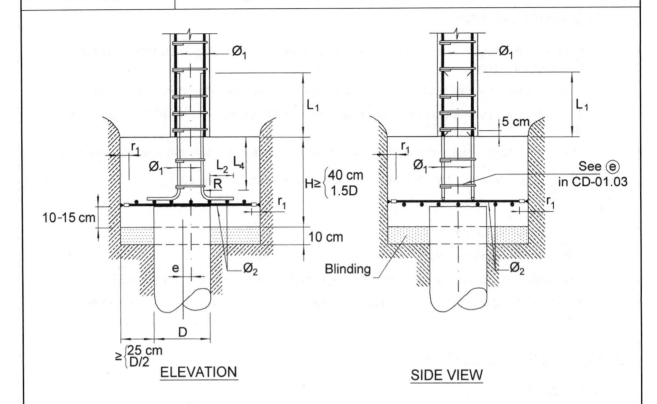

ELEVATION

SIDE VIEW

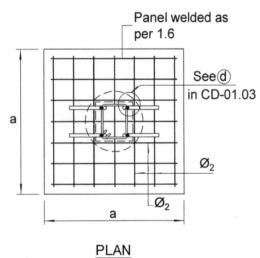

PLAN

AR. Maximum eccentricities

AR. e_{max} ⎰ 5 cm (Intense site supervision)
10 cm (Standard site supervision)
15 cm (Limited site supervision)

(e_{max} is the maximum eccentricity in any direction)

NOTE: See CD-01.30

1. RECOMMENDATIONS

1. In the following notes, caps are always assumed to be cast into forms.

2. The reinforcement panel rests on the pile surface at the cutoff level.

3. $r_1 = 2.5$ cm.

4. Length L_2 shall suffice to tie the ϕ_1 starter bars securely to two ϕ_2 transverse bars. (It may not be less than 2s, where s is the space between ϕ_2 bars.)

5. The starter bar ties are supports whose sole purpose is to keep the starter bar assembly firmly in place while the concrete is cast in the cap. The starter bar assembly is to be tied at all cross points. These are not the same ties as in the column.

6. The top of the cap is smoothed with a float or power float except in the contact area with the future column, where a rough surface such as is generated by the vibrator is needed.

7. In soft soils, the 25 cm over the level of the future blinding should not be excavated until shortly before pouring the concrete to ensure that the soil is not softened by rainfall immediately before placement.

8. The cap subgrade must be compacted before the blinding is poured.

9. The blinding is floated or smoothed with a power float. Its standard 15-cm thickness may be varied to absorb tolerances in foundation subgrade levelling.

10. Before casting the concrete, the joint surface should be cleaned and pressure hosed. The concrete should not be cast until the surface dries.

11. See 1.2 and 1.3 for descriptions of how to tie bars and place spacers.

2. STATUTORY LEGISLATION

EC2 (5), EN 1536 (24).

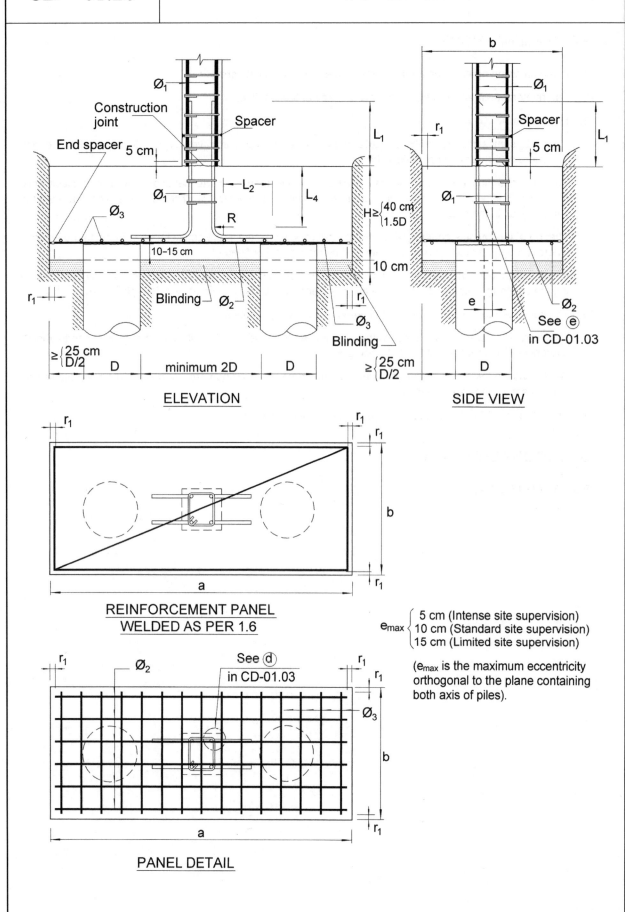

ELEVATION

SIDE VIEW

REINFORCEMENT PANEL
WELDED AS PER 1.6

$$e_{max} \begin{cases} 5 \text{ cm (Intense site supervision)} \\ 10 \text{ cm (Standard site supervision)} \\ 15 \text{ cm (Limited site supervision)} \end{cases}$$

(e_{max} is the maximum eccentricity orthogonal to the plane containing both axis of piles).

PANEL DETAIL

1. RECOMMENDATIONS

1. In the following notes, caps are always assumed to be cast into forms.

2. The reinforcement lattice rests on the pile surface at the cutoff level.

3. **AR.** The starter bars are diameter ϕ_1 bars. L_4 should be $\geq \ell_{bd}$, but length ℓ_{bd} is defined in EC2 for the least favourable case. In this case the side covers for the ϕ_1 bars are very large. See (12), p. 69. A safe value would be $\ell'_{bd} = \frac{2}{3}\ell_{bd}$. If $L_4 < \frac{2}{3}\ell_{bd}$, often a more suitable solution than increasing the depth of the cap is to use two starter bars for every ϕ_1 bar in the wall. The sum of the cross-sections of these two bars should not be less than the ϕ_1 bar cross-section, but their diameter should be such that $\frac{2}{3}\ell_{bd} < L_4$, where ℓ_{bd} is the anchorage length. The arrangement detail is as in alternative (d). This rule can be applied wherever the bar cover is $\geq 10\,\phi$ but not under 10 cm.

4. Length L_2 shall suffice to tie the ϕ_1 starter bars securely to two ϕ_3 transverse bars. (It may not be less than 2s, where s is the space between ϕ_3 bars.)

5. The starter bar ties are supports whose sole purpose is to keep the starter bar assembly firmly in place while the concrete is cast in the cap. The starter bar assembly is to be tied at all cross points. These are not the same ties as in the column.

6. The top of the cap is smoothed with a float or power float except in the contact area with the future column, where a rough surface such as is generated by the vibrator is needed.

7. In soft soils, the 25 cm over the level of the future blinding should not be excavated until shortly before pouring the concrete to ensure that the soil is not softened by rainfall immediately before placement.

8. The cap subgrade must be compacted before the blinding is poured.

9. The blinding is floated or smoothed with a power float. Its standard 15-cm thickness may be varied to absorb tolerances in foundation subgrade levelling.

10. Before casting the concrete, the joint surface should be cleaned and pressure hosed. The concrete should not be cast until the surface dries.

11. See 1.2 and 1.3 for descriptions of how to tie bars and place spacers.

12. **AR.** In large caps, top and side reinforcing grids may be advisable to prevent shrinkage and thermal contraction cracking.

2. STATUTORY LEGISLATION

See EC2 (5), EN 1536 (24).

THREE-PILE CAP

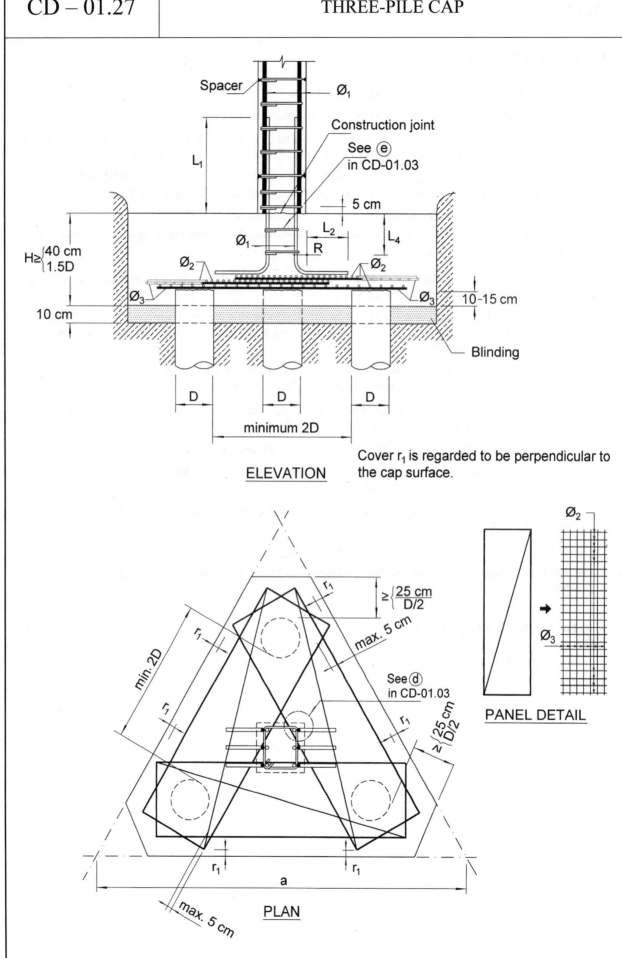

ELEVATION

Cover r_1 is regarded to be perpendicular to the cap surface.

PLAN

PANEL DETAIL

1. RECOMMENDATIONS

1. In the following notes, caps are always assumed to be cast into forms.

2. The reinforcement panels rest on the pile surface at the cutoff level.

3. $r_1 = 2.5$ cm.

4. **AR.** The starter bars are diameter ϕ_1 bars. L_4 should be $\geq \ell_{bd}$, but length ℓ_{bd} is defined in EC-2 for the least favourable case. In this case the side covers for the ϕ_1 bars are very large. See (12), p. 69. A safe value would be $\ell'_{bd} = \frac{2}{3}\ell_{bd}$. If $L_4 < \frac{2}{3}\ell_{bd}$, often a more suitable solution than increasing the depth of the footing is to use two starter bars for every ϕ_1 bar in the wall. The sum of the cross-sections of these two bars should not be less than the ϕ_1 bar cross-section, but their diameter should be such that $\frac{2}{3}\ell_{bd} < L_4$, where ℓ_{bd} is the anchorage length. The arrangement detail is as in alternative (d). This rule can be applied wherever the bar cover is $\geq 10\,\phi$ but not under 10 cm.

5. The L_2 length on the starter bars should suffice to tie them to two ϕ_2 reinforcing bars. (It may not be less than 2s, where s is the space between ϕ_2 bars.)

6. The starter bar ties are supports whose sole purpose is to keep the starter bar assembly firmly in place while the concrete is cast in the cap. The starter bar assembly is to be tied at all cross points. These are not the same ties as in the column.

7. The top of the cap is smoothed with a float or power float except in the contact area with the future column, where a rough surface such as is generated by the vibrator is needed.

8. In soft soils, the 25 cm over the level of the future blinding should not be excavated until shortly before pouring the concrete to ensure that the soil is not softened by rainfall immediately before placement.

9. The cap subgrade must be compacted before the blinding is poured.

10. The blinding is floated or smoothed with a power float. Its standard 15-cm thickness may be varied to absorb tolerances in foundation subgrade levelling.

11. Before casting the concrete, the joint surface should be cleaned and pressure hosed. The concrete should not be cast until the surface dries.

12. See 1.2 and 1.3 for descriptions of how to tie bars and place spacers.

13. **AR.** In large caps, top and side reinforcing grids may be advisable to prevent shrinkage and thermal contraction cracking.

2. STATUTORY LEGISLATION

See EC2 (5), EN 1536 (24).

FOUR-PILE CAP

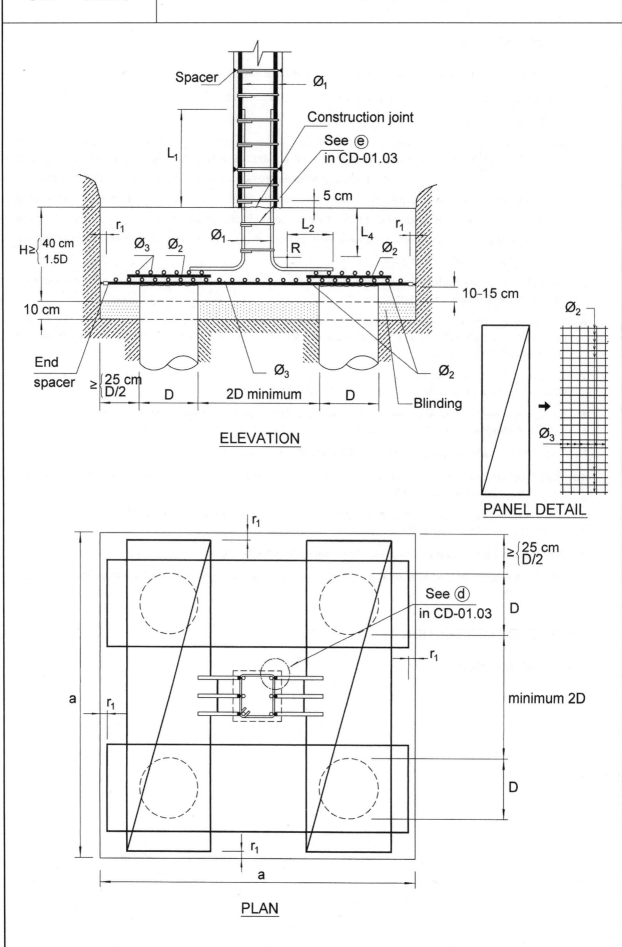

ELEVATION

PANEL DETAIL

PLAN

1. RECOMMENDATIONS

1. In the following notes, caps are always assumed to be cast into forms.

2. The reinforcement grid rests on the pile surface at the cutoff level.

3. $r_1 = 2.5$ cm.

4. **AR.** The starter bars are diameter ϕ_1 bars. L_4 should be $\geq \ell_{bd}$, but length ℓ_{bd} is defined in EC-2 for the least favourable case. In this case the side covers for the ϕ_1 bars are very large. See (12), p. 69. A safe value would be $\ell'_{bd} = \frac{2}{3}\ell_{bd}$. If $L_4 < \frac{2}{3}\ell_{bd}$, often a more suitable solution than increasing the depth of the footing is to use two starter bars for every ϕ_1 bar in the wall. The sum of the cross-sections of these two bars should not be less than the ϕ_1 bar cross-section, but their diameter should be such that $\frac{2}{3}\ell_{bd} < L_4$, where ℓ_{bd} is the anchorage length. The arrangement detail is as in alternative (d). This rule can be applied wherever the bar cover is $\geq 10\,\phi$ but not under 10 cm.

5. The L_2 length on the starter bars should suffice to tie them to two ϕ_2 reinforcing bars. (It may not be less than 2s, where s is the space between ϕ_2 bars.)

6. The starter bar ties are supports whose sole purpose is to keep the starter bar assembly firmly in place while the concrete is cast in the cap. The starter bar assembly is to be tied at all cross points. These are not the same ties as in the column.

7. The top of the cap is smoothed with a float or power float except in the contact area with the future column, where a rough surface such as is generated by the vibrator is needed.

8. In soft soils, the 25 cm over the level of the future blinding should not be excavated until shortly before pouring the concrete to ensure that the soil is not softened by rainfall immediately before placement.

9. The cap subgrade must be compacted before the blinding is poured.

10. The blinding is floated or smoothed with a power float. Its standard 15-cm thickness may be varied to absorb tolerances in foundation subgrade levelling.

11. Before casting the concrete, the joint surface should be cleaned and pressure hosed. The concrete should not be cast until the surface dries.

12. See 1.2 and 1.3 for descriptions of how to tie bars and place spacers.

13. **AR.** In large caps, top and side reinforcing may be advisable to prevent shrinkage and thermal contraction cracking.

2. STATUTORY LEGISLATION

See EC2 (5), EN 1536 (24).

GROUP PILE CAP (N > 4)

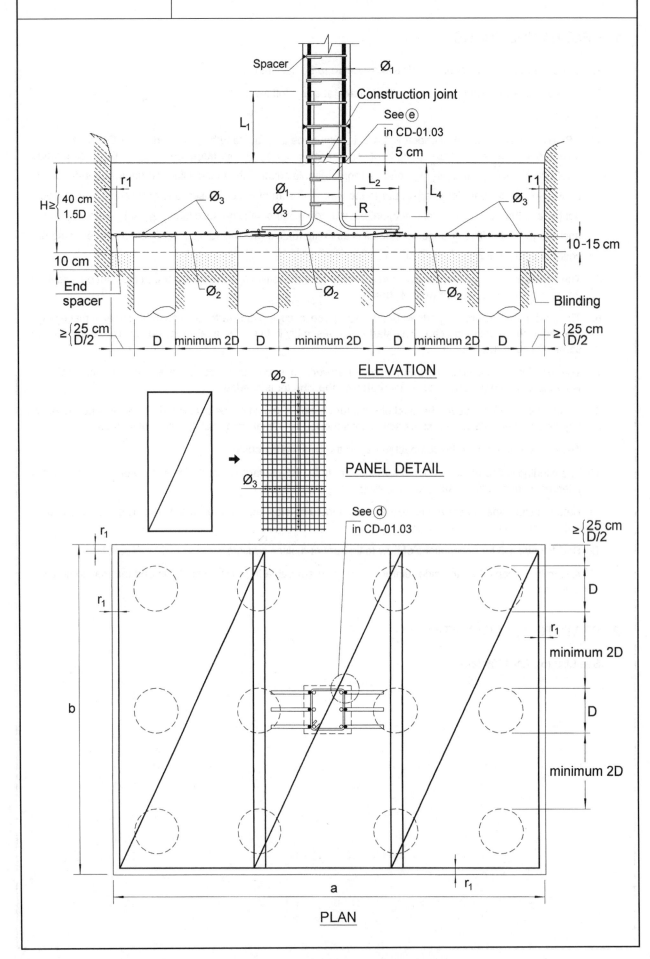

ELEVATION

PANEL DETAIL

PLAN

1. RECOMMENDATIONS

1. In the following notes, caps are always assumed to be cast into forms.

2. The reinforcement grid rests on the pile surface at the cutoff level.

3. $r_1 = 2.5$ cm.

4. **AR.** The starter bars are diameter ϕ_1 bars. L_4 should be $\geq \ell_{bd}$, but length ℓ_{bd} is defined in EC-2 for the least favourable case. In this case the side covers for the ϕ_1 bars are very large. See (12), p. 69. A safe value would be $\ell'_{bd} = \dfrac{2}{3}\ell_{bd}$. If $L_4 < \dfrac{2}{3}\ell_{bd}$, often a more suitable solution than increasing the depth of the footing is to use two starter bars for every ϕ_1 bar in the wall. The sum of the cross-sections of these two bars should not be less than the ϕ_1 bar cross-section, but their diameter should be such that $\dfrac{2}{3}\ell_{bd} < L_4$, where ℓ_{bd} is the anchorage length. The arrangement detail is as in alternative (d). This rule can be applied wherever the bar cover is $\geq 10\,\phi$ but not under 10 cm.

5. The L_2 length on the starter bars should suffice to tie them to two ϕ_3 reinforcing bars. (It may not be less than 2s, where s is the space between ϕ_3 bars.)

6. The starter bar ties are supports whose sole purpose is to keep the starter bar assembly firmly in place while the concrete is cast in the cap. The starter bar assembly is to be tied at all cross points. These are not the same ties as in the column.

7. The top of the cap is smoothed with a float or power float except in the contact area with the future column, where a rough surface such as is generated by the vibrator is needed.

8. In soft soils, the 25 cm over the level of the future blinding should not be excavated until shortly before pouring the concrete to ensure that the soil is not softened by rainfall immediately before placement.

9. The cap subgrade must be compacted before the blinding is poured.

10. The blinding is floated or smoothed with a power float. Its standard 15-cm thickness may be varied to absorb tolerances in foundation subgrade levelling.

11. Before casting the concrete, the joint surface should be cleaned and pressure hosed. The concrete should not be cast until the surface dries.

12. See 1.2 and 1.3 for descriptions of how to tie bars and place spacers.

13. In large caps, top and side reinforcing grids may be advisable to prevent shrinkage and thermal contraction cracking.

2. STATUTORY LEGISLATION

This type of pile is not addressed in EC2 (5).

CD – 01.30	CENTRING BEAM FOR ONE- OR TWO-PILE CAPS

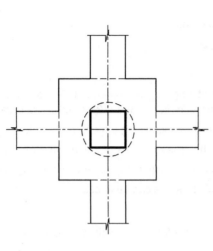

ONE-PILE CAP - PLAN

TWO-PILE CAP - PLAN

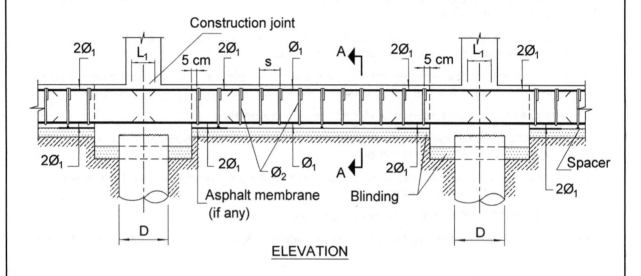

ELEVATION

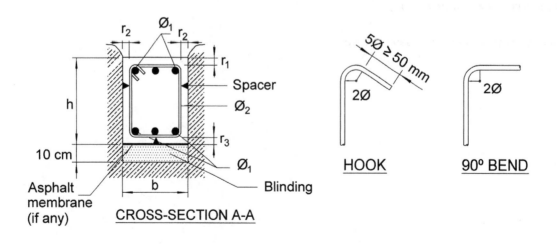

CROSS-SECTION A-A

HOOK

90° BEND

1. RECOMMENDATIONS

1. The beam is assumed to be cast in formwork.

2. $r_1 = 2.5$ cm.

3. $r_2 = 2.5$ cm.

4. $r_3 = 2.5$ cm.

5. L_1 is the ϕ_1 bar anchorage length.

6. The bottom of the excavation is compacted before pouring the blinding.

7. The top of the beam is smoothed with a float or power float.

8. The blinding is floated or smoothed with a power float. Its standard 10-cm thickness may be varied to absorb tolerances in foundation subgrade levelling.

9. Where the footing is to be poured on soils that constitute an aggressive medium, it should be protected by an asphalt membrane as specified.

10. See 1.2 and 1.3 for descriptions of how to tie bars and place spacers.

11. Care should be taken to avoid beam breakage when compacting fill over centring beams.

12. This beam is built to resist the action effects induced by accidental eccentricity. See CD – 01.25.

2. STATUTORY LEGISLATION

None in place.

3. RECOMMENDED ALTERNATIVE CODES

See (7).

4. SPECIFIC REFERENCES

See Section 12 in EC2 (5).

Group 02

**Retaining walls
and
basement walls**

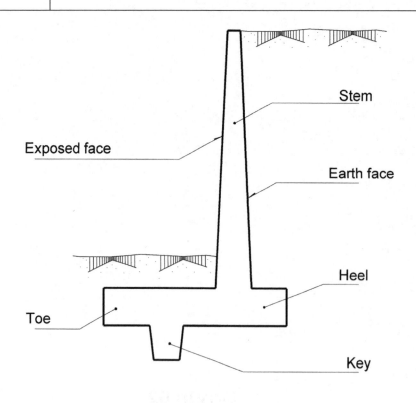

Stem

Exposed face

Earth face

Heel

Toe

Key

1. RECOMMENDATIONS

None.

2. LEGISLATION

EC2 makes no reference to this member.

It is mentioned in EC7 (26), but essentially in connection with geotechnical aspects.

3. SPECIFIC REFERENCES

See (23) and (27).

CANTILEVER RETAINING WALLS.
FOOTING

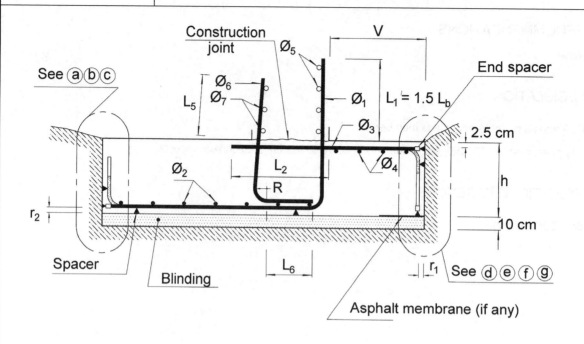

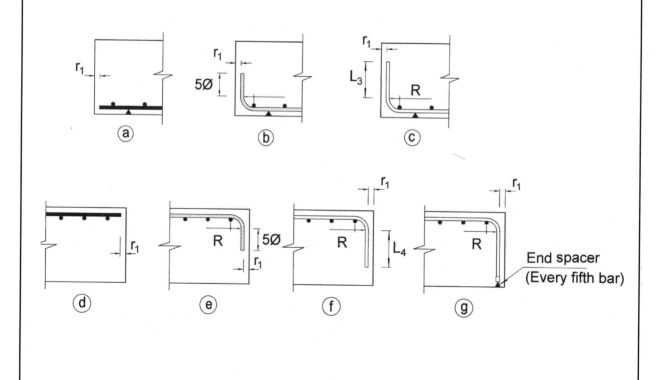

1. RECOMMENDATIONS

1. See 1.5 to determine when to use anchorage type a, b or c on the ϕ_1 reinforcing bars.

2. The use of anchor type d, e, f or g in ϕ_3 bars should be determined on the grounds of the anchorage calculations. See 1.5.

3. Lower values can be obtained applying table 8.3 F EC-2. Length L_1 is 2.0 times the ϕ_1 bar lap length (see EC2, 8.7.3).

4. L_5 is the ϕ_6 bar lap length.

5. $r_1 = 7.5$ cm if the concrete is poured directly against the soil.

6. $r_2 = 2.5$ cm but not smaller than ϕ_1.

7. If the ϕ_3 bar diameter is not large and cantilever V is not overly long, the bars may be tied to the ϕ_1 and ϕ_6 bars to hold them in place, in which case the bar spacing must be coordinated accordingly. Otherwise, special chairs must be used or solution (g) must be implemented.

8. To keep the ϕ_1 starter bars in place during concrete pouring, they should be tied to the ϕ_5 bars. Similarly, the ϕ_6 bars should be tied to the ϕ_7 bars and, advisably, to the ϕ_3 bars to ensure stiffness while the foundations are poured.

9. The top of the foundations is smoothed with a float or power float except in the contact area with the future vertical member (construction joint), where a rough surface such as is generated by the vibrator is needed. Before casting the concrete to form the vertical member, the foundation surface should be cleaned and moistened. The new concrete should not be cast until the surface of the existing concrete dries.

10. In soft soils, the top 25 cm underneath the blinding concrete should not be excavated until shortly before pouring the concrete to ensure that the soil is not softened by rainfall immediately before casting.

11. The subgrade should be compacted before pouring the blinding.

12. This blinding is floated or smoothed with a power float. Its 10-cm depth can be varied to absorb the levelling tolerances at the bottom of the foundations.

13. In humid soils that constitute an aggressive medium, an asphalt membrane should be used as specified to protect the concrete.

14. See 1.2 and 1.3 for descriptions of how to tie bars and place spacers.

2. SPECIFIC REFERENCES

See Chapter 5 in (23) and (27).

CANTILEVER RETAINING WALLS.
STEM

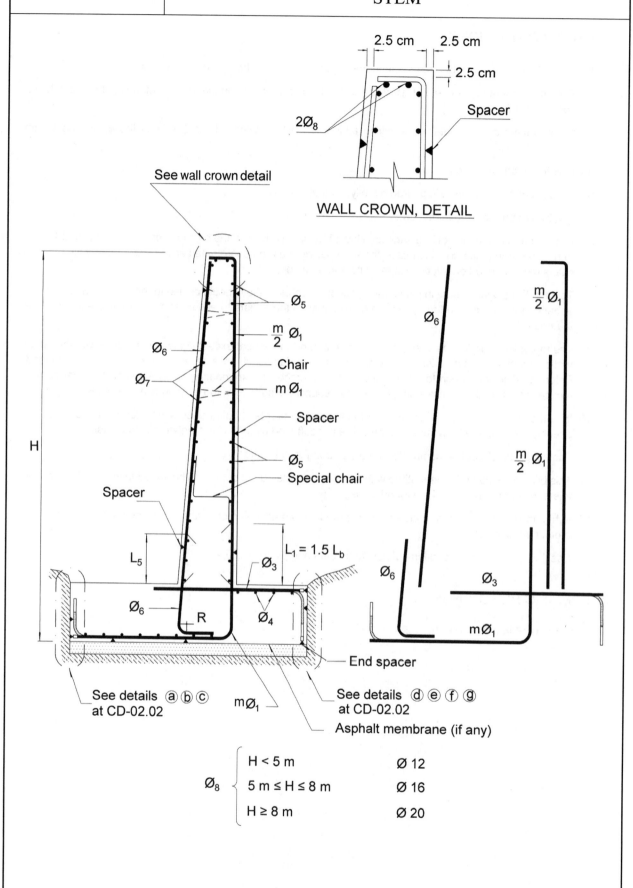

WALL CROWN, DETAIL

$$\varnothing_8 \begin{cases} H < 5\ \text{m} & \varnothing\ 12 \\ 5\ \text{m} \le H \le 8\ \text{m} & \varnothing\ 16 \\ H \ge 8\ \text{m} & \varnothing\ 20 \end{cases}$$

110

1. RECOMMENDATIONS

1. Length L_1 is 1.5 times the ϕ_1 bar lap length (see EC2, 8.7.3).

2. L_5 is the ϕ_6 bar lap length.

3. r_3 (cover of ϕ_6) is 2.5 cm but not smaller than ϕ_6.

4. r_4 (cover of ϕ_1) is 2.5 cm but may not smaller than ϕ_1 if formwork is placed against the soil. $r_4 = 7.5$ cm if the concrete is poured directly against the soil.

5. In the upper part of the wall, the two grids can be separated with chairs. In high or very high walls, special chairs must be used.

6. See 1.2 and 1.3 for descriptions of how to tie bars and place spacers.

7. The ϕ_8 reinforcing bars are used to control drying shrinkage and thermal contraction at the top of the wall.

2. SPECIFIC REFERENCES

See Chapter 5, (23) and (27).

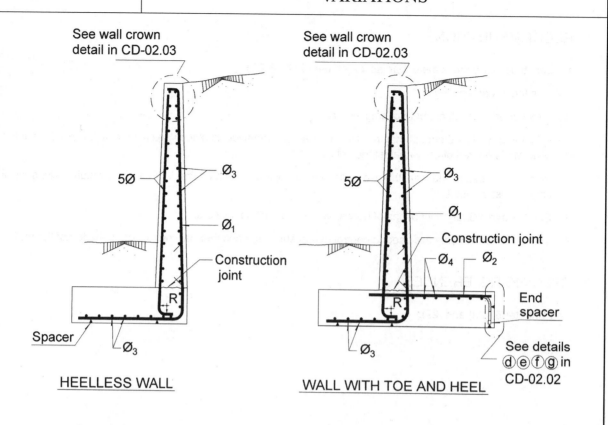

HEELLESS WALL

WALL WITH TOE AND HEEL

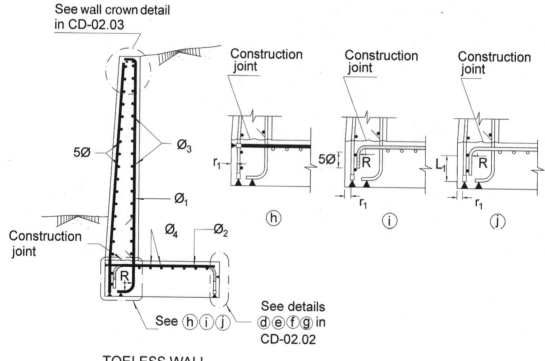

TOELESS WALL

1. RECOMMENDATIONS

1. The Recommendations set out in CD – 02.02 and CD – 02.03 are applicable to these variations.

2. For toeless walls, the total length of the ϕ_1 bar that is embedded in the foundations must be greater than or equal to its anchorage length.

2. STATUTORY LEGISLATION

None in place. EC7 (26) is applicable for calculating thrust only.

3. SPECIFIC REFERENCES

See Chapter 5 in (23) and (27).

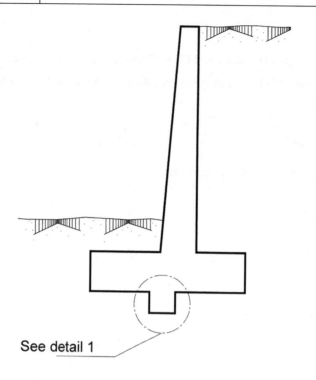

See detail 1

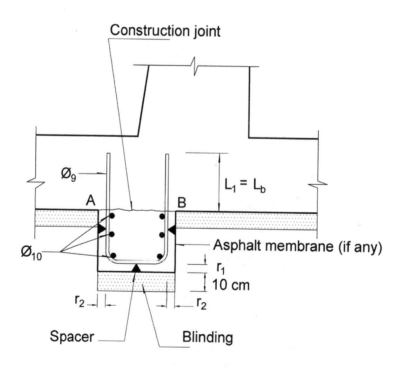

Construction joint

$\emptyset_9$

A   B

$L_1 = L_b$

$\emptyset_{10}$

Asphalt membrane (if any)

r_1

10 cm

r_2 r_2

Spacer   Blinding

DETAIL 1

1. RECOMMENDATIONS

1. L_1 is the ϕ_9 bar lap length.

2. Cover r_1 should be 2.5 cm but not smaller than ϕ_9.

3. Cover r_2 should be 7.5 cm.

4. In soft soils, the top 25 cm underneath the blinding concrete should not be excavated until shortly before pouring the concrete to ensure that the soil is not softened by rainfall immediately before casting.

5. The subgrade should be compacted before pouring the blinding.

6. The blinding is floated or smoothed with a power float. Its 10 cm depth can be varied to absorb the levelling tolerances at the bottom of the foundations.

7. A rough surface such as is generated by vibration is needed at construction joint AB, which should not be smoothed. Before casting the concrete to form the vertical member, the foundation surface should be carefully cleaned and moistened. The new concrete should not be cast until the surface of the existing concrete dries.

8. See 1.2 and 1.3 for descriptions of how to tie bars and place spacers.

2. STATUTORY LEGISLATION

None in place. EC7 (26) is applicable for calculating thrust only.

3. SPECIFIC REFERENCES

See Chapter 5 in (23) and (27).

CANTILEVER RETAINING WALLS.
CONSTRUCTION JOINTS IN FOOTINGS

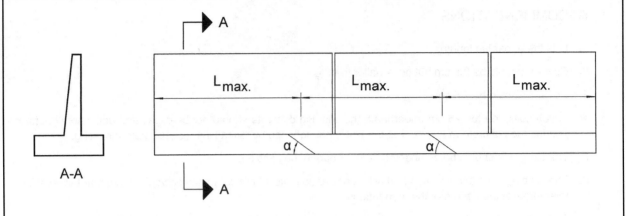

A-A

MAXIMUM DISTANCE

CLIMATE	$L_{max.}$ BY SEASON	
	WARM	COLD
DRY	12 m	16 m
HUMID	16 m	22 m

NOTE: The joint may adopt the natural slope of the vibrated concrete
($\alpha \approx 25°$) or be formed with welded-wire mesh.

1. RECOMMENDATIONS

1. With the use of galvanised welded-wire mesh as 'formwork', the resulting joint can be positioned vertically.

2. In this case, the welded-wire mesh should be tied to the reinforcement perimeter and stiffened with pieces of off cuts steel. (See CD – 02.10.)

2. STATUTORY LEGISLATION

None in place.

3. RECOMMENDED ALTERNATIVE CODES

None in place.

4. SPECIFIC REFERENCES

See Chapter 14 in (23) and (27).

CD – 02.07	CANTILEVER RETAINING WALLS. VERTICAL CONTRACTION JOINTS IN THE STEM

WALL HEIGHT	RECOMMENDED DISTANCE BETWEEN VERTICAL CONTRACTION JOINTS
H ≤ 2.4m	3 H
2.40 < H ≤ 3.6m	2 H
H > 3.6m	H (*)

(*) **AR**. The maximum distance should not exceed 7.5m

1. RECOMMENDATIONS

1. See CD – 02.09 for details.
2. The distances specified entail observance of the minimum shrinkage reinforcement ratios and curing rules.

2. STATUTORY LEGISLATION

EC2 makes no reference to this subject.

3. RECOMMENDED ALTERNATIVE CODES

See ACI (29).

4. SPECIFIC REFERENCES

See (23) and Chapter 14 in (27).

	CANTILEVER RETAINING WALLS.
CD – 02.08	HORIZONTAL CONSTRUCTION JOINTS IN ARCHITECTURAL CONCRETE

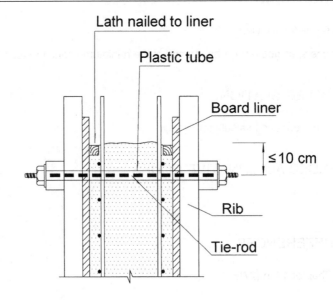

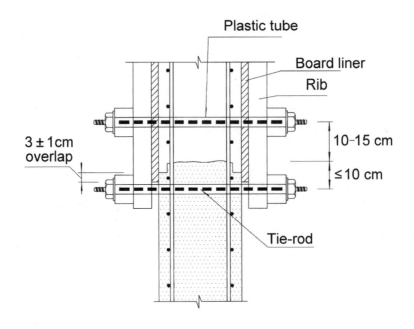

1. RECOMMENDATIONS

1. The holes in the plastic tube may be:

   (a) plugged with plastic stoppers;

   (b) filled with mortar, except the 3 or 4 cm closest to the exposed surfaces, to be able to identify the holes;

   (c) filled flush with mortar.

2. If solution 1 (c) is adopted, a tentative mortar should be prepared by mixing around two parts of the cement used in the works with about one part of white cement for a perfect long-term blend with the wall colour. (The proportions should be adjusted as needed by trial and error.)

2. STATUTORY LEGISLATION

None in place.

3. RECOMMENDED ALTERNATIVE CODES

See ACI (29).

4. SPECIFIC REFERENCES

See Chapter 14 in (27).

CANTILEVER RETAINING WALLS.
VERTICAL CONTRACTION JOINTS

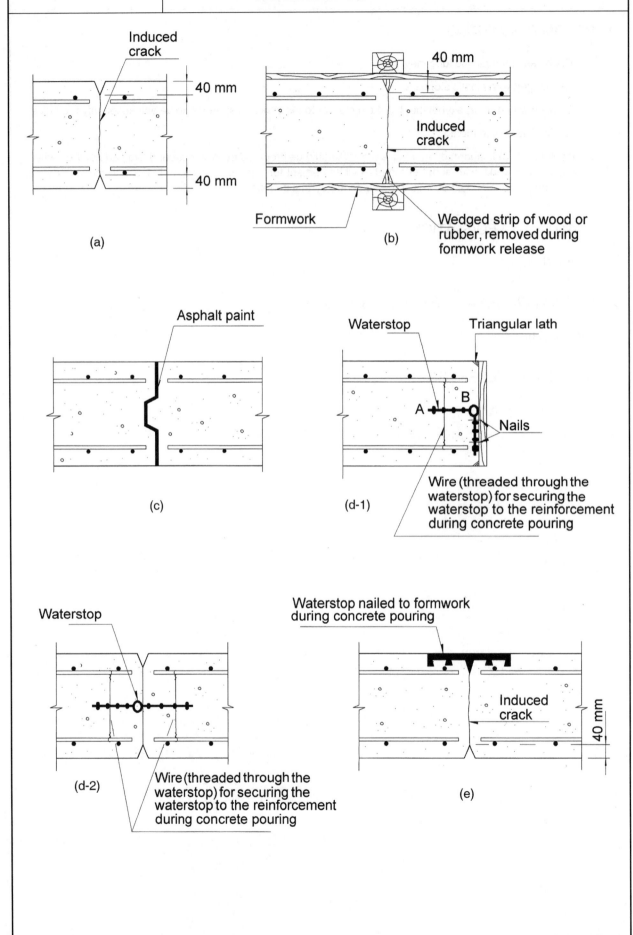

Induced crack

40 mm

40 mm

(a)

40 mm

Induced crack

Formwork

Wedged strip of wood or rubber, removed during formwork release

(b)

Asphalt paint

(c)

Waterstop

Triangular lath

A B

Nails

Wire (threaded through the waterstop) for securing the waterstop to the reinforcement during concrete pouring

(d-1)

Waterstop

(d-2)

Wire (threaded through the waterstop) for securing the waterstop to the reinforcement during concrete pouring

Waterstop nailed to formwork during concrete pouring

Induced crack

40 mm

(e)

1. RECOMMENDATIONS

1. Joints (a) and (b) can transmit stress. Joint (c) can only transmit horizontal bending stress. Joints (a) and (b) may require corrosion protection in the area around the reinforcing bar that crosses the crack if significant amounts of water seep in through the outer side of the member.

2. Joints (a), (b) and (c) are scantly waterproof.

3. Joints (d) and (e) are fully waterproof.

4. Vigorous vibration must be conducted with extreme care around waterstops.

5. The waterstop should penetrate at least 30 cm into the wall foundation. Welding must be performed very meticulously in long splices.

6. The bevelled and sunken edges in joints (a), (b) and (e) must reduce the wall section by at least 25 per cent to ensure that the crack will form.

2. STATUTORY LEGISLATION

See ACI (29).

3. SPECIFIC REFERENCES

See Chapter 14 in (27).

CANTILEVER RETAINING WALLS.
EXPANSION JOINTS

EXPANSION JOINTS ARE IMPERATIVE IN ALL SOIL TYPE OR
FOUNDATION SUPPORT LEVEL CHANGES.
MAXIMUM INTER-JOINT SPACING, ≤ 25 m.
THEY ARE ALSO NECESSARY IN CHANGES OF DIRECTION,
WHERE THE FILL IS IN THE CONCAVE AREA (CASES (a) AND
(b)) ②.

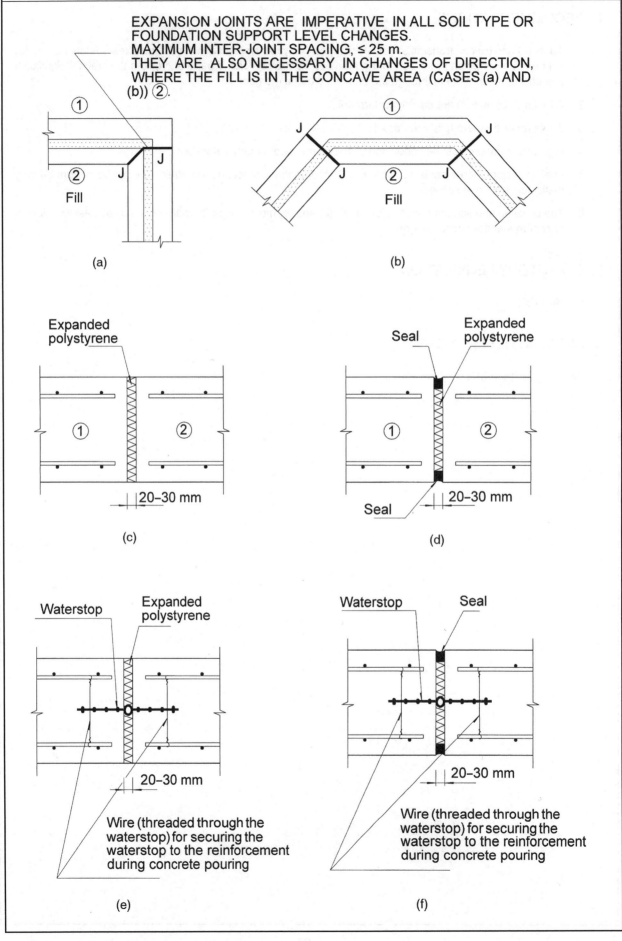

1. RECOMMENDATIONS

1. The failure to make J-J joints in (a) and (b) places horizontal tensile stress in the stems and may induce torsion in the footings. In this case, the members affected should be designed accordingly.

2. Expanded polystyrene may be replaced by any readily compressed material.

3. The seal for joint (d) should rest on the polystyrene.

4. Vigorous vibration must be conducted with extreme care around waterstops.

5. The waterstop should penetrate at least 30 cm into the wall foundation. Welding must be performed very meticulously in long splices.

2. STATUTORY LEGISLATION

None in place.

3. RECOMMENDED ALTERNATIVE CODES

See ACI (29).

4. SPECIFIC REFERENCES

See Chapter 14 in (27).

CANTILEVER RETAINING WALLS.
FILL AND DRAINAGE

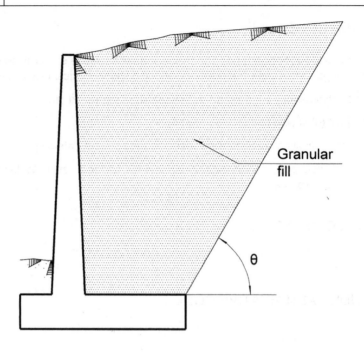

Granular fill

θ

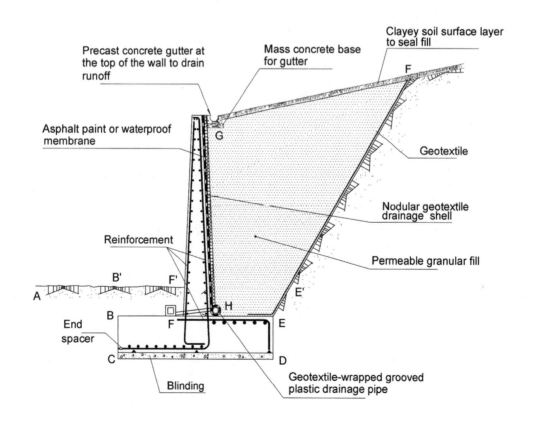

Precast concrete gutter at the top of the wall to drain runoff

Mass concrete base for gutter

Clayey soil surface layer to seal fill

F

Asphalt paint or waterproof membrane

G

Geotextile

Nodular geotextile drainage shell

Permeable granular fill

Reinforcement

B'   F'

A

E'

B

H

F   E

End spacer

C   D

Blinding

Geotextile-wrapped grooved plastic drainage pipe

1. RECOMMENDATIONS

1. Angle θ must be calculated.

2. Granular fill should be used to ensure effective drainage. Otherwise, the thrust cannot be assessed using the usual Rankine and Coulomb theories on earth pressure.

3. For water proofing the wall there are two systems, (b), (c). In solution (b), waterproofing consists of a mere coat of asphalt paint. In (c), a waterproofing membrane is used.

4. Alternative (c) is the solution of choice when watertightness must be guaranteed.

2. STATUTORY LEGISLATION

None in place.

3. SPECIFIC REFERENCES

See (27).

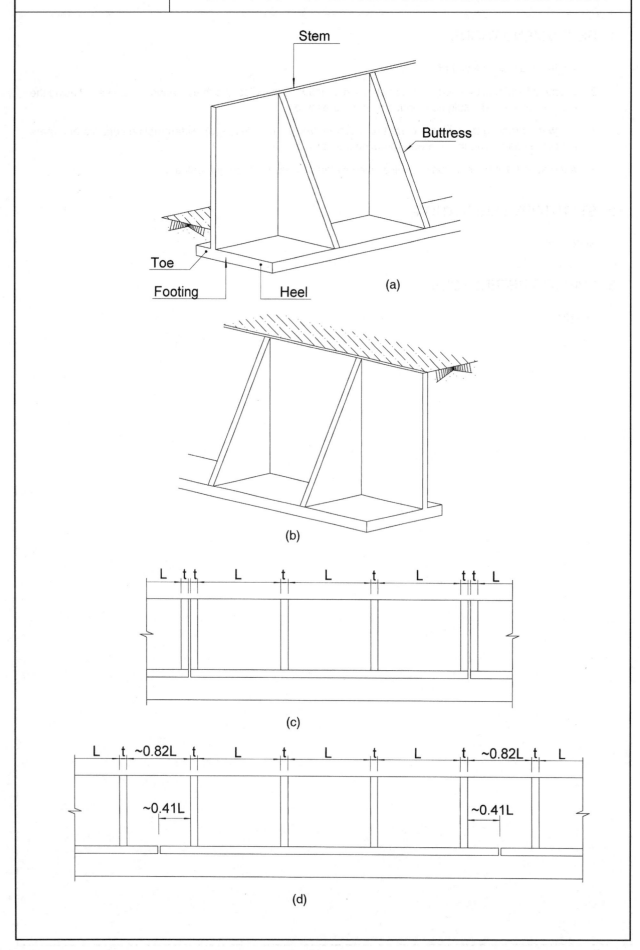

(a)

(b)

(c)

(d)

1. RECOMMENDATIONS

1. Solution (a) is the standard solution.

2. While distributing the buttress as in (c) means duplication along the expansion joint, the advantage to this approach is that the formwork is the same for all the cells. Solution (d) calls for different formwork for the cells alongside joints.

2. STATUTORY LEGISLATION

None in place.

3. RECOMMENDED ALTERNATIVE CODES

None in place.

4. SPECIFIC REFERENCES

See Chapter 8 in (27).

BUTTRESS WALLS.
FOOTING

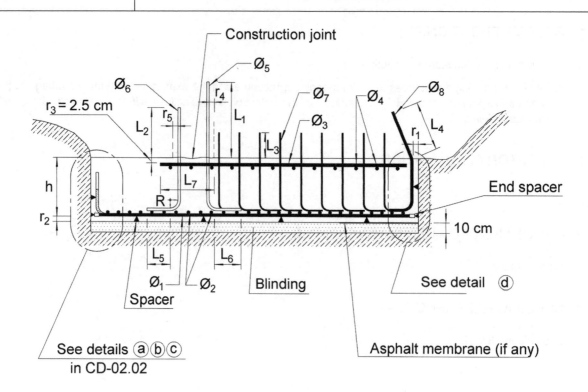

Construction joint

$\emptyset_5$

$\emptyset_6$

$r_3 = 2.5$ cm

$\emptyset_7$

$\emptyset_4$

$\emptyset_8$

r_4

L_2

r_5

L_1

$\emptyset_3$

L_3

r_1

L_4

L_7

End spacer

h

R

r_2

10 cm

L_5

L_6

See detail ⓓ

$\emptyset_1$

$\emptyset_2$

Blinding

Spacer

See details ⓐⓑⓒ
in CD-02.02

Asphalt membrane (if any)

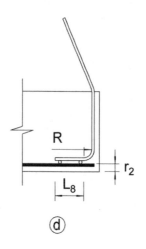

R

r_2

L_8

ⓓ

1. RECOMMENDATIONS

1. L_1 is 1.5 times the ϕ_5 bar lap length.

2. L_2 is the ϕ_8 bar lap length.

3. L_3 is the ϕ_7 bar lap length. The length that the ϕ_7 bars are embedded in the foundations should not be smaller than their anchorage length.

4. L_4 is 1.5 times the ϕ_8 bar lap length.

5. Lengths L_5 and L_6 should not be under 2 s, where s is the space between the ϕ_2 bars, to ensure that the ϕ_5 and ϕ_6 starter bars are tied securely to the grid at the bottom of the foundations. Analogously, the preceding arrangements are applicable to the ϕ_7 starter bars and the minimum value of L_8.

6. L_7 is the ϕ_3 bar lap length.

7. Cover r_1 should be 7.5 cm if the concrete is poured directly against the soil.

8. Cover r_2 should be 2.5 cm but not smaller than ϕ_1.

9. $r_4 = 2.5$ cm $\geq \phi_5$.

10. $r_5 = 2.5$ cm $\geq \phi_6$.

11. To keep the ϕ_5, ϕ_6 and ϕ_7 starter bars in place during concreting, they should be tied to the ϕ_3 or ϕ_4 bars.

12. See 1.5 to determine when to use anchor type a, b or c on the ϕ_1 reinforcing bars.

13. In soft soils, the top 25 cm underneath the blinding concrete should not be excavated until shortly before pouring the concrete to ensure that the soil is not softened by rainfall immediately before casting.

14. The subgrade should be compacted before pouring the blinding.

15. The blinding is floated or smoothed with a power float. Its 10-cm depth can be varied to absorb the levelling tolerances at the bottom of the foundations.

16. The top of the foundations is smoothed with a float or power float except in the contact area with the future vertical member (construction joint), where a rough surface such as is generated by the vibrator is needed. Before casting the concrete to form the vertical member, the foundation surface should be cleaned and moistened. The new concrete should not be cast until the surface of the existing concrete dries.

17. Where the footing is to be poured against soils that constitute an aggressive medium, it should be protected by an asphalt membrane as specified.

18. See 1.2 and 1.3 for descriptions of how to tie bars and place spacers.

2. STATUTORY LEGISLATION

None in place.

3. SPECIFIC REFERENCES

See Chapter 8 in (27).

BUTTRESS WALLS.
STEM

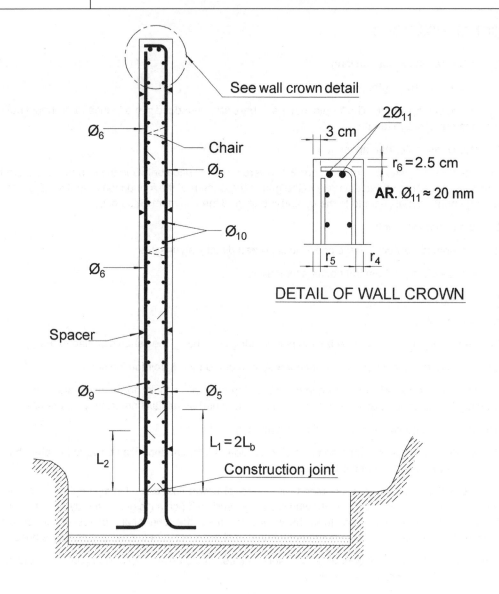

See wall crown detail

$\emptyset_6$

Chair

$\emptyset_5$

$\emptyset_{10}$

$\emptyset_6$

Spacer

$\emptyset_9$

$\emptyset_5$

$L_1 = 2L_b$

L_2

Construction joint

3 cm

$2\emptyset_{11}$

$r_6 = 2.5$ cm

AR. $\emptyset_{11} \approx 20$ mm

r_5 | r_4

DETAIL OF WALL CROWN

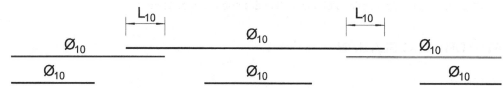

L_{10}

L_{10}

$\emptyset_{10}$

$\emptyset_{10}$

$\emptyset_{10}$

$\emptyset_{10}$

$\emptyset_{10}$

$\emptyset_{10}$

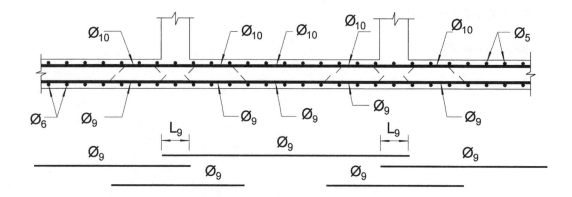

$\emptyset_{10}$

$\emptyset_{10}$

$\emptyset_{10}$

$\emptyset_{10}$

$\emptyset_{10}$

$\emptyset_5$

$\emptyset_6$

$\emptyset_9$

$\emptyset_9$

$\emptyset_9$

$\emptyset_9$

$\emptyset_9$

L_9

L_9

$\emptyset_9$

$\emptyset_9$

$\emptyset_9$

$\emptyset_9$

$\emptyset_9$

1. RECOMMENDATIONS

1. A.R. L_1 is 2.0 times the ϕ_5 bar lap length, L_b. Lower values can be obtained applying table 8.3 of EC2.

2. L_2 is the ϕ_6 bar lap length.

3. L_9 is the ϕ_9 bar lap length.

4. L_{10} is the ϕ_{10} bar lap length.

5. $r_4 = 2.5$ cm $\geq \phi_5$.

6. $r_5 = 2.5$ cm $\geq \phi_6$.

7. See 1.2 and 1.3 for descriptions of how to tie bars and place spacers.

8. In exposed concrete walls, the solution for horizontal construction joints is as specified in CD – 02.08.

9. Vertical contraction joints are to be made further to CD – 02.07, types (a) or (b) in CD – 02.09, or CD – 02.21. They will, in any case, be positioned midway between buttresses.

2. STATUTORY LEGISLATION

None in place.

3. SPECIFIC REFERENCES

See Chapter 8 in (27).

BUTTRESS WALLS.
INSIDE BUTTRESS

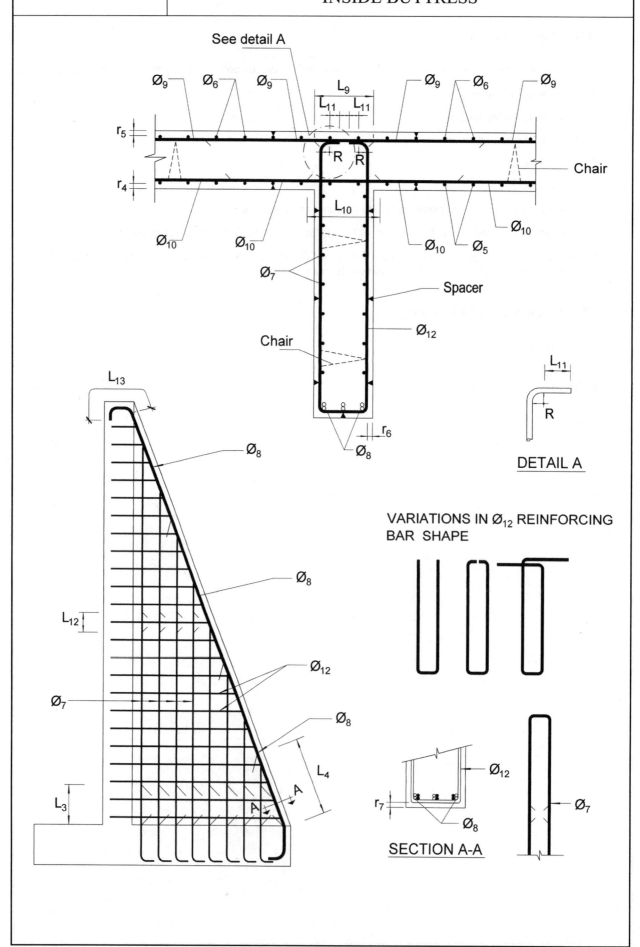

See detail A

DETAIL A

VARIATIONS IN $\emptyset_{12}$ REINFORCING BAR SHAPE

SECTION A-A

1. RECOMMENDATIONS

1. L_9 is the ϕ_9 bar lap length.

2. L_{10} is the ϕ_{10} bar lap length.

3. L_{11} is deduced from the ϕ_{12} stirrup anchorage. See variations in arrangement. The ϕ_{12} anchorage length is counted from its point of entry in the wall.

4. L_4 is 1.5 times the ϕ_8 bar lap length. (Note that the lap length may be conditioned by the use of paired or bundled bars.)

5. The ϕ_8 bars rest directly on the top of the foundations.

6. In thick stirrups, the bend radius must not be overly sharp.

7. L_{12} is the ϕ_7 bar lap length.

8. L_{13} is the ϕ_8 bar lap length.

9. $r_4 = 2.5$ cm $\geq \phi_5$.

10. $r_5 = 2.5$ cm $\geq \phi_6$.

11. $r_6 = 2.5$ cm $\geq \phi_{12}$.

12. $r_7 = 2.5$ cm $\geq \phi_e$ - ϕ_{12}, where ϕ_e is the equivalent diameter for bundled bars, if used. ($\phi_e = \phi_8$ if no paired or bundled bars are used.)

13. In exposed concrete walls, the solution for horizontal construction joints is as specified in CD – 02.08.

14. See 1.2 and 1.3 for descriptions of how to tie bars and place spacers.

2. STATUTORY LEGISLATION

None in place.

3. SPECIFIC REFERENCES

See Chapter 8 in (27).

BUTTRESS WALLS.
END BUTTRESS

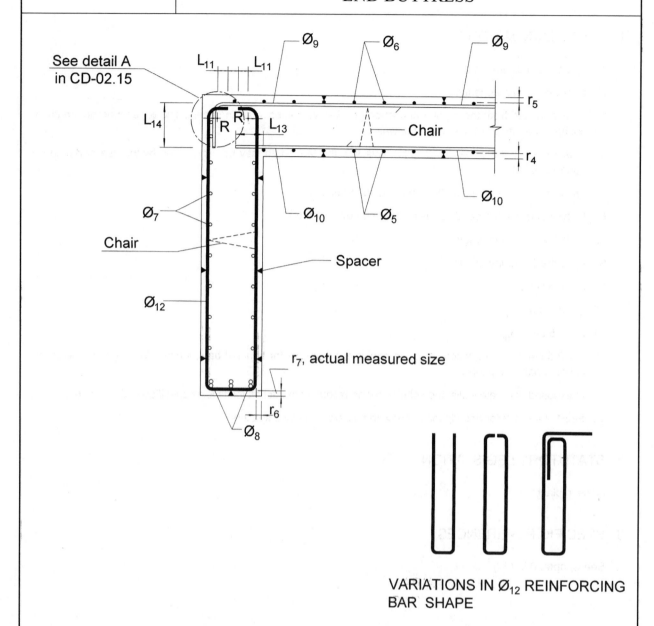

See detail A in CD-02.15

Chair

Chair

Spacer

r_7, actual measured size

VARIATIONS IN $\varnothing_{12}$ REINFORCING BAR SHAPE

1. RECOMMENDATIONS

All the Recommendations in CD – 02.15 are applicable, in addition to the following.

1. L_{14} is deduced from the ϕ_9 anchorage length.

2. L_{13} is deduced from the ϕ_{10} anchorage length.

3. In exposed concrete walls, the solution for horizontal construction joints is as specified in CD – 02.08.

2. STATUTORY LEGISLATION

None in place.

3. SPECIFIC REFERENCES

See Chapter 8 in (27).

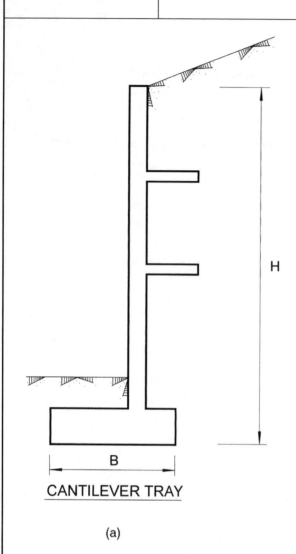

B

CANTILEVER TRAY

(a)

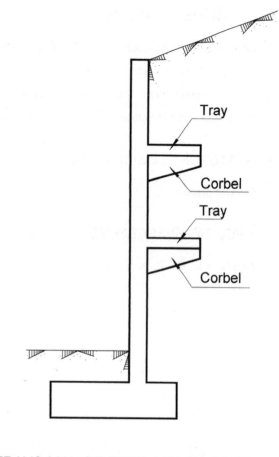

Tray

Corbel

Tray

Corbel

TRAYS MAY BE PRECAST OR CAST
IN PLACE OVER CORBELS

(b)

NOTES

1. RECOMMENDATIONS

None.

2. STATUTORY LEGISLATION

None in place.

3. SPECIFIC REFERENCES

See Chapter 9 in (27).

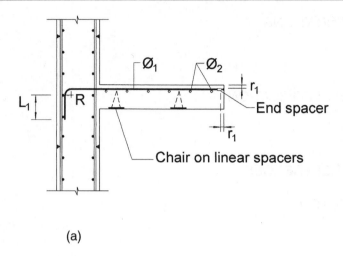

(a)

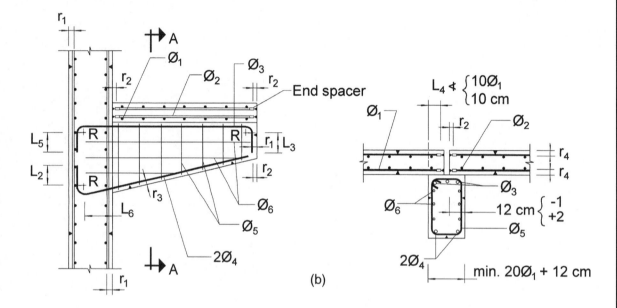

(b)

SECTION A-A

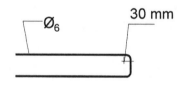

1. RECOMMENDATIONS

1. $r_1 = 2.5$ cm $\geq \phi_1$.

2. $r_2 = 2,5$ cm.

3. $r_3 = r_4 = 2.5$ cm.

4. L_1 is deduced from the ϕ_1 bar anchorage length.

5. L_5 is deduced from the ϕ_3 bar anchorage length.

6. L_2 should be s, where s is the spacing between the outer side horizontal bars, to ensure its attachment to these bars. This should also be the minimum value for L_5.

7. L_3 is deduced from the ϕ_3 bar anchorage conditions. (See Recommendation 6 in CD – 01.10.)

8. L_6 is the ϕ_5 bar lap length.

9. At its narrowest, the corbel must be wide enough to support the trays and absorb any tolerances.

10. In solution (a), the trays may be cast *in situ* in formwork or over the corbels, in turn poured against the fill on the outer side of the wall.

11. Solution (b) calls for cast-in-place corbels and precast trays.

2. STATUTORY LEGISLATION

None in place.

3. SPECIFIC REFERENCES

See Chapter 9 in (27).

BASEMENT WALLS.
FACADE FOOTING

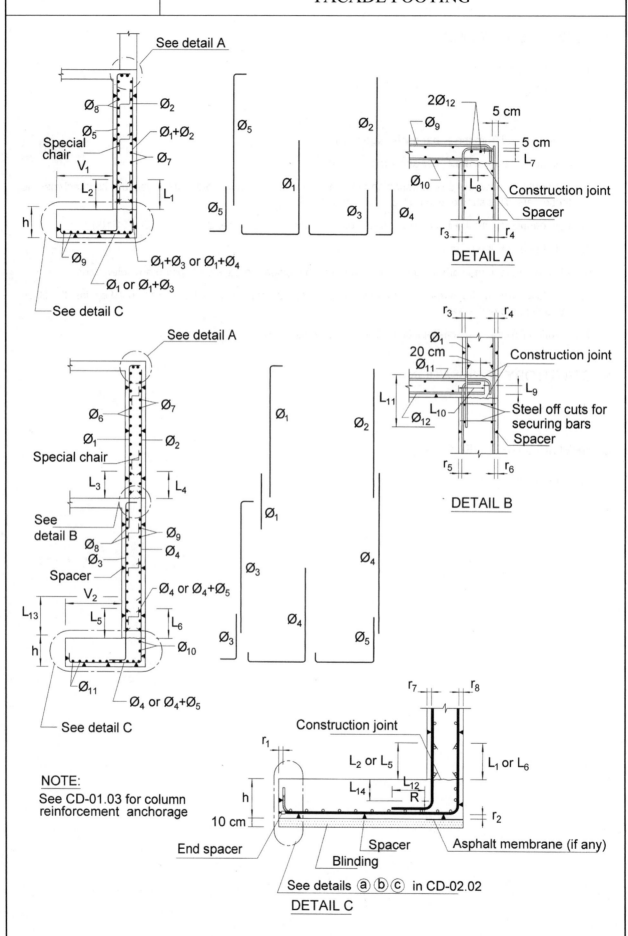

See detail A

$\emptyset_8$

$\emptyset_2$

$\emptyset_5$

$\emptyset_1+\emptyset_2$

Special chair

$\emptyset_7$

V_1

L_2

L_1

h

$\emptyset_9$

$\emptyset_1+\emptyset_3$ or $\emptyset_1+\emptyset_4$

$\emptyset_1$ or $\emptyset_1+\emptyset_3$

See detail C

$\emptyset_5$

$\emptyset_1$

$\emptyset_5$

$\emptyset_2$

$\emptyset_3$

$\emptyset_4$

$2\emptyset_{12}$

$\emptyset_9$

5 cm

5 cm

L_7

$\emptyset_{10}$

L_8

Construction joint

Spacer

r_3

r_4

DETAIL A

See detail A

$\emptyset_6$

$\emptyset_7$

$\emptyset_1$

$\emptyset_2$

Special chair

L_3

L_4

See detail B

$\emptyset_8$

$\emptyset_9$

$\emptyset_3$

$\emptyset_4$

Spacer

V_2

$\emptyset_4$ or $\emptyset_4+\emptyset_5$

L_{13}

L_5

L_6

h

$\emptyset_{10}$

$\emptyset_{11}$

$\emptyset_4$ or $\emptyset_4+\emptyset_5$

See detail C

$\emptyset_1$

$\emptyset_1$

$\emptyset_3$

$\emptyset_3$

$\emptyset_2$

$\emptyset_4$

$\emptyset_4$

$\emptyset_5$

r_3

r_4

$\emptyset_1$

20 cm

$\emptyset_{11}$

Construction joint

L_{11}

L_9

$\emptyset_{12}$

L_{10}

Steel off cuts for securing bars

Spacer

r_5

r_6

DETAIL B

NOTE:
See CD-01.03 for column reinforcement anchorage

r_7

r_8

Construction joint

r_1

L_2 or L_5

L_1 or L_6

h

L_{14}

L_{12}

R

r_2

10 cm

End spacer

Spacer

Asphalt membrane (if any)

Blinding

See details (a) (b) (c) in CD-02.02

DETAIL C

1. RECOMMENDATIONS

NOTE: In the remarks below, the same designations have sometimes been used for bars and covers whether one or two basements are involved. In each case, they should be interpreted as required for the wall to which they refer.

1. $r_1 = 7.5$ cm if the concrete for the footing is poured directly against the soil.

   $r_2 = 2.5$ cm $\begin{cases} \geq \phi_1 \text{ and } \geq \phi_3 \text{ (one basement)} \\ \geq \phi_4 \text{ and } \geq \phi_5 \text{ (two basements)} \end{cases}$ $\qquad r_6 = 2.5$ cm $\geq \phi_4$

   $r_3 = 2.5$ cm $\geq \begin{cases} \phi_5 \text{ (one basement)} \\ \phi_1 \text{ (two basements)} \end{cases}$ $\qquad r_7 = \begin{cases} r_3 \text{ (one basement)} \\ r_5 \text{ (two basements)} \end{cases}$

   $r_4 = 2.5$ cm $\geq \phi_2$ (one and two basements) $\qquad r_8 = \begin{cases} r_4 \text{ (one basement)} \\ r_6 \text{ (two basements)} \end{cases}$

   $r_5 = 2.5$ cm $\geq \phi_3$ (two basements)

2. The horizontal part of the starter bar anchorage area should be tied to hold these bars in place during concrete pouring. A diagonal made of off cuts may be required in addition to the chairs to stiffen the assembly. For L_{14}, see CD – 01.03, recommendation 3.

3. See 1.5 to determine when to use anchor type a, b or c for bars ϕ_1, ϕ_3, ϕ_4 and ϕ_5.

4. L_1 is the lap length of the thicker of the ϕ_2 or ϕ_3 bars, or of the thicker of the ϕ_2 or ϕ_4 bars.

5. L_2 is the ϕ_5 bar lap length.

6. L_3 is the ϕ_1 bar lap length.

7. L_4 is the lap length of the thicker of the ϕ_2 or ϕ_4 bars.

8. L_5 is the ϕ_3 bar lap length.

9. L_6 is the ϕ_4 bar lap length.

10. Note that, according to the calculations for a wall in a single basement, the foundations may either be reinforced only with ϕ_1 bars or also require ϕ_3 bars, in which case no ϕ_4 starter bars are used.

11. L_{11} is the ϕ_1 bar lap length (detail B).

12. L_{12} should be at least $2s$ long, where s is the spacing between the transverse bars to which the respective bar is tied.

13. In detail B, if the two walls in the two basements differ in thickness, they should be tied with bar cutting off cuts.

14. In soft soils, the top 25 cm underneath the blinding concrete should not be excavated until shortly before pouring the concrete to ensure that the soil is not softened by rainfall immediately before casting.

15. The subgrade should be compacted before pouring the blinding.

16. The blinding is floated or smoothed with a power float. Its 10-cm depth can be varied to absorb the levelling tolerances at the bottom of the foundations.

17. The top of the foundations is smoothed with a float or power float except in the contact area with the future vertical member (construction joint), where a rough surface such as is generated by the vibrator is needed. Before casting the concrete to form the vertical member, the foundation surface should be cleaned and moistened. The new concrete should not be cast until the surface of the existing concrete dries.

18. Where the concrete is to be poured against soils that constitute an aggressive medium, it should be protected by an asphalt membrane as specified.

19. In walls enclosing two basements, a rough surface such as is generated by vibration is needed for the horizontal joints. Before pouring the new concrete, the joints should be cleaned and moistened and the surface allowed to dry before proceeding.

20. See 1.2 and 1.3 for descriptions of how to tie bars and place spacers.

2. STATUTORY LEGISLATION

None in place.

3. SPECIFIC REFERENCES

See Chapter 10 in (27).

BASEMENT WALL.
CENTRED FOOTING

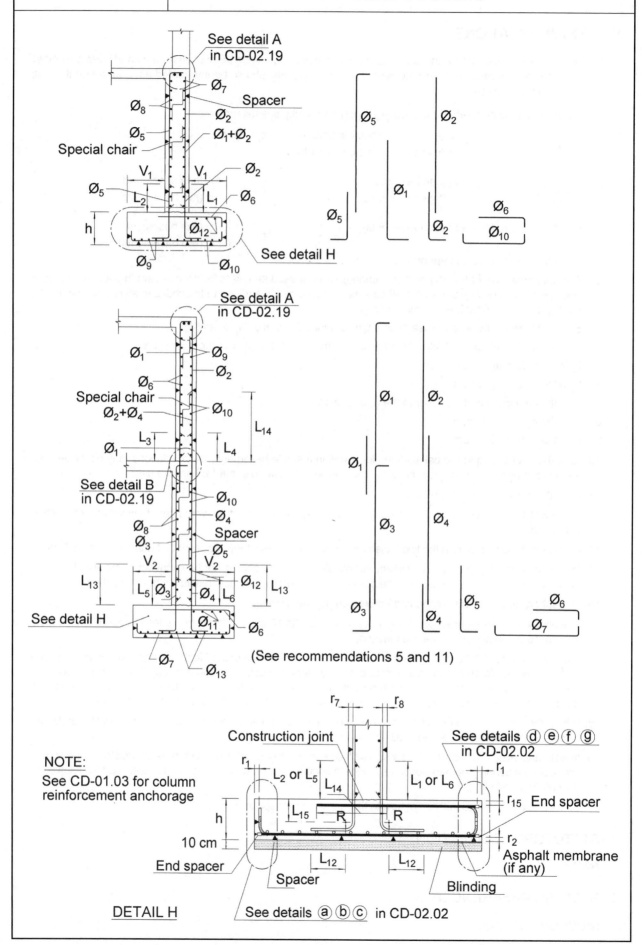

See detail A
in CD-02.19

$\varnothing_7$

Spacer

$\varnothing_8$

$\varnothing_2$

$\varnothing_5$

$\varnothing_1 + \varnothing_2$

Special chair

V_1 V_1 $\varnothing_2$

$\varnothing_5$

L_2 L_1 $\varnothing_6$

h $\varnothing_{12}$

$\varnothing_9$ $\varnothing_{10}$

See detail H

$\varnothing_5$

$\varnothing_2$

$\varnothing_1$

$\varnothing_5$

$\varnothing_2$

$\varnothing_6$

$\varnothing_{10}$

See detail A
in CD-02.19

$\varnothing_1$ $\varnothing_9$

$\varnothing_6$ $\varnothing_2$

Special chair

$\varnothing_2 + \varnothing_4$ $\varnothing_{10}$

L_3 L_{14}

$\varnothing_1$ L_4

See detail B
in CD-02.19

$\varnothing_{10}$

$\varnothing_4$

$\varnothing_8$

$\varnothing_3$ Spacer

$\varnothing_5$

V_2 V_2

L_{13} $\varnothing_{12}$ L_{13}

L_5 $\varnothing_3$ $\varnothing_4$ L_6

See detail H

$\varnothing_{11}$ $\varnothing_6$

$\varnothing_7$ $\varnothing_{13}$

(See recommendations 5 and 11)

$\varnothing_1$ $\varnothing_2$

$\varnothing_1$

$\varnothing_3$ $\varnothing_4$

$\varnothing_3$ $\varnothing_4$ $\varnothing_5$

$\varnothing_6$

$\varnothing_7$

NOTE:
See CD-01.03 for column
reinforcement anchorage

r_7 r_8

Construction joint

See details ⓓ ⓔ ⓕ ⓖ
in CD-02.02

r_1

L_2 or L_5

L_{14}

L_1 or L_6

r_1

r_{15} End spacer

L_{15} R R

h

10 cm

r_2

Asphalt membrane
(if any)

End spacer

L_{12} L_{12}

Spacer

Blinding

DETAIL H

See details ⓐ ⓑ ⓒ in CD-02.02

1. RECOMMENDATIONS

1. $r_1 = 7.5$ cm if the concrete for the footing is poured directly against the soil.

$r_2 = 2.5$ cm $\begin{cases} \geq \phi_{10} \text{ (one basement)} \\ \geq \phi_7 \text{ (two basements)} \end{cases}$ $\qquad r_6 = 2.5$ cm $\geq \phi_4$

$r_3 = 2.5$ cm $\geq \begin{cases} \phi_5 \text{ (one basement)} \\ \phi_1 \text{ (two basements)} \end{cases}$ $\qquad r_7 = \begin{cases} r_3 \text{ (one basement)} \\ r_5 \text{ (two basements)} \end{cases}$

$r_4 = 2.5$ cm $\geq \phi_2$ (one and two basements) $\qquad r_8 = \begin{cases} r_4 \text{ (one basement)} \\ r_6 \text{ (two basements)} \end{cases}$

$r_5 = 2.5$ cm $\geq \phi_3$ (two basements)

2. The horizontal part of the starter bar anchorage area should be tied to hold these bars in place during concrete pouring. A diagonal made of off cuts may also be required to stiffen the assembly, in addition to chairs. For L_{15}, see CD – 01.03, recommendation 3.

3. See 1.5 to determine when to use anchor type a, b or c in ϕ_1 or ϕ_{10} reinforcing bars.

4. L_1 is the ϕ_2 bar lap length.

5. L_2 is the ϕ_5 bar lap length.

6. L_3 is the ϕ_1 bar lap length.

7. L_4 is the lap length of the thicker of the ϕ_2 or ϕ_4 bars.

8. L_5 is the ϕ_3 bar lap length.

9. L_6 is the ϕ_4 bar lap length.

10. L_{11} is the ϕ_1 bar lap length (detail B).

11. L_{12} should be at least 2 s long, where s is the spacing between the transverse bars to which the respective bar is tied.

12. In detail B, if the two walls in the two basements differ in thickness, they should be tied with bar cutting off cuts.

13. L_{13} is the length of in the wall of two basements bars anchorage length.

14. In soft soils, the top 25 cm underneath the blinding concrete should not be excavated until shortly before pouring the concrete to ensure that the soil is not softened by rainfall immediately before casting.

15. The subgrade should be compacted before pouring the blinding.

16. The blinding is floated or smoothed with a power float. Its 10-cm depth can be varied to absorb the levelling tolerances at the bottom of the foundations.

17. The top of the foundations is smoothed with a float or power float except in the contact area with the future vertical member (construction joint), where a rough surface such as is generated by the vibrator is needed. Before pouring the concrete to form the vertical member, the foundation surface should be cleaned and moistened. The new concrete should not be cast until the surface of the existing concrete dries.

18. Where the concrete is to be poured against soils that constitute an aggressive medium, it should be protected by an asphalt membrane as specified.

19. In walls enclosing two basements, a rough surface such as is generated by vibration is needed at the horizontal joints. Before pouring the new concrete, the joints should be cleaned and moistened and the surface allowed to dry before proceeding.

20. See 1.2 and 1.3 for descriptions of how to tie bars and place spacers.

2. STATUTORY LEGISLATION

None in place.

3. SPECIFIC REFERENCES

See Chapter 10 in (27).

CD – 02.21	BASEMENT WALL. VERTICAL CONTRACTION JOINT

(a) VERTICAL CONTRACTION JOINT

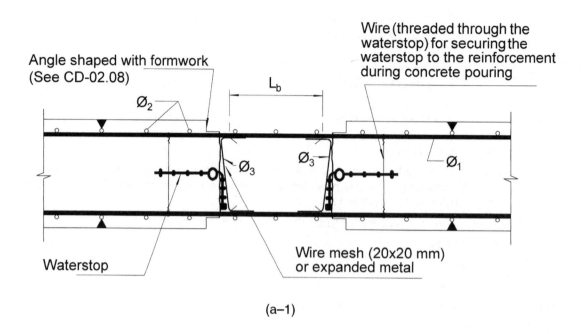

Wire (threaded through the waterstop) for securing the waterstop to the reinforcement during concrete pouring

Angle shaped with formwork (See CD-02.08)

$\emptyset_2$

L_b

$\emptyset_3$

$\emptyset_3$

$\emptyset_1$

Waterstop

Wire mesh (20x20 mm) or expanded metal

(a–1)

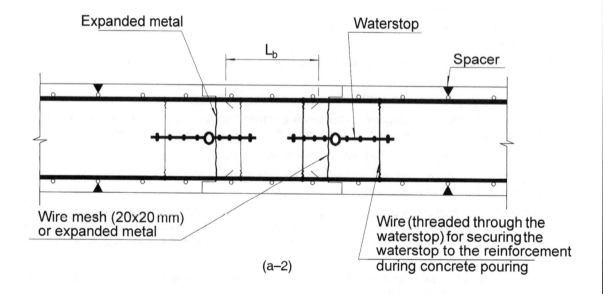

Expanded metal

Waterstop

Spacer

L_b

Wire mesh (20x20 mm) or expanded metal

Wire (threaded through the waterstop) for securing the waterstop to the reinforcement during concrete pouring

(a–2)

(b) EXPANSION JOINT (See CD-02.10). The specifications for expansion joints in basement walls are the same as in retaining walls

1. RECOMMENDATIONS

1. The solution depicted refers to basement walls that are also foundations, i.e. subjected to vertical bending stress. In any other case, the solutions set out in CD – 02.09 are simpler and easier.

2. L_b is the ϕ_1 bar lap length.

3. Galvanised expanded metal or small-mesh welded-wire 'formwork' must be tied to the grids and stiffened with off cuts from ϕ_3 bars, tied to the ϕ_1 bars. Off cuts 12 mm in diameter and spaced at 50 cm suffice and may be withdrawn before the concrete is poured.

4. Vigorous vibration around waterstops must be conducted with extreme care.

5. The joint must be left open for two days in winter and three in summer in cold, humid environments; and three days in winter and five in summer in warm, dry climates.

2. STATUTORY LEGISLATION

None in place.

3. SPECIFIC REFERENCES

See Chapter 10 in (27).

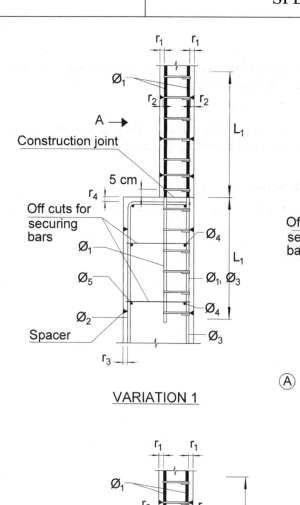

VARIATION 1

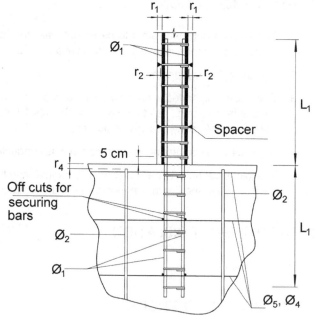

Ⓐ

VIEW FROM A

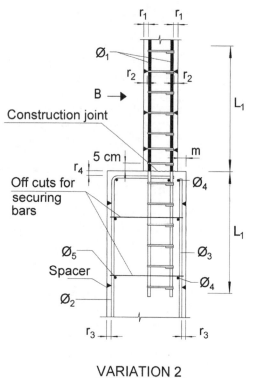

VARIATION 2

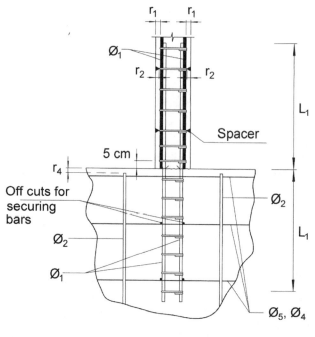

Ⓑ

VIEW FROM B

148

1. RECOMMENDATIONS

1. If the column width is the same or narrower than the wall width, the bond is as shown in detail A. Variation 1 affords a straightforward solution. Variation 2 is an alternative, in which the column is set back by a certain length from the outer side of the wall (3/4 cm generally suffices) to accommodate the starter bars. (Solutions involving starter bar deviation may also be adopted. See CD – 03.04, for instance.) (The horizontal reinforcement may have to be strengthened due to the concentration of loads in the column.)

2. The ϕ_1 starter bars are tied to two straight pieces of off cuts. Care must be taken to ensure that the tips are covered.

3. The starter bars must have stirrups of the same diameter and spacing as in the column unless the side cover of concrete over all the bars is ten times their diameter, in which case two assembly stirrups suffice.

4. L_1 is the ϕ_1 bar lap length.

5. If the column protrudes from the wall, its reinforcing bars and respective stirrups must reach into the foundations (see detail G). L_2 must be at least 2 s, where s is the spacing between distribution bars in the foundations.

6. See 1.5 to determine when to use anchor type a, b or c in the ϕ_1 reinforcing bars.

7. The reinforcement is arranged at wall corners as shown in details E and F. L_5 is the ϕ_8 bar lap length.

8. Further to calculations, L_6, which is initially the ϕ_6 bar lap length, may need to be lengthened to withstand horizontal bending.

9. See CD – 02.11 for recommendations on fillers, waterproofing and drainage.

2. STATUTORY LEGISLATION

None in place.

3. SPECIFIC REFERENCES

See Chapter 10 in (27).

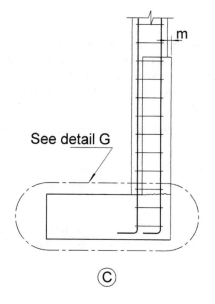

See detail G

C

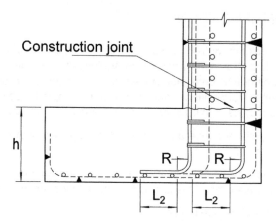

Construction joint

h

R R

L_2 L_2

DETAIL G (Reinforcement in column)

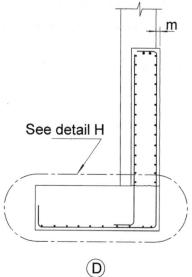

See detail H

D

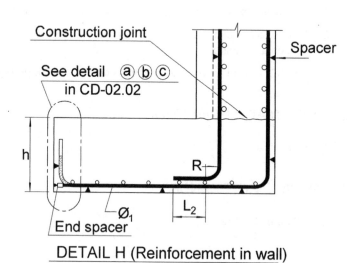

Construction joint

See detail ⓐ ⓑ ⓒ
in CD-02.02

Spacer

h

R

$\emptyset_1$

L_2

End spacer

DETAIL H (Reinforcement in wall)

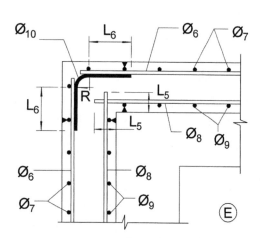

$\emptyset_{10}$ L_6 $\emptyset_6$ $\emptyset_7$

R

L_6 L_5

L_5

$\emptyset_8$ $\emptyset_9$

$\emptyset_6$ $\emptyset_8$

$\emptyset_7$ $\emptyset_9$

E

REINFORCEMENT IN CORNER
OF WALL

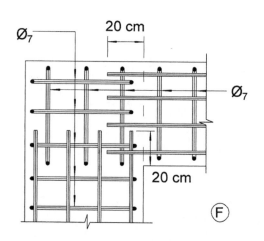

20 cm

$\emptyset_7$

$\emptyset_7$

20 cm

F

REINFORCEMENT IN FOOTING

1. RECOMMENDATIONS

1. If the column width is the same or narrower than the wall width, the bond is as shown in detail A. Variation 1 affords a straightforward solution. Variation 2 is an alternative, in which the column is set back by a certain length from the outer side of the wall (3/4 cm generally suffices) to accommodate the starter bars. (Solutions involving starter bar deviation may also be adopted. See CD – 03.04, for instance.) (The horizontal reinforcement may have to be strengthened due to the concentration of loads in the column.)

2. The ϕ_1 starter bars are tied to two straight pieces of off cuts. Care must be taken to ensure that the tips are covered.

3. The starter bars must have stirrups of the same diameter and spacing as in the column unless the side cover of concrete over all the bars is ten times their diameter, in which case two assembly stirrups suffice.

4. L_1 is the ϕ_1 bar lap length.

5. If the column protrudes from the wall, its reinforcing bars and respective stirrups must reach into the foundations (see detail G). L_2 must be at least 2 s, where s is the spacing between distribution bars in the foundations.

6. See 1.5 to determine when to use anchor type a, b or c in the ϕ_1 reinforcing bars.

7. The reinforcement is arranged at wall corners as shown in details E and F. L_5 is the ϕ_8 bar lap length.

8. Further to calculations, L_6, which is initially the ϕ_6 bar lap length, may need to be lengthened to withstand horizontal bending.

9. See CD – 02.11 for Recommendations on fillers, waterproofing and drainage.

2. STATUTORY LEGISLATION

None in place.

3. SPECIFIC REFERENCES

See Chapter 10 in (27).

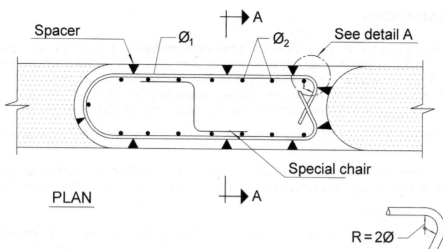

Spacer $\varnothing_1$ $\varnothing_2$ See detail A

Special chair

PLAN

$R = 2\varnothing$

$5\varnothing \geq 50\,mm$

DETAIL A

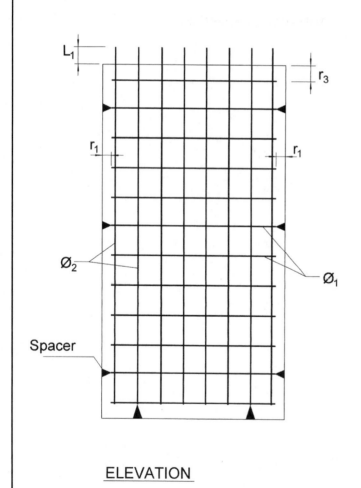

L_1

r_3

r_1 r_1

$\varnothing_2$ $\varnothing_1$

Spacer

ELEVATION

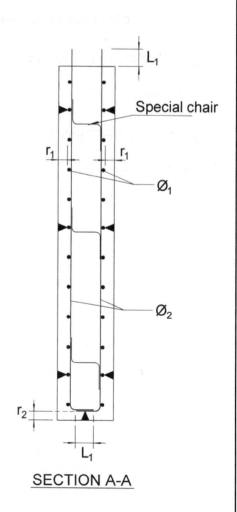

L_1

Special chair

r_1 r_1

$\varnothing_1$

$\varnothing_2$

r_2

L_1

SECTION A-A

1. RECOMMENDATIONS

1. $r_1 = 7.5$ cm.
2. $r_2 = 7.5$ cm.
3. Use revolving wheel spacers, positioned on the horizontal reinforcement.
4. L_1 is the ϕ_2 bar lap length.
5. The length of L_2 can be greater for constructional reasons.
6. A rough surface such as is generated by vibration is needed at the top.
7. All reinforcement ties should be welded.
8. The special chairs should be welded to the ϕ_2 bars.
9. See 1.4 for welding details.

2. STATUTORY LEGISLATION

None in place.

3. SPECIFIC REFERENCES

See Chapter 11 in (27).

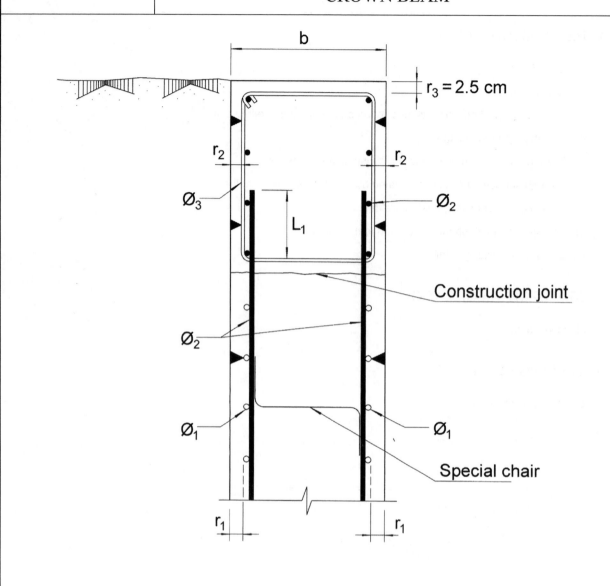

b

$r_3 = 2.5$ cm

r_2 r_2

$\emptyset_3$ $\emptyset_2$

L_1

Construction joint

$\emptyset_2$

$\emptyset_1$ $\emptyset_1$

Special chair

r_1 r_1

1. RECOMMENDATIONS

1. For the diaphragm wall, see CD – 02.23.

2. The construction joint surface is cleaned and moistened before pouring the concrete for the beam. The beam is cast when the surface of the joint dries.

3. The top of the beam is smoothed with a float or power float.

4. If b > 65 cm, use multiple stirrups.

2. STATUTORY LEGISLATION

None in place.

3. SPECIFIC REFERENCES

See Chapter 11 in (27).

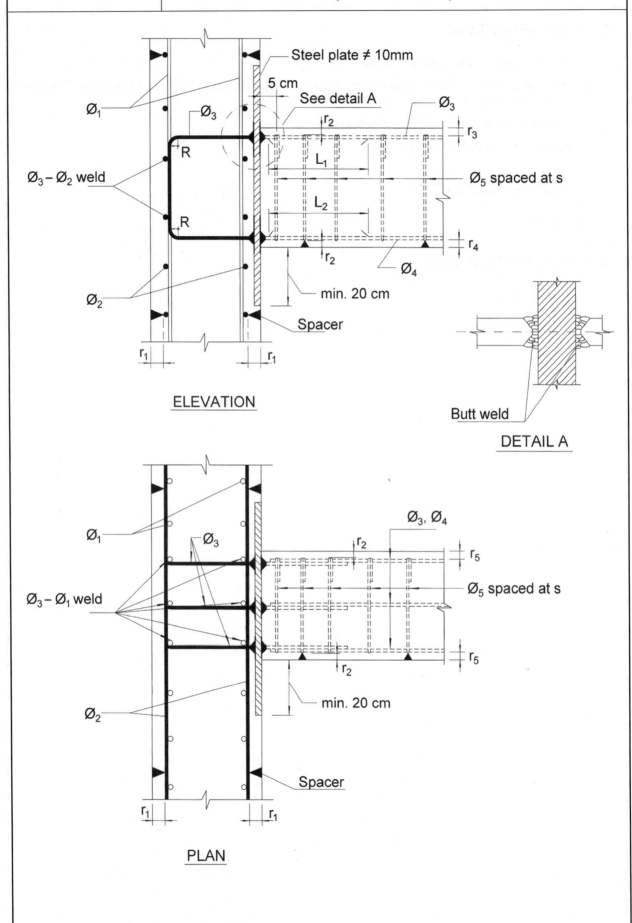

| CD – 02.25 | DIAPHRAGM WALL–BEAM BOND (VARIATION 1) |

Steel plate ≠ 10mm

5 cm

See detail A

ELEVATION

DETAIL A

Butt weld

min. 20 cm

Spacer

PLAN

1. RECOMMENDATIONS

1. See Recommendations in CD – 02.23.

2. $r_1 = 7.5$ cm; $r_2 = 2.5$ cm; $r_3 = 2.5$ cm; $r_4 = 2.5$ cm; $r_5 = 2.5$ cm.

3. See 1.4 for welding details.

4. Particular care is needed when welding ϕ_3 to ϕ_2 and ϕ_1 bars.

5. Ensure plate attachment stiffness. In any event, the design for the plate–beam bond must make allowance for plate sliding and shifting when the diaphragm wall concrete is poured.

6. L_1 is the ϕ_3 bar lap length.

7. L_2 is the ϕ_4 lap length.

8. See 1.2 and 1.3 for descriptions of how to tie bars and place spacers.

2. STATUTORY LEGISLATION

None in place.

3. SPECIFIC REFERENCES

See Chapter 11 in (27).

CD – 02.26	DIAPHRAGM WALL–BEAM BOND (VARIATION 2)

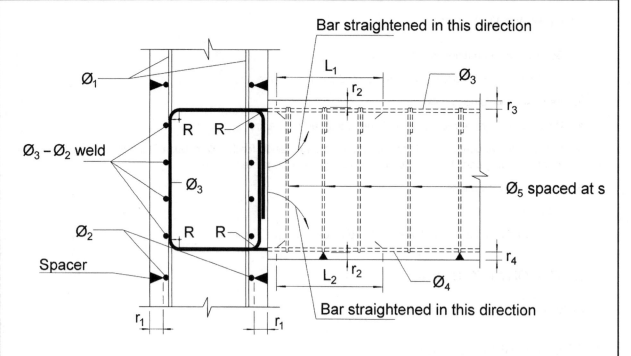

ELEVATION

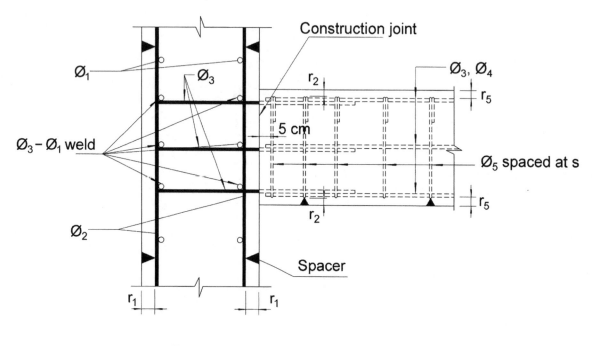

PLAN

1. RECOMMENDATIONS

1. See Recommendations in CD – 02.23.

2. $r_1 = 7.5$ cm; $r_2 = 2.5$ cm; $r_3 = 2.5$ cm; $r_4 = 2.5$ cm; $r_5 = 2.5$ cm.

3. See 1.4 for welding details.

4. The ϕ_3 bars designed to be straightened are to be welded to the ϕ_1 and ϕ_2 bars and must be left uncovered at the inner side of the diaphragm wall.

5. Radius R should be the standard bend radius.

6. Ensure plate attachment stiffness. In any event, the design for the plate–beam bond must make allowance for plate sliding and shifting when the diaphragm wall concrete is poured.

7. L_1 is the ϕ_3 bar lap length.

8. L_2 is the ϕ_4 lap length.

9. See 1.2 and 1.3 for descriptions of how to tie bars and place spacers.

2. STATUTORY LEGISLATION

None in place.

3. SPECIFIC REFERENCES

See Chapter 11 in (27).

DIAPHRAGM WALL–BEAM BOND
(VARIATION 3)

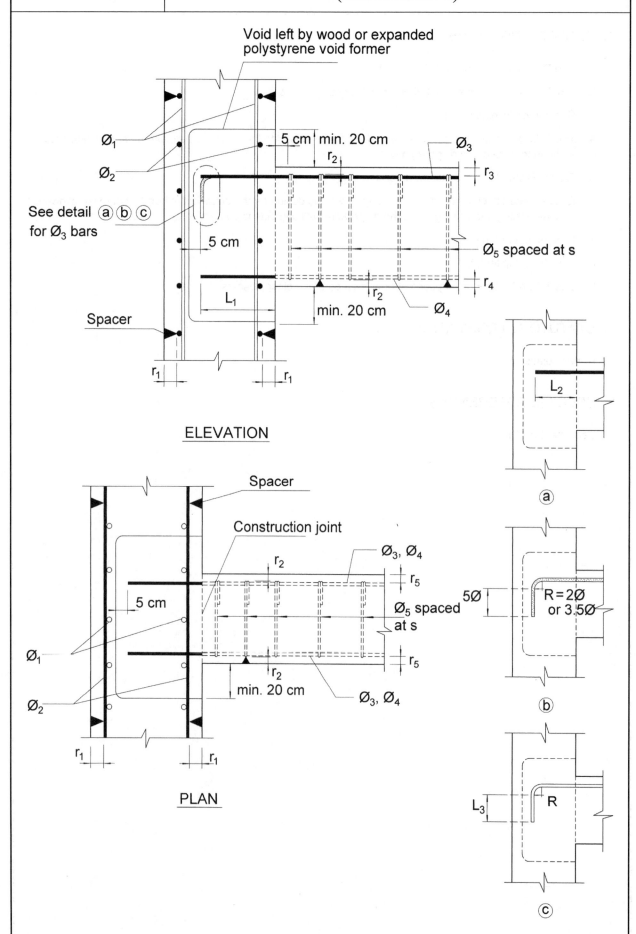

Void left by wood or expanded polystyrene void former

$\emptyset_1$

$\emptyset_2$

See detail (a)(b)(c) for $\emptyset_3$ bars

5 cm

Spacer

r_1

5 cm | min. 20 cm

r_2

$\emptyset_3$

r_3

$\emptyset_5$ spaced at s

r_4

r_2

$\emptyset_4$

L_1

min. 20 cm

r_1

ELEVATION

Spacer

Construction joint

$\emptyset_3, \emptyset_4$

r_2

r_5

5 cm

$\emptyset_5$ spaced at s

$\emptyset_1$

r_5

$\emptyset_2$

r_2

min. 20 cm

$\emptyset_3, \emptyset_4$

r_1

r_1

PLAN

L_2

(a)

$5\emptyset$

$R = 2\emptyset$ or $3.5\emptyset$

(b)

L_3

R

(c)

1. RECOMMENDATIONS

1. See Recommendations in CD – 02.23.

2. $r_1 = 7.5$ cm ; $r_2 = 2.5$ cm ; $r_3 \geq \phi_3$; $r_4 \geq \phi_4$; $r_5 \geq \phi_3$; $\phi_5 \geq \phi_4$.

3. The void formers must be carefully tied to the diaphragm wall cage. In any event, the design for the plate–beam bond must make allowance for void formers and shifting when the diaphragm wall concrete is poured.

4. Before casting the concrete to form the vertical member, the concrete around the void area should be cleaned and moistened. The new concrete should not be cast until the surface of the existing concrete dries.

5. Special care must be taken during concrete pouring and compacting in the void area.

6. See 1.2 and 1.3 for descriptions of how to tie bars and place spacers.

2. STATUTORY LEGISLATION

None in place.

3. SPECIFIC REFERENCES

See Chapter 11 in (27).

DIAPHRAGM WALL–SLAB BOND
(VARIATION 1)

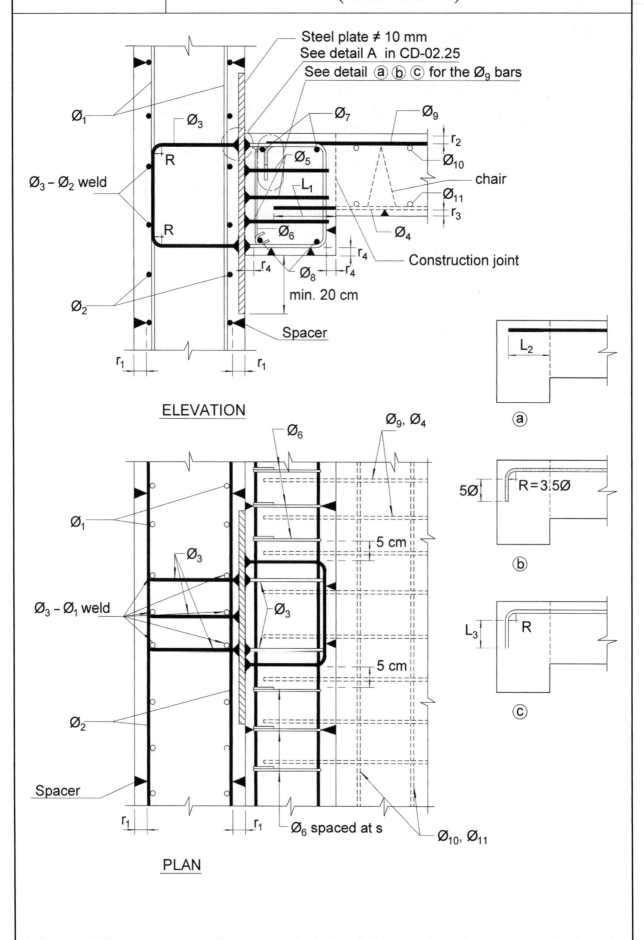

Steel plate ≠ 10 mm
See detail A in CD-02.25
See detail ⓐ ⓑ ⓒ for the Ø₉ bars

$\varnothing_1$

$\varnothing_3$

$\varnothing_7$

$\varnothing_9$

R

$\varnothing_5$

r_2

$\varnothing_{10}$

$\varnothing_3 - \varnothing_2$ weld

L_1

chair

$\varnothing_{11}$

r_3

R

$\varnothing_6$

$\varnothing_4$

r_4

Construction joint

r_4

$\varnothing_8$

r_4

$\varnothing_2$

min. 20 cm

Spacer

r_1

r_1

ELEVATION

$\varnothing_6$

$\varnothing_9, \varnothing_4$

$\varnothing_1$

5 cm

$\varnothing_3$

$\varnothing_3 - \varnothing_1$ weld

$\varnothing_3$

5 cm

$\varnothing_2$

Spacer

r_1

r_1

$\varnothing_6$ spaced at s

$\varnothing_{10}, \varnothing_{11}$

PLAN

L_2

ⓐ

5Ø

$R = 3.5\varnothing$

ⓑ

L_3

R

ⓒ

1. RECOMMENDATIONS

1. See the Recommendations in CD – 02.23 for the diaphragm wall.

2. $r_1 = 7.5$ cm ; $r_2 = 2.5$ cm $\geq \phi_9$.

3. $r_3 = 2.5$ cm $\geq \phi_4$.

4. $r_4 = 2.5$ cm.

5. See 1.4 for welding details.

6. See 1.2 and 1.3 for descriptions of how to tie bars and place spacers.

2. STATUTORY LEGISLATION

None in place.

3. SPECIFIC REFERENCES

See Chapter 11 in (27).

DIAPHRAGM WALL–SLAB BOND
(VARIATION 2)

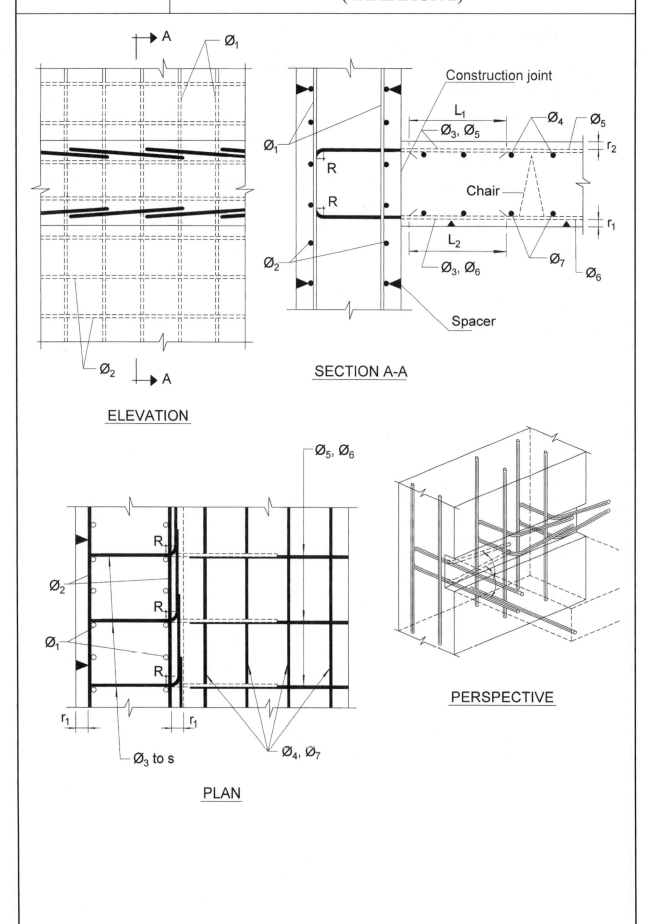

ELEVATION

SECTION A-A

PLAN

PERSPECTIVE

1. RECOMMENDATIONS

1. See the Recommendations in CD – 02.23 for the diaphragm wall.

2. For the bend radius R see Table T-1.1.

3. The ϕ_3 bars are welded to the vertical diaphragm wall reinforcement. They must be uncovered on the inner side, to be readily visible with superficial chipping.

4. See 1.4 for welding details.

5. r_1 = 7.5 cm.

6. r_2 = 2.5 cm.

7. L_1 is the lap length of the thicker of the ϕ_3 or ϕ_5 bars.

8. L_2 is the lap length of the thicker of the ϕ_3 or ϕ_6 bars.

9. See 1.2 and 1.3 for descriptions of how to tie bars and place spacers.

2. STATUTORY LEGISLATION

None in place.

3. SPECIFIC REFERENCES

See Chapter 11 in (27).

WALLS. DRAINAGE CHAMBER AND
CHANNEL IN DIAPHRAGM WALL

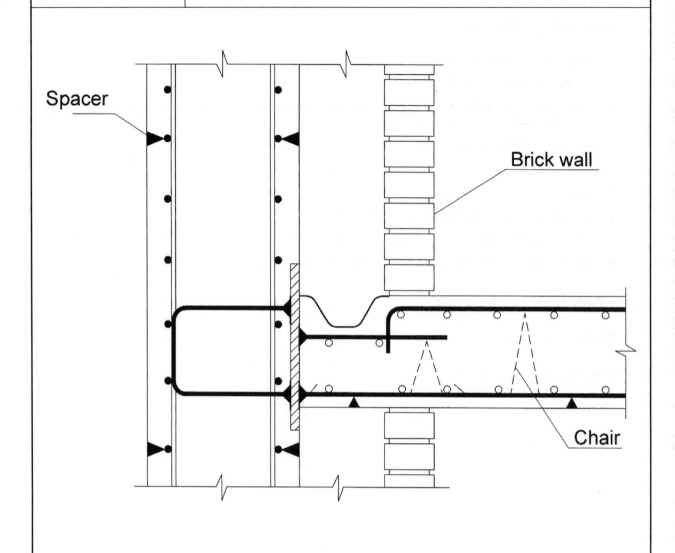

Spacer

Brick wall

Chair

1. RECOMMENDATIONS

1. This is the solution to be used to prevent water from seeping across the diaphragm wall.

2. The detail shows a diaphragm wall–slab bond analogous to the bond described in CD – 02.28. The system can be adapted to any other circumstance.

3. The channel should slope toward the downpipes.

4. Openings must be left in the brick wall to allow water seeping across the diaphragm wall to evaporate.

2. STATUTORY LEGISLATION

None in place.

3. SPECIFIC REFERENCES

See Chapter 11 in (27).

Group 03

Columns and joints

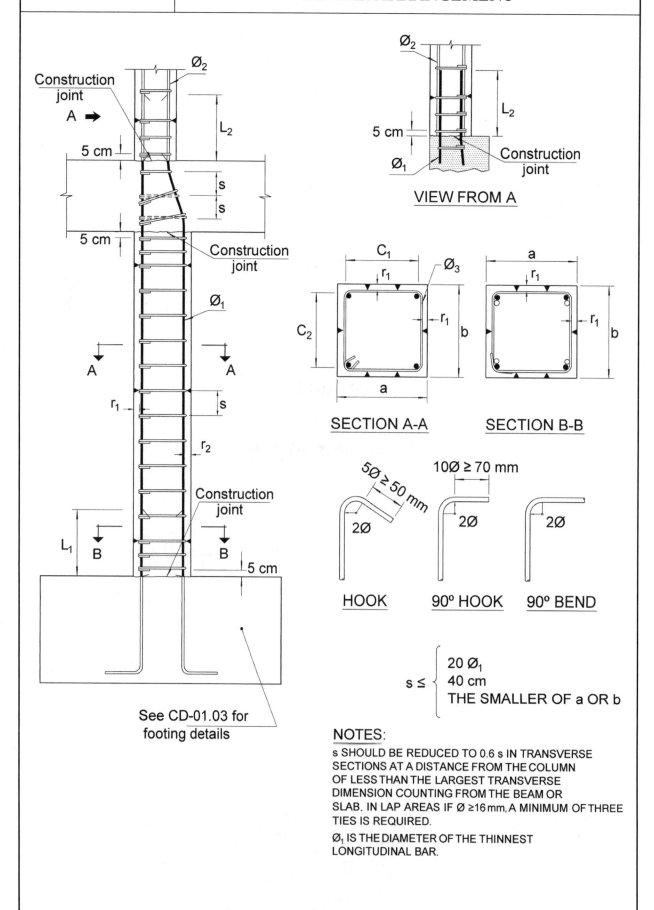

| CD – 03.01 | COLUMNS SPRINGING FROM THE FOOTING. TIE BAR ARRANGEMENT |

VIEW FROM A

SECTION A-A   SECTION B-B

HOOK   90° HOOK   90° BEND

$$s \le \begin{cases} 20\ \emptyset_1 \\ 40\ cm \\ \text{THE SMALLER OF a OR b} \end{cases}$$

See CD-01.03 for footing details

NOTES:

s SHOULD BE REDUCED TO 0.6 s IN TRANSVERSE SECTIONS AT A DISTANCE FROM THE COLUMN OF LESS THAN THE LARGEST TRANSVERSE DIMENSION COUNTING FROM THE BEAM OR SLAB. IN LAP AREAS IF Ø ≥16 mm, A MINIMUM OF THREE TIES IS REQUIRED.

$\emptyset_1$ IS THE DIAMETER OF THE THINNEST LONGITUDINAL BAR.

| CD – 03.01 | NOTES |

1. RECOMMENDATIONS

1. L_1 is the ϕ_1 bar lap length. L_2 is the ϕ_2 bar lap length.

2. See Tables T-1.1 and T-1.2 for bend radii.

3. Cover r_1 = 2.5 cm.

4. Cover r_2 may not be less than ϕ_1.

5. The tie bars are bent in the starter bars and hooked in all others.

6. As a rule, the tie bars in the upper and lower lengths of the column can be used in the joints if appropriately slanted (solid line). Otherwise, special tie bars are needed (dashed line).

7. See CD – 03.16 for tie bar shapes and bar layouts.

8. See 1.2 for bar tying procedures and 1.3 for spacer placement.

9. Before pouring the concrete for the column, the bottom joint should be cleaned and moistened, and the new concrete should not be cast until the existing concrete surface dries.

2. STATUTORY LEGISLATION

See EC2 (5).

3. RECOMMENDED ALTERNATIVE CODES

See ACI 318-08 (22) and (15).

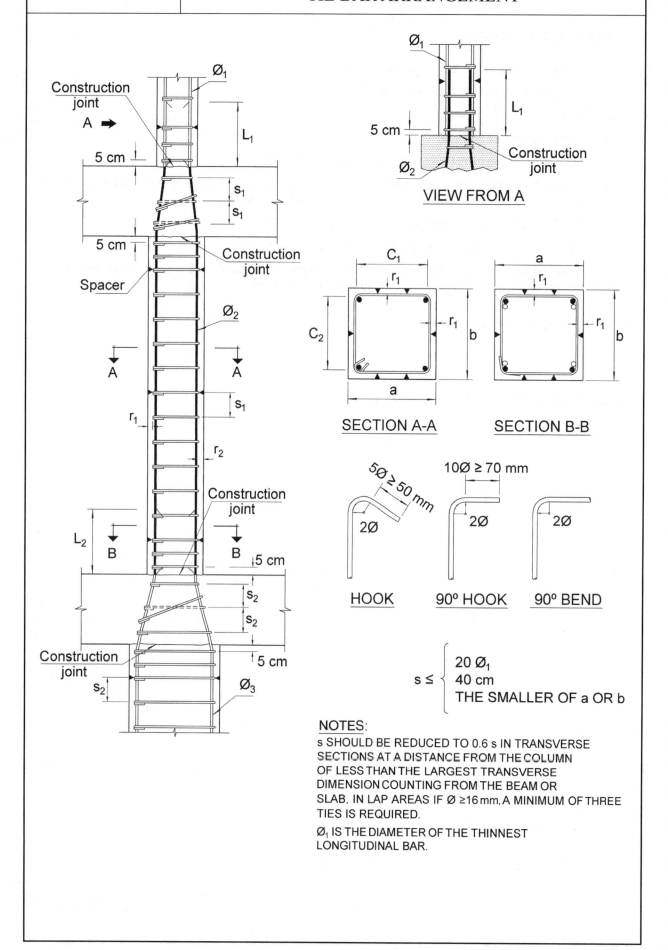

COLUMNS IN INTERMEDIATE STOREYS.
TIE BAR ARRANGEMENT

Construction joint

$\varnothing_1$

A →

5 cm

L_1

s_1

s_1

5 cm

Construction joint

Spacer

$\varnothing_2$

A   A

s_1

r_1

r_2

Construction joint

L_2

B   B

5 cm

s_2

s_2

5 cm

Construction joint

s_2

$\varnothing_3$

$\varnothing_1$

L_1

5 cm

$\varnothing_2$

Construction joint

VIEW FROM A

SECTION A-A

C_1

r_1

C_2

r_1

a

b

SECTION B-B

a

r_1

r_1

b

HOOK

$5\varnothing \geq 50$ mm

$2\varnothing$

90° HOOK

$10\varnothing \geq 70$ mm

$2\varnothing$

90° BEND

$2\varnothing$

$$s \leq \begin{cases} 20\,\varnothing_1 \\ 40 \text{ cm} \\ \text{THE SMALLER OF } a \text{ OR } b \end{cases}$$

NOTES:

s SHOULD BE REDUCED TO 0.6 s IN TRANSVERSE SECTIONS AT A DISTANCE FROM THE COLUMN OF LESS THAN THE LARGEST TRANSVERSE DIMENSION COUNTING FROM THE BEAM OR SLAB. IN LAP AREAS IF $\varnothing \geq 16$ mm, A MINIMUM OF THREE TIES IS REQUIRED.

$\varnothing_1$ IS THE DIAMETER OF THE THINNEST LONGITUDINAL BAR.

1. RECOMMENDATIONS

1. L_1 is the ϕ_1 bar lap length. L_2 is the ϕ_2 bar lap length.

2. See Tables T-1.1 and T-1.2 for bend radii.

3. Cover $r_1 = 2.5$ cm.

4. Cover $r_2 \geq \phi_2$.

5. The tie bars are bent in the starter bars and hooked in all others.

6. As a rule, the tie bars in the upper and lower lengths of the column can be used in the joints if appropriately slanted (solid line). Otherwise, special tie bars are needed (dashed line).

7. See CD – 03.16 for tie bar shapes and bar layouts.

8. See 1.2 for bar tying procedures and 1.3 for spacer placement.

9. Before pouring the concrete for the column, the bottom joint should be cleaned and moistened, and the new concrete should not be cast until the existing concrete surface dries.

2. STATUTORY LEGISLATION

See EC2 (5).

3. RECOMMENDED ALTERNATIVE CODES

See ACI 318-08 (22) and (15).

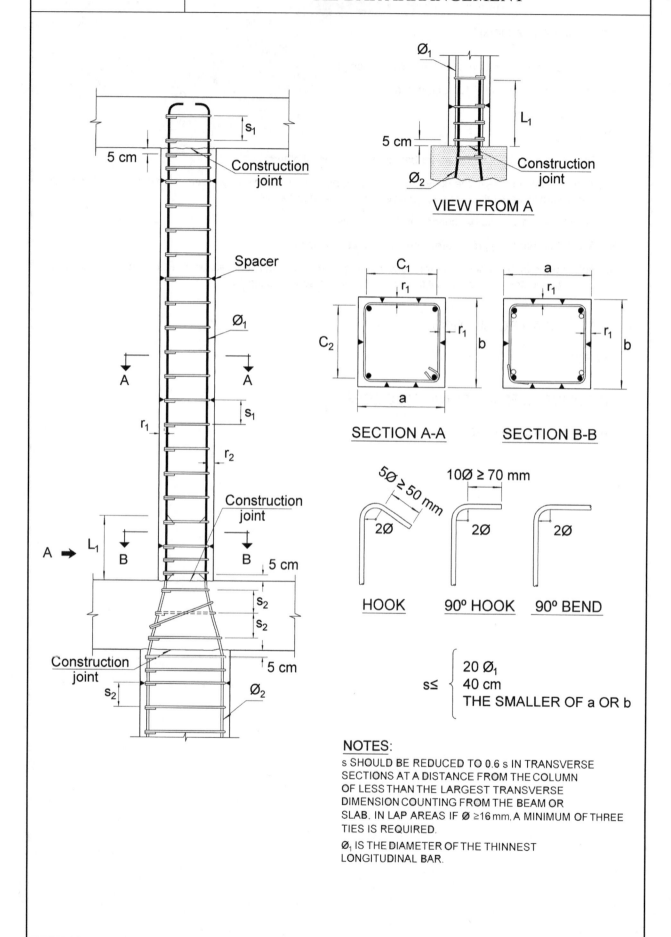

CD – 03.03

COLUMNS IN TOP STOREY.
TIE BAR ARRANGEMENT

s_1

5 cm

Construction joint

$\emptyset_1$

L_1

5 cm

$\emptyset_2$

Construction joint

VIEW FROM A

Spacer

$\emptyset_1$

A — A

s_1

r_1

r_2

Construction joint

$A \rightarrow$ L_1

B

B

5 cm

Construction joint

s_2

$\emptyset_2$

s_2

s_2

5 cm

C_1

r_1

C_2

r_1

b

a

SECTION A-A

a

r_1

r_1

b

SECTION B-B

$5\emptyset \geq 50$ mm

$2\emptyset$

HOOK

$10\emptyset \geq 70$ mm

$2\emptyset$

90° HOOK

$2\emptyset$

90° BEND

$$s \leq \begin{cases} 20\ \emptyset_1 \\ 40\ \text{cm} \\ \text{THE SMALLER OF a OR b} \end{cases}$$

NOTES:

s SHOULD BE REDUCED TO 0.6 s IN TRANSVERSE SECTIONS AT A DISTANCE FROM THE COLUMN OF LESS THAN THE LARGEST TRANSVERSE DIMENSION COUNTING FROM THE BEAM OR SLAB. IN LAP AREAS IF $\emptyset \geq 16$ mm, A MINIMUM OF THREE TIES IS REQUIRED.

$\emptyset_1$ IS THE DIAMETER OF THE THINNEST LONGITUDINAL BAR.

1. RECOMMENDATIONS

1. L_1 is the ϕ_1 bar lap length.

2. See Tables T-1.1 and T-1.2 for bend radii.

3. Cover $r_1 = 2.5$ cm.

4. Cover $r_2 \geq \phi_1$.

5. The tie bars are bent in the starter bars and hooked in all others.

6. As a rule, the tie bars in the upper and lower lengths of the column can be used in the joints if appropriately slanted (solid line). Otherwise, special tie bars are needed (dashed line).

7. See CD – 03.16 for tie bar shapes and bar layouts.

8. See 1.2 for bar tying procedures and 1.3 for spacer placement.

9. Before pouring the concrete for the column, the bottom joint should be cleaned and moistened, and the new concrete should not be cast until the existing concrete surface dries. Analogously, the top joint should be cleaned and moistened before pouring the concrete for the lintel.

2. STATUTORY LEGISLATION

See EC2 (5).

3. RECOMMENDED ALTERNATIVE CODES

See ACI 318-08 (22) and (15).

INTERMEDIATE JOINT IN EDGE COLUMNS
(VARIATION 1)

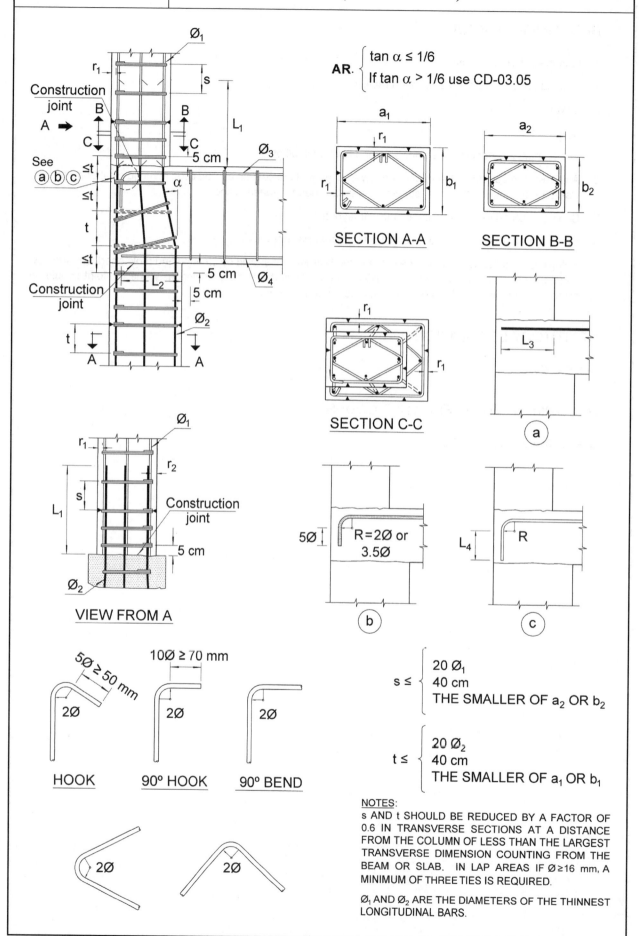

AR. $\tan \alpha \leq 1/6$
If $\tan \alpha > 1/6$ use CD-03.05

SECTION A-A

SECTION B-B

SECTION C-C

a

b

c

VIEW FROM A

$5\emptyset \geq 50$ mm
$2\emptyset$
HOOK

$10\emptyset \geq 70$ mm
$2\emptyset$
90° HOOK

$2\emptyset$
90° BEND

$2\emptyset$

$2\emptyset$

$s \leq \begin{cases} 20\,\emptyset_1 \\ 40 \text{ cm} \\ \text{THE SMALLER OF } a_2 \text{ OR } b_2 \end{cases}$

$t \leq \begin{cases} 20\,\emptyset_2 \\ 40 \text{ cm} \\ \text{THE SMALLER OF } a_1 \text{ OR } b_1 \end{cases}$

NOTES:
s AND t SHOULD BE REDUCED BY A FACTOR OF 0.6 IN TRANSVERSE SECTIONS AT A DISTANCE FROM THE COLUMN OF LESS THAN THE LARGEST TRANSVERSE DIMENSION COUNTING FROM THE BEAM OR SLAB. IN LAP AREAS IF $\emptyset \geq 16$ mm, A MINIMUM OF THREE TIES IS REQUIRED.

$\emptyset_1$ AND $\emptyset_2$ ARE THE DIAMETERS OF THE THINNEST LONGITUDINAL BARS.

1. RECOMMENDATIONS

1. L_1 is the ϕ_1 bar lap length.

2. See Tables T-1.1 and T-1.2 for bend radii.

3. Cover $r_1 = 2.5$ cm.

4. Cover $r_2 \geq \phi_2$.

5. To calculate lengths L_3 and L_4, see EC2, anchorage lengths.

6. The tie bars are bent in the starter bars and hooked in all others.

7. As a rule, the tie bars in the upper and lower lengths of the column can be used in the joints if appropriately slanted (solid line). Otherwise, special tie bars are needed (dashed line).

8. See CD – 03.16 for tie bar shapes and bar layouts.

9. See 1.2 for bar tying procedures and 1.3 for spacer placement.

10. Angle α is understood to be the actual measured magnitude. According to EC2, tie bars need not be calculated if $\tan \alpha \leq \dfrac{1}{12}$. ACI 318-08 simply recommends $\tan \alpha \leq \dfrac{1}{6}$.

2. STATUTORY LEGISLATION

See EC2 (5).

3. RECOMMENDED ALTERNATIVE CODES

See ACI 318-08 (22) and (15).

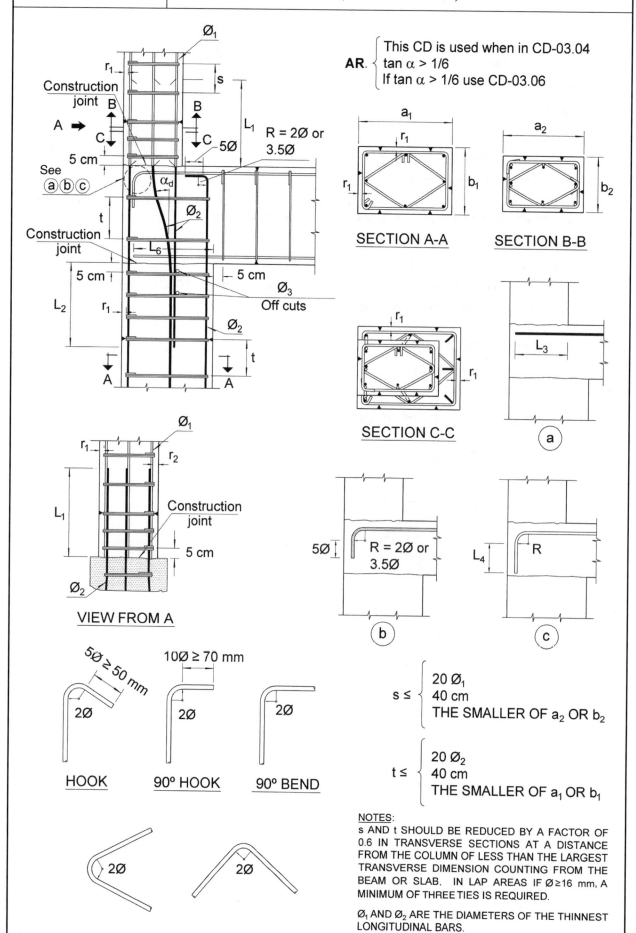

CD – 03.05

INTERMEDIATE JOINT IN EDGE COLUMNS (VARIATION 2)

Construction joint

$\varnothing_1$

r_1

A →

B

L_1

R = 2Ø or 3.5Ø

5Ø

5 cm

See
ⓐⓑⓒ

α_d

$\varnothing_2$

t

Construction joint

L_6

5 cm

5 cm

$\varnothing_3$

Off cuts

L_2

r_1

$\varnothing_2$

t

AR. This CD is used when in CD-03.04
tan α > 1/6
If tan α > 1/6 use CD-03.06

a_1

r_1

r_1

b_1

SECTION A-A

a_2

b_2

SECTION B-B

r_1

r_1

SECTION C-C

L_3

ⓐ

$\varnothing_1$

r_1

r_2

L_1

Construction joint

5 cm

$\varnothing_2$

VIEW FROM A

5Ø R = 2Ø or 3.5Ø

ⓑ

L_4

R

ⓒ

5Ø ≥ 50 mm
2Ø
HOOK

10Ø ≥ 70 mm
2Ø
90º HOOK

2Ø
90º BEND

$s \leq$ { 20 $\varnothing_1$
40 cm
THE SMALLER OF a_2 OR b_2 }

$t \leq$ { 20 $\varnothing_2$
40 cm
THE SMALLER OF a_1 OR b_1 }

2Ø

2Ø

NOTES:
s AND t SHOULD BE REDUCED BY A FACTOR OF 0.6 IN TRANSVERSE SECTIONS AT A DISTANCE FROM THE COLUMN OF LESS THAN THE LARGEST TRANSVERSE DIMENSION COUNTING FROM THE BEAM OR SLAB. IN LAP AREAS IF Ø≥16 mm, A MINIMUM OF THREE TIES IS REQUIRED.

$\varnothing_1$ AND $\varnothing_2$ ARE THE DIAMETERS OF THE THINNEST LONGITUDINAL BARS.

1. RECOMMENDATIONS

1. L_1 is the ϕ_1 bar lap length. L_2 is the ϕ_2 bar lap length.

2. See Tables T-1.1 and T-1.2 for bend radii.

3. Cover $r_1 = 2.5$ cm.

4. Cover $r_2 \geq \phi_2$.

5. To calculate lengths L_3 and L_4, see EC2, anchorage lengths.

6. The tie bars are bent in the starter bars and hooked in all others.

7. For tie bars in the joint, see the respective details.

8. See CD – 03.16 for tie bar shapes and bar layouts.

9. See 1.2 for bar tying procedures and 1.3 for spacer placement.

10. Angle α is understood to be the actual measured magnitude.

2. STATUTORY LEGISLATION

See EC2 (5).

3. RECOMMENDED ALTERNATIVE CODES

See ACI 318-08 (22) and (15).

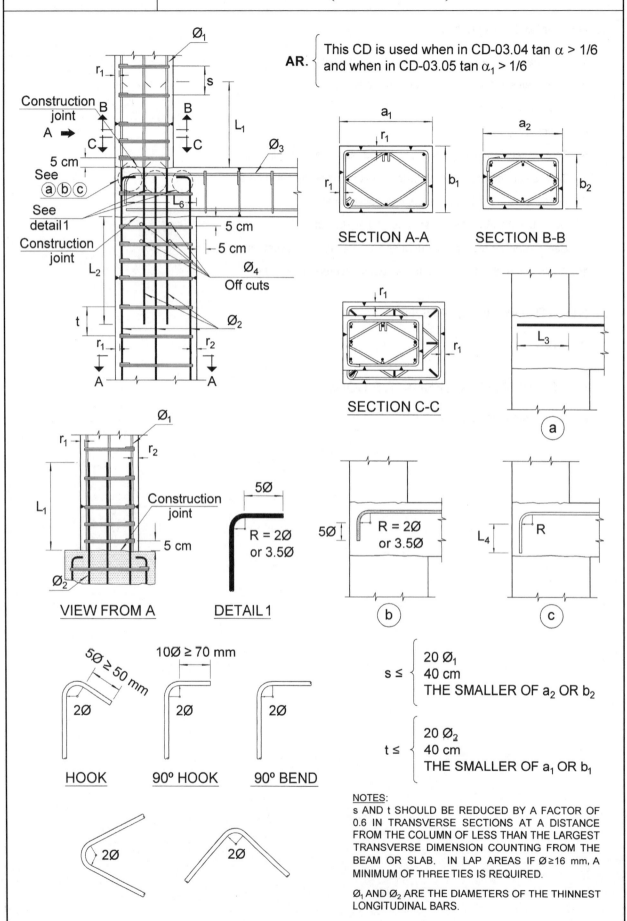

| CD – 03.06 | INTERMEDIATE JOINT IN EDGE COLUMNS (VARIATION 3) |

AR. $\begin{cases}$ This CD is used when in CD-03.04 tan α > 1/6 and when in CD-03.05 tan α_1 > 1/6 $\end{cases}$

SECTION A-A

SECTION B-B

SECTION C-C

VIEW FROM A

DETAIL 1

a

b

c

HOOK

90° HOOK

90° BEND

$s \leq \begin{cases} 20\ \emptyset_1 \\ 40\ cm \\ \text{THE SMALLER OF } a_2 \text{ OR } b_2 \end{cases}$

$t \leq \begin{cases} 20\ \emptyset_2 \\ 40\ cm \\ \text{THE SMALLER OF } a_1 \text{ OR } b_1 \end{cases}$

NOTES:
s AND t SHOULD BE REDUCED BY A FACTOR OF 0.6 IN TRANSVERSE SECTIONS AT A DISTANCE FROM THE COLUMN OF LESS THAN THE LARGEST TRANSVERSE DIMENSION COUNTING FROM THE BEAM OR SLAB. IN LAP AREAS IF $\emptyset \geq 16$ mm, A MINIMUM OF THREE TIES IS REQUIRED.

$\emptyset_1$ AND $\emptyset_2$ ARE THE DIAMETERS OF THE THINNEST LONGITUDINAL BARS.

1. RECOMMENDATIONS

1. L_1 is the ϕ_1 bar lap length. L_2 is the ϕ_2 bar lap length.
2. See Tables T-1.1 and T-1.2 for bend radii.
3. Cover $r_1 = 2.5$ cm.
4. The tie bars are bent in the starter bars and hooked in all others.
5. For tie bars in the joint, see the respective details.
6. Cover $r_2 \geq \phi_2$.
7. To calculate lengths L_3 and L_4, see EC2, anchorage lengths.
8. See 1.2 for bar tying procedures and 1.3 for spacer placement.

2. STATUTORY LEGISLATION

See EC2 (5).

3. RECOMMENDED ALTERNATIVE CODES

See ACI 318-08 (22) and (15).

INTERMEDIATE CORNER JOINT
(VARIATION 1)

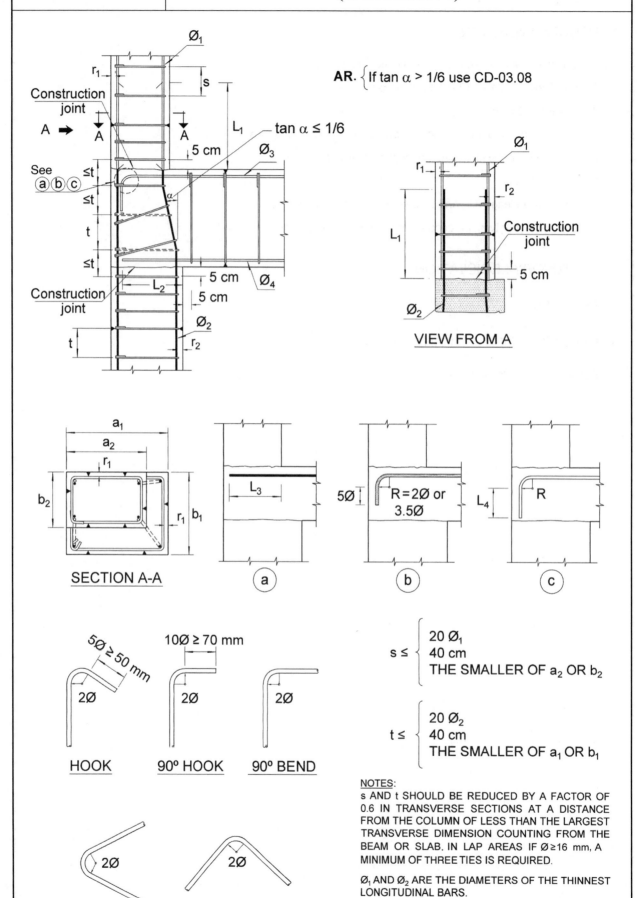

AR. $\{$ If tan α > 1/6 use CD-03.08

tan α ≤ 1/6

VIEW FROM A

SECTION A-A

(a)

50 $R = 2\emptyset$ or 3.5$\emptyset$

(b)

(c)

HOOK
5$\emptyset$ ≥ 50 mm
2$\emptyset$

90° HOOK
10$\emptyset$ ≥ 70 mm
2$\emptyset$

90° BEND
2$\emptyset$

2$\emptyset$

2$\emptyset$

$s \le \begin{cases} 20\ \emptyset_1 \\ 40\ \text{cm} \\ \text{THE SMALLER OF } a_2 \text{ OR } b_2 \end{cases}$

$t \le \begin{cases} 20\ \emptyset_2 \\ 40\ \text{cm} \\ \text{THE SMALLER OF } a_1 \text{ OR } b_1 \end{cases}$

NOTES:
s AND t SHOULD BE REDUCED BY A FACTOR OF 0.6 IN TRANSVERSE SECTIONS AT A DISTANCE FROM THE COLUMN OF LESS THAN THE LARGEST TRANSVERSE DIMENSION COUNTING FROM THE BEAM OR SLAB. IN LAP AREAS IF $\emptyset \ge 16$ mm, A MINIMUM OF THREE TIES IS REQUIRED.

$\emptyset_1$ AND $\emptyset_2$ ARE THE DIAMETERS OF THE THINNEST LONGITUDINAL BARS.

182

1. RECOMMENDATIONS

1. L_1 is the ϕ_1 bar lap length.

2. See Tables T-1.1 and T-1.2 for bend radii.

3. Cover $r_1 = 2.5$ cm.

4. Cover $r_2 \geq \phi_2$.

5. To calculate lengths L_3 and L_4, see EC2, anchorage lengths.

6. The tie bars are bent in the starter bars and hooked in all others.

7. As a rule, the tie bars in the upper and lower lengths of the column can be used in the joints if appropriately slanted (solid line). Otherwise, special tie bars are needed (dashed line).

8. See CD – 03.16 for tie bar shapes and bar layouts.

9. See 1.2 for bar tying procedures and 1.3 for spacer placement.

10. Angle α is understood to be the actual measured magnitude.

2. STATUTORY LEGISLATION

See EC2 (5).

3. RECOMMENDED ALTERNATIVE CODES

See ACI 318-08 (22) and (15).

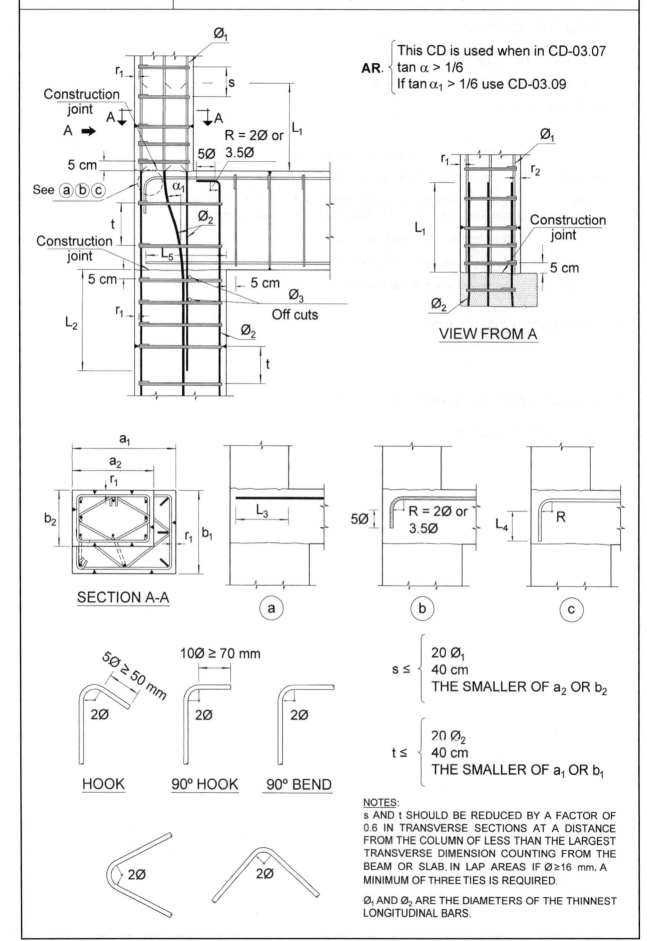

CD – 03.08

INTERMEDIATE CORNER JOINT (VARIATION 2)

AR. This CD is used when in CD-03.07
tan α > 1/6
If tan α_1 > 1/6 use CD-03.09

VIEW FROM A

SECTION A-A

(a) (b) (c)

HOOK   90° HOOK   90° BEND

$s \le$ 20 $\emptyset_1$
40 cm
THE SMALLER OF a_2 OR b_2

$t \le$ 20 $\emptyset_2$
40 cm
THE SMALLER OF a_1 OR b_1

NOTES:
s AND t SHOULD BE REDUCED BY A FACTOR OF 0.6 IN TRANSVERSE SECTIONS AT A DISTANCE FROM THE COLUMN OF LESS THAN THE LARGEST TRANSVERSE DIMENSION COUNTING FROM THE BEAM OR SLAB. IN LAP AREAS IF Ø ≥16 mm, A MINIMUM OF THREE TIES IS REQUIRED.

$\emptyset_1$ AND $\emptyset_2$ ARE THE DIAMETERS OF THE THINNEST LONGITUDINAL BARS.

1. RECOMMENDATIONS

1. The limitation on angle α is understood to be the actual spatial magnitude. The main bars are on different planes in this arrangement. Reinforcement schedules should be formulated with care. The arrangement specified does not reduce bending strength in the lap area.

2. L_1 is the ϕ_1 bar lap length. L_2 is the ϕ_2 bar lap length.

3. See Tables T-1.1 and T-1.2 for bend radii.

4. Cover $r_1 = 2.5$ cm.

5. Cover $r_2 \geq \phi_2$.

6. To calculate lengths L_3 and L_4, see EC2, anchorage lengths.

7. The tie bars are bent in the starter bars and hooked in all others.

8. For tie bars in the joint, see the respective details.

9. See CD – 03.16 for tie bar shapes and bar layouts.

10. See 1.2 for bar tying procedures and 1.3 for spacer placement.

2. STATUTORY LEGISLATION

See EC2 (5).

3. RECOMMENDED ALTERNATIVE CODES

See ACI 318-08 (22) and (15).

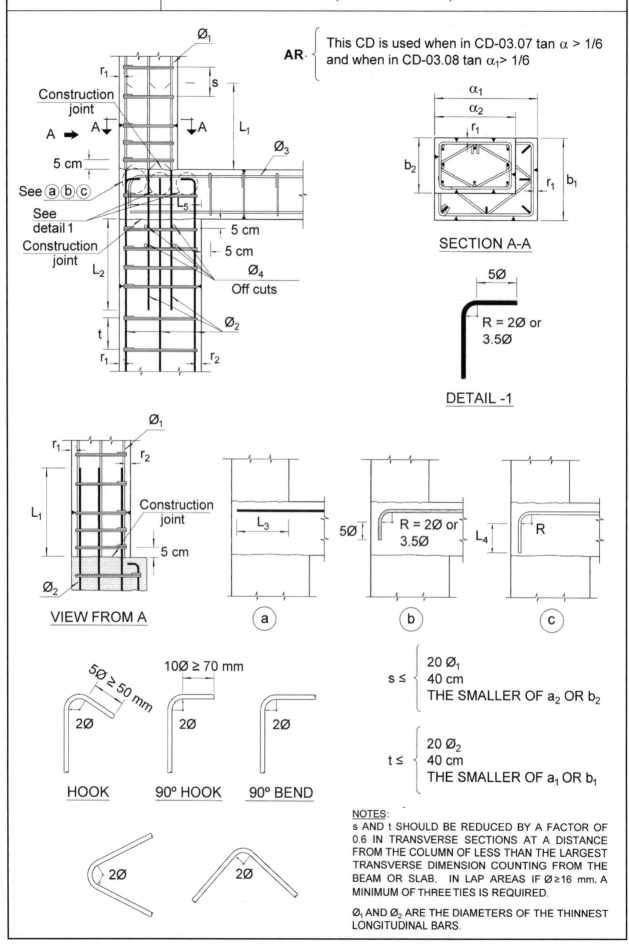

CD – 03.09

INTERMEDIATE CORNER JOINT (VARIATION 3)

AR. This CD is used when in CD-03.07 tan α > 1/6 and when in CD-03.08 tan α_1 > 1/6

Construction joint

See (a)(b)(c)

See detail 1

Construction joint

Off cuts

SECTION A-A

R = 2Ø or 3.5Ø

DETAIL -1

VIEW FROM A

(a)

(b) R = 2Ø or 3.5Ø

(c)

HOOK 5Ø ≥ 50 mm

90° HOOK 10Ø ≥ 70 mm

90° BEND

$s \leq$ $\begin{cases} 20\,Ø_1 \\ 40\ cm \\ \text{THE SMALLER OF } a_2 \text{ OR } b_2 \end{cases}$

$t \leq$ $\begin{cases} 20\,Ø_2 \\ 40\ cm \\ \text{THE SMALLER OF } a_1 \text{ OR } b_1 \end{cases}$

NOTES:
s AND t SHOULD BE REDUCED BY A FACTOR OF 0.6 IN TRANSVERSE SECTIONS AT A DISTANCE FROM THE COLUMN OF LESS THAN THE LARGEST TRANSVERSE DIMENSION COUNTING FROM THE BEAM OR SLAB. IN LAP AREAS IF Ø ≥ 16 mm, A MINIMUM OF THREE TIES IS REQUIRED.

Ø$_1$ AND Ø$_2$ ARE THE DIAMETERS OF THE THINNEST LONGITUDINAL BARS.

CD – 03.09	NOTES

1. RECOMMENDATIONS

1. L_1 is the ϕ_1 bar lap length. L_2 is the ϕ_2 bar lap length.
2. See Tables T-1.1 and T-1.2 for bend radii.
3. Cover $r_1 = 2.5$ cm.
4. Cover $r_2 \geq \phi_2$.
5. To calculate lengths L_4 and L_5, see EC2, anchorage lengths.
6. The tie bars are bent in the starter bars and hooked in all others.
7. For tie bars in the joint, see the respective details.
8. See CD – 03.16 for tie bar shapes and bar layouts.
9. See 1.2 for bar tying procedures and 1.3 for spacer placement.

2. STATUTORY LEGISLATION

See EC2 (5).

3. RECOMMENDED ALTERNATIVE CODES

See ACI 318-08 (22) and (15).

CD – 03.10	FACADE OR CORNER JOINT ON LAST STOREY

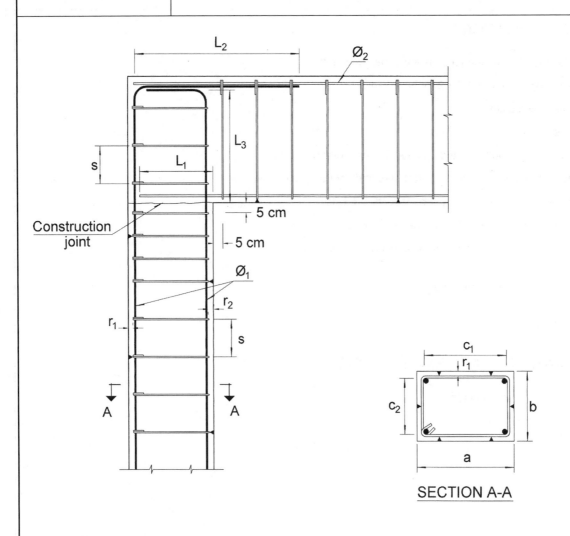

SECTION A-A

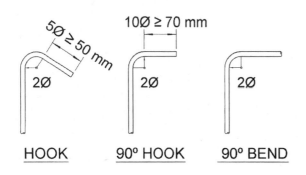

HOOK   90° HOOK   90° BEND

NOTES:
s SHOULD BE REDUCED BY A FACTOR OF 0.6 IN TRANSVERSE
SECTIONS AT A DISTANCE FROM THE COLUMN
OF LESS THAN THE LARGEST TRANSVERSE
DIMENSION COUNTING FROM THE BEAM OR
SLAB. IN LAP AREAS IF $\emptyset \geq 16$ mm, A MINIMUM OF THREE
TIES IS REQUIRED.
$\emptyset_1$ IS THE DIAMETERS OF THE THINNEST
LONGITUDINAL BARS.

$$s \leq \begin{cases} 20\ \emptyset_1 \\ 40\ \text{cm} \\ \text{THE SMALLER OF a OR b} \end{cases}$$

1. RECOMMENDATIONS

1. L_2 is the lap length of the thicker of the ϕ_1 or ϕ_2 bars.

2. See Tables T-1.1 and T-1.2 for bend radii.

3. Cover $r_1 = 2.5$ cm.

4. Cover $r_2 \geq \phi_1$.

5. To calculate length L_1, see EC2, anchorage lengths.

6. L_3 should not be shorter than the ϕ_1 bar anchorage length.

7. See CD – 03.16 for tie bar shapes and bar layouts.

8. See 1.2 for bar tying procedures and 1.3 for spacer placement.

2. STATUTORY LEGISLATION

See EC2 (5).

3. RECOMMENDED ALTERNATIVE CODES

See ACI 318-08 (22) and (15).

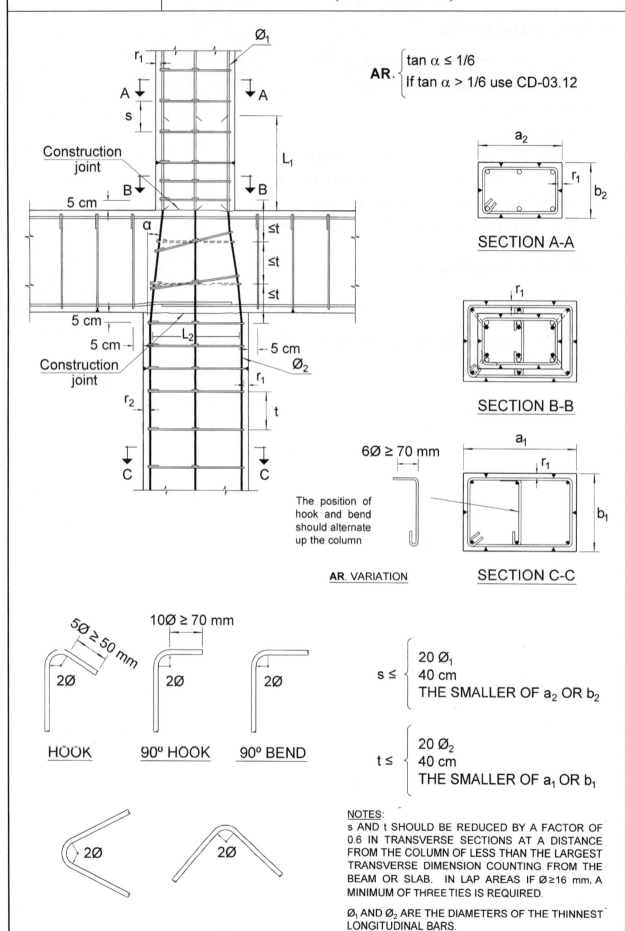

CD – 03.11

INSIDE JOINT IN INTERMEDIATE STOREYS (VARIATION 1)

$$AR. \begin{cases} \tan \alpha \le 1/6 \\ \text{If } \tan \alpha > 1/6 \text{ use CD-03.12} \end{cases}$$

SECTION A-A

SECTION B-B

$6\emptyset \ge 70$ mm

The position of hook and bend should alternate up the column

AR. VARIATION

SECTION C-C

$5\emptyset \ge 50$ mm

$10\emptyset \ge 70$ mm

$2\emptyset$

$2\emptyset$

$2\emptyset$

HOOK   90° HOOK   90° BEND

$2\emptyset$ $2\emptyset$

$$s \le \begin{cases} 20 \, \emptyset_1 \\ 40 \text{ cm} \\ \text{THE SMALLER OF } a_2 \text{ OR } b_2 \end{cases}$$

$$t \le \begin{cases} 20 \, \emptyset_2 \\ 40 \text{ cm} \\ \text{THE SMALLER OF } a_1 \text{ OR } b_1 \end{cases}$$

NOTES:
s AND t SHOULD BE REDUCED BY A FACTOR OF 0.6 IN TRANSVERSE SECTIONS AT A DISTANCE FROM THE COLUMN OF LESS THAN THE LARGEST TRANSVERSE DIMENSION COUNTING FROM THE BEAM OR SLAB. IN LAP AREAS IF Ø≥16 mm, A MINIMUM OF THREE TIES IS REQUIRED.

$\emptyset_1$ AND $\emptyset_2$ ARE THE DIAMETERS OF THE THINNEST LONGITUDINAL BARS.

CD – 03.11	NOTES

1. RECOMMENDATIONS

1. L_1 is the ϕ_1 bar lap length.

2. See Tables T-1.1 and T-1.2 for bend radii.

3. Cover $r_1 = 2.5$ cm.

4. Cover $r_2 \geq \phi_2$.

5. L_2 is the lap length.

6. The tie bars are bent in the starter bars and hooked in all others.

7. As a rule, the tie bars in the upper and lower lengths of the column can be used in the joints if appropriately slanted (solid line). Otherwise, special tie bars are needed (dashed line).

8. See CD – 03.16 for tie bar shapes and bar layouts.

9. See 1.2 for bar tying procedures and 1.3 for spacer placement.

10. Angle α is understood to be the actual measured magnitude.

11. The variation shown in section C-C, taken from ACI 318-08 (22) greatly simplifies placing of the reinforcement. It may not be used in areas subject to seismic risk.

2. STATUTORY LEGISLATION

See EC2 (5).

3. RECOMMENDED ALTERNATIVE CODES

See ACI 318-08 (22) and (15).

INSIDE JOINT IN INTERMEDIATE STOREYS (VARIATION 2)

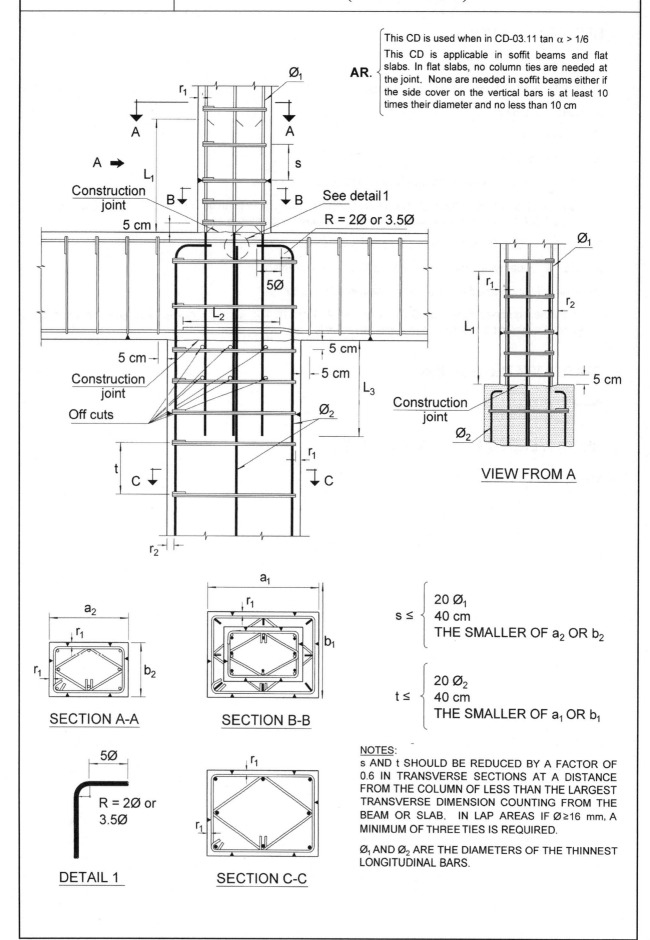

AR. This CD is used when in CD-03.11 tan α > 1/6

This CD is applicable in soffit beams and flat slabs. In flat slabs, no column ties are needed at the joint. None are needed in soffit beams either if the side cover on the vertical bars is at least 10 times their diameter and no less than 10 cm

See detail 1

R = 2Ø or 3.5Ø

VIEW FROM A

SECTION A-A

SECTION B-B

$$s \le \begin{cases} 20\,\varnothing_1 \\ 40 \text{ cm} \\ \text{THE SMALLER OF } a_2 \text{ OR } b_2 \end{cases}$$

$$t \le \begin{cases} 20\,\varnothing_2 \\ 40 \text{ cm} \\ \text{THE SMALLER OF } a_1 \text{ OR } b_1 \end{cases}$$

DETAIL 1

R = 2Ø or 3.5Ø

SECTION C-C

NOTES:
s AND t SHOULD BE REDUCED BY A FACTOR OF 0.6 IN TRANSVERSE SECTIONS AT A DISTANCE FROM THE COLUMN OF LESS THAN THE LARGEST TRANSVERSE DIMENSION COUNTING FROM THE BEAM OR SLAB. IN LAP AREAS IF Ø≥16 mm, A MINIMUM OF THREE TIES IS REQUIRED.

Ø₁ AND Ø₂ ARE THE DIAMETERS OF THE THINNEST LONGITUDINAL BARS.

1. RECOMMENDATIONS

1. L_1 is the ϕ_1 bar lap length. L_3 is the ϕ_2 bar lap length.

2. See Tables T-1.1 and T-1.2 for bend radii.

3. Cover $r_1 = 2.5$ cm.

4. Cover $r_2 \geq \phi_2$.

5. L_2 is the lap length.

6. The tie bars are bent in the starter bars and hooked in all others.

7. For tie bars in the joint, see the respective details.

8. See CD – 03.16 for tie bar shapes and bar layouts.

9. See 1.2 for bar tying procedures and 1.3 for spacer placement.

2. STATUTORY LEGISLATION

See EC2 (5).

3. RECOMMENDED ALTERNATIVE CODES

See ACI 318-08 (22) and (15).

CD – 03.13	INTERMEDIATE JOINT IN CIRCULAR COLUMNS

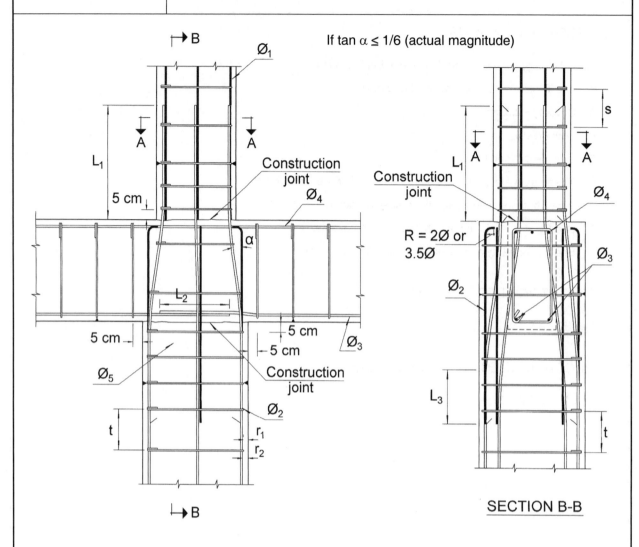

If tan $\alpha \leq 1/6$ (actual magnitude)

Construction joint

SECTION B-B

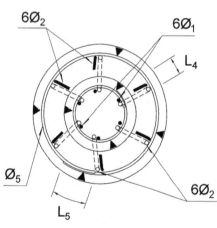

SECTION A-A

1. RECOMMENDATIONS

1. L_1 is the ϕ_1 bar lap length. L_2 is the ϕ_3 bar lap length.

2. See Tables T-1.1 and T-1.2 for bend radii.

3. Cover $r_1 = 2.5$ cm.

4. The tie bars are bent in the starter bars and hooked in all others.

5. L_3 is the ϕ_2 bar lap length.

6. The joint calls for special tie bars.

7. See 1.2 for bar tying procedures and 1.3 for spacer placement.

8. Angle α is understood to be the actual measured magnitude.

2. STATUTORY LEGISLATION

See EC2 (5).

3. RECOMMENDED ALTERNATIVE CODES

See ACI 318-08 (22) and (15).

TRANSITION FROM CIRCULAR TO RECTANGULAR COLUMNS

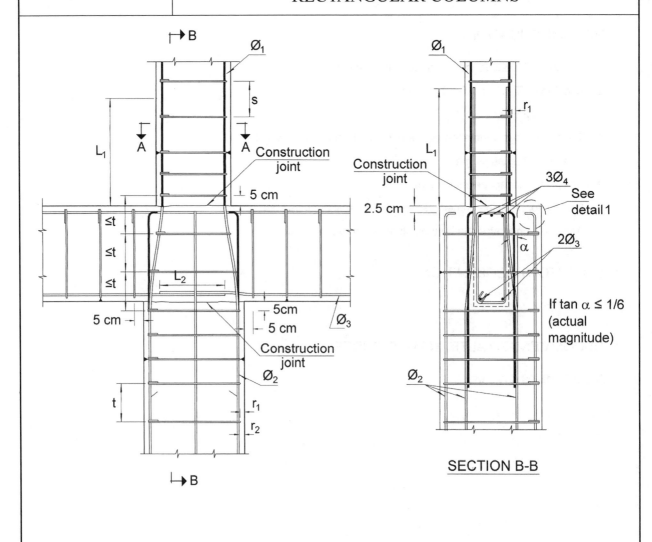

$\varnothing_1$

B

s

A

A

L_1

Construction joint

5 cm

$\leq t$

$\leq t$

$\leq t$

L_2

5 cm

5cm

5 cm

Construction joint

$\varnothing_2$

$\varnothing_3$

t

r_1

r_2

B

$\varnothing_1$

r_1

L_1

Construction joint

$3\varnothing_4$

See detail 1

2.5 cm

α

$2\varnothing_3$

If tan $\alpha \leq 1/6$ (actual magnitude)

$\varnothing_2$

SECTION B-B

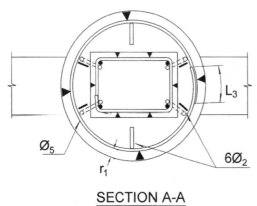

L_3

$\varnothing_5$

r_1

$6\varnothing_2$

SECTION A-A

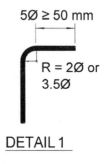

$5\varnothing \geq 50$ mm

R = $2\varnothing$ or $3.5\varnothing$

DETAIL 1

1. RECOMMENDATIONS

1. L_1 is the ϕ_1 bar lap length. L_2 is the ϕ_3 bar lap length.

2. See Tables T-1.1 and T-1.2 for bend radii.

3. Cover $r_1 = 2.5$ cm.

4. The tie bars are bent in the starter bars and hooked in all others.

5. L_3 is the ϕ_5 bar lap length.

6. The joint calls for special tie bars.

7. See 1.2 for bar tying procedures and 1.3 for spacer placement.

8. Angle α is understood to be the actual measured magnitude.

2. STATUTORY LEGISLATION

See EC2 (5).

3. RECOMMENDED ALTERNATIVE CODES

See ACI 318-08 (22) and (15).

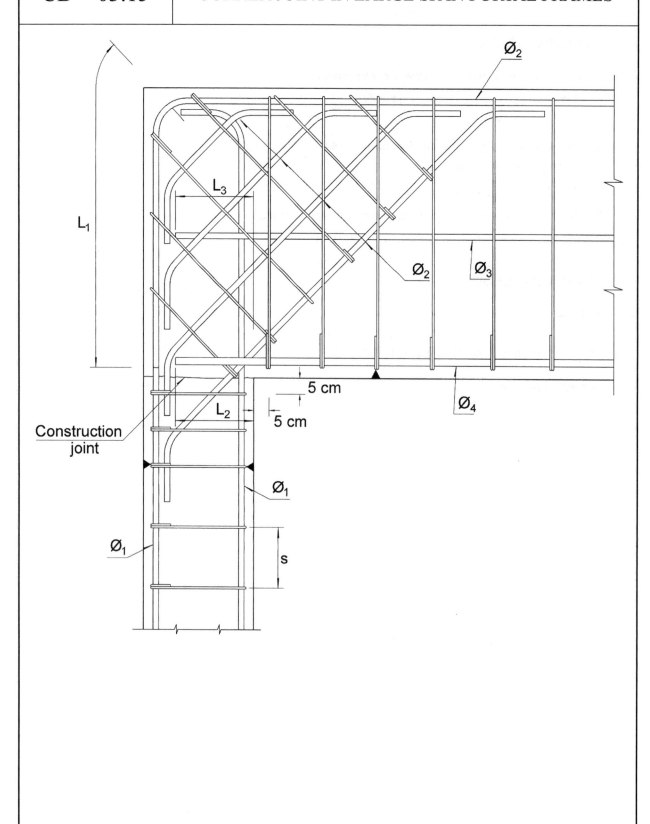

1. RECOMMENDATIONS

1. To calculate length L_2, see EC2, anchorage lengths.

2. L_1 is the lap length of the thicker of the ϕ_1 or ϕ_2 bars.

3. Bar ϕ_3 reinforces the web. It has a conventional anchorage (L_3). An arrangement calling for 20 $\phi_3 \geq 15$ cm is suggested.

2. STATUTORY LEGISLATION

None in place.

3. RECOMMENDED ALTERNATIVE CODES

None in place.

BAR ARRANGEMENT AND SHAPES OF TIES IN COLUMNS

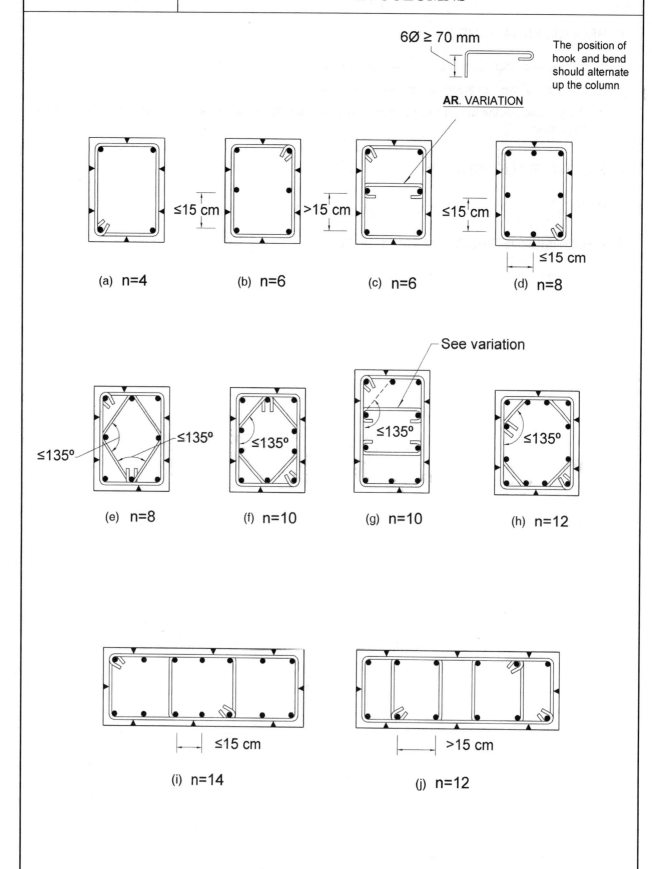

6Ø ≥ 70 mm

AR. VARIATION

The position of hook and bend should alternate up the column

≤15 cm   >15 cm   ≤15 cm   ≤15 cm

(a) n=4   (b) n=6   (c) n=6   (d) n=8

See variation

≤135°   ≤135°   ≤135°   ≤135°

(e) n=8   (f) n=10   (g) n=10   (h) n=12

≤15 cm   >15 cm

(i) n=14   (j) n=12

1. RECOMMENDATIONS

1. See CD – 03.01, CD – 03.02 and CD – 03.03 for the distribution of tie bars along the column, cover and spacers, respectively.

2. The maximum spacing between longitudinal reinforcing bars is 35 cm (not specified in EC2).

3. The tie bars should brace the longitudinal reinforcement at an angle of $\leq 135^\circ$ (details (e), (f), (g) and (h)) (not specified in EC2).

4. **AR.** Only every other bar need be braced if they are spaced at more than 15 cm (details (c), (j)). This provision is laid down in ACI-318-08 (22).

5. The general principle is that the smallest possible number of bars should be used (see CD – 03.17 on possible bundling).

6. In long columns, the use of thick bars is advisable, with preference given to detail (i) over detail (j) arrangements.

7. Rather than very thick tie bars, the use of bundled ties is recommended.

8. When two or more ties are used, different shapes may be jointed or staggered at 10 ϕ, provided ties of the same type are spaced within the maximum allowable distance. This second solution is preferable when, following the preceding recommendation, double ties are used rather than four-ties.

9. Complex tie bar arrangements lead to segregation during concrete pouring and consequently to lower concrete strength.

2. STATUTORY LEGISLATION

This subject is scantly dealt with in EC2 (5).

3. RECOMMENDED ALTERNATIVE CODES

See ACI 318-08 (22) and (15).

BUNDLED BAR ARRANGEMENTS

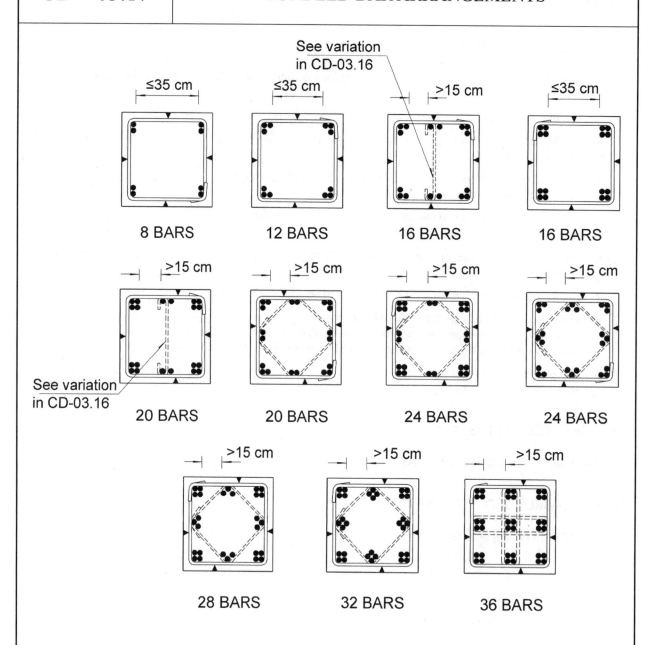

See variation
in CD-03.16

≤35 cm ≤35 cm >15 cm ≤35 cm

8 BARS **12 BARS** **16 BARS** **16 BARS**

>15 cm >15 cm >15 cm >15 cm

See variation
in CD-03.16

20 BARS **20 BARS** **24 BARS** **24 BARS**

>15 cm >15 cm >15 cm

28 BARS **32 BARS** **36 BARS**

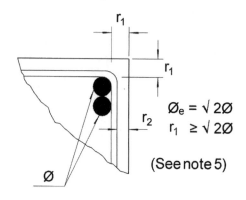

$$\emptyset_e = \sqrt{2}\emptyset$$
$$r_1 \geq \sqrt{2}\emptyset$$

(See note 5)

1. RECOMMENDATIONS

1. In 20-ϕ tie bar spacing, where ϕ is the diameter of the main reinforcement, ϕ is defined to be the diameter of the individual bars comprising the bundle. Note that the maximum spacing between bundles is 35 cm.

2. For tie bar diameter, $\phi > \dfrac{1}{4}\phi_e$, where ϕ_e is the equivalent diameter of the thickest bundle (see Table T-1.8 for equivalent diameters for bundles).

3. Rather than very thick tie bars, the use of bundled pairs of tie bars is recommended.

4. When two or more tie bars are used, different combinations (such as the 24-bar solution shown opposite) may be bundled or staggered at 24 ϕ, provided ties of the same type are spaced within the maximum allowable distance. This second solution is preferable when, following the preceding recommendation, double tie bars are used rather than four-tie bundles.

5. Note that the bundle cover may not be smaller than the equivalent diameter, but must be measured to the actual physical position of the bars in the bundle.

6. See 1.2 for bar tying procedures and 1.3 for spacer placement.

2. STATUTORY LEGISLATION

See EC2 (5).

3. RECOMMENDED ALTERNATIVE CODES

See ACI 318-08 (22) and (15).

ARRANGEMENT OF LAPS IN COLUMNS WITH BUNDLED BARS

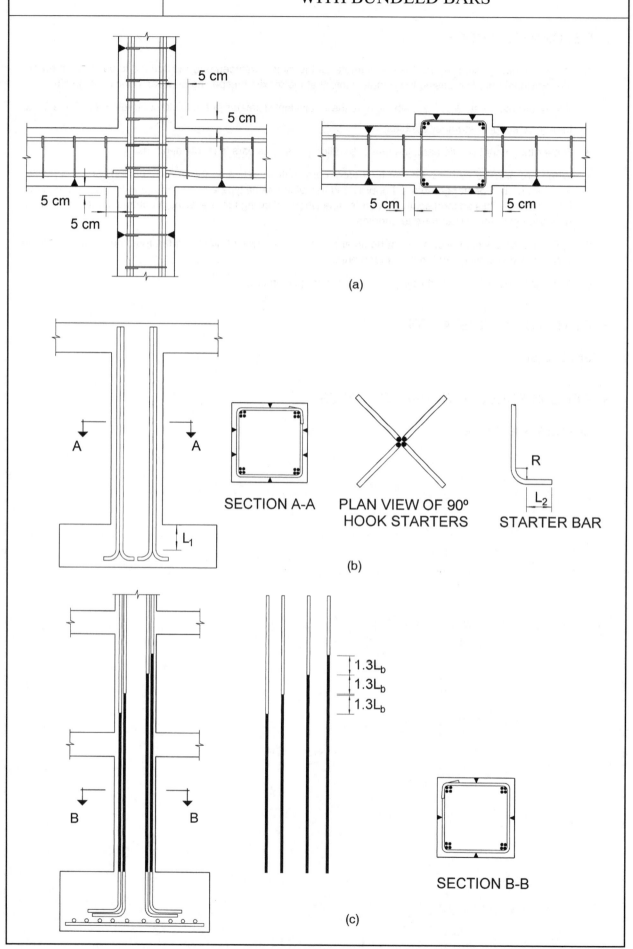

5 cm

5 cm

5 cm

5 cm

5 cm

5 cm

(a)

A   A

L_1

SECTION A-A

PLAN VIEW OF 90°
HOOK STARTERS

R

L_2

STARTER BAR

(b)

$1.3L_b$
$1.3L_b$
$1.3L_b$

B   B

SECTION B-B

(c)

1. RECOMMENDATIONS

1. The column over-width with respect to the beam should be carefully studied to ensure that the reinforcing bars do not interfere with one another at beam–column crossings (detail (a)).

2. In one-storey columns, the four bars in a bundle may all be active (detail (b)).

3. L_1 should be greater than or equal to the group anchorage length ($L_1 = 1.3\,L_b$, where L_b is the anchorage length of the individual bars).

4. L_2 should be less than double the spacing between bars in the footing grid.

5. Since no more than four starter bars can be accommodated in several-storey columns, the most suitable solution is to butt splice (but not weld) the bundles from the second storey up, which entails assuming only three bars for strength calculations. While the fourth bar is merely a splice bar, it is more practical to place it along the entire length of the bundle than in the splice area only.

2. STATUTORY LEGISLATION

See EC2.

3. RECOMMENDED ALTERNATIVE CODES

See ACI 318-08 (22) and (15).

EDGE SCHEDULE.
COLUMN SCHEDULE

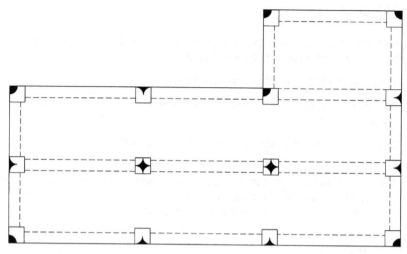

PLAN OF STRUCTURE

 Column with reference at vertex B and sides AB and BC

 Column with reference at side AB and centreline

Vertical lines to be maintained in all the height of the building

 Column with reference at centreline

COLUMN	STOREY						
	Ground	1st	2nd	3rd	4th	5th	6th
1-3-5-9							
2-4							
6-8-10-12							
13-15-20				35 x 25 6Ø20 - B Ties Ø6mm to 20 cm			
14-16-17-18 -19							

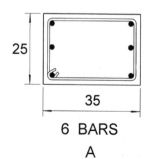

25   35

6 BARS

A

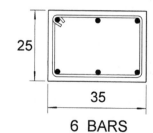

25   35

6 BARS

B

1. RECOMMENDATIONS

1. With the schedule shown, a clear indication can be given of the loads transmitted on each storey.

2. Note that cases ABC and AB appear not only in corner and edge columns, but in many other circumstances (such as courtyards, stairwells or machinery shafts, to name but a few).

3. The position of the reinforcing bars in each column is specified by an 'A' or 'B' in the respective cell on the column schedule.

   A diagram to distinguish between A and B should be included in a note attached to the schedule.

2. STATUTORY LEGISLATION

None in place.

3. RECOMMENDED ALTERNATIVE CODES

None in place.

Group 04

Walls subjected to axial loads

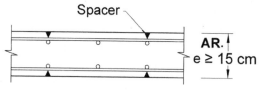

HORIZONTAL SECTION
(a) MINIMUM THICKNESS

$AR.$ $e \geq 15$ cm

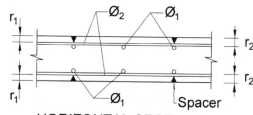

HORIZONTAL SECTION
(b) COVER

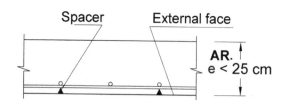

Spacer   External face

HORIZONTAL SECTION
(c) WALLS REINFORCED ON ONE SIDE ONLY

$AR.$ $e < 25$ cm

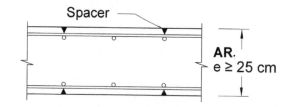

Spacer

HORIZONTAL SECTION
(d) WALLS WITH $e \geq 25$ cm (REINFORCEMENT MANDATORY ON BOTH SIDES)

$AR.$ $e \geq 25$ cm

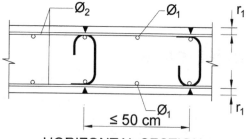

≤ 50 cm

HORIZONTAL SECTION

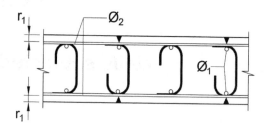

HORIZONTAL SECTION

a) On the horizontal plane the cross-ties should be placed alternately on each side with the 90° hooks.
b) The 90° hooks should be placed alternately in the vertical direction.

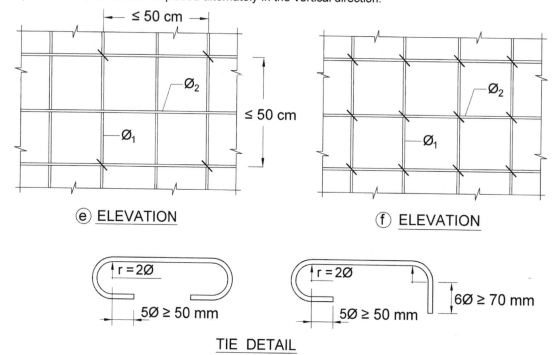

≤ 50 cm

≤ 50 cm

(e) ELEVATION

(f) ELEVATION

$r = 2\emptyset$

$5\emptyset \geq 50$ mm

$r = 2\emptyset$

$5\emptyset \geq 50$ mm

$6\emptyset \geq 70$ mm

TIE DETAIL

1. RECOMMENDATIONS

1. For ease of pouring and to prevent segregation, the minimum recommended thickness is 15 cm. Nonetheless, particular care is required when pouring walls with e < 25 cm.

2. $r_1 = 2.5$ cm; $r_2 \geq \phi_2$.

3. **AR.** Structural calculations permitting, where e < 25 cm the reinforcement may be placed on one side only (the outer side if the wall is a facade). Where e $\geq$ 25 cm, reinforcement is needed on both sides. The required drying shrinkage and thermal contraction ratios must be observed. See details (c) and (d).

4. If the geometric ratio of the ϕ_1 vertical reinforcement is ≤ 2 per cent, $\phi \leq 16$ mm, or if welded-wire mesh reinforcement is used with a cover of $2\,\phi$, no tie bars are required. Otherwise, four ties per m² are required.

5. **AR.** If the conditions specified in recommendation 4 are not met, and:

   (a) $\phi_1 \leq 12$ mm, tie bars must be spaced, horizontally and vertically, at no more than 50 cm; or

   (b) $\phi > 12$ mm, and in all cases where vertical compression reinforcement is needed, tie bars must be placed at all intersections and spaced vertically at no more than $20\,\phi_1$.

6. See CD – 02.08 for horizontal construction joints in exposed concrete.

7. If vertical contraction joints are required, see CD – 02.09 and in particular, CD – 02.21.

8. See 1.2 and 1.3 for descriptions of how to tie bars and place spacers.

9. Recommendations 3 and 5 are taken from ACI 318-08 (22).

2. STATUTORY LEGISLATION

See EC2 (5).

3. RECOMMENDED ALTERNATIVE CODES

See ACI 318-08 (22), which deals with the subject more extensively than EC2 (5).

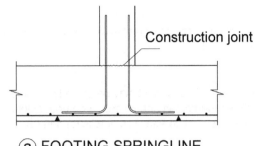

(a) FOOTING SPRINGLINE

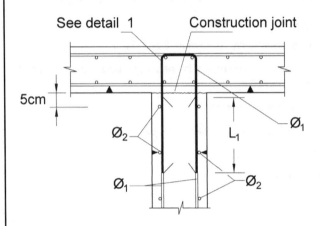

(c) BOND TO ROOF SLAB

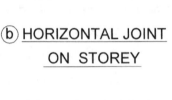

(b) HORIZONTAL JOINT ON STOREY

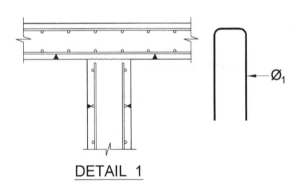

DETAIL 1

DETAIL 2

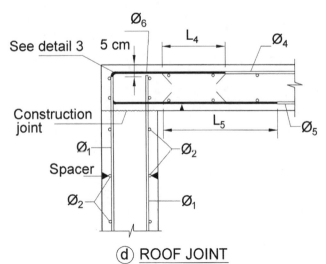

(d) ROOF JOINT

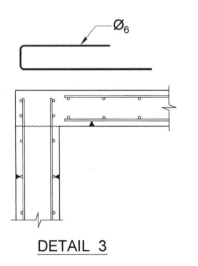

DETAIL 3

1. RECOMMENDATIONS

1. For cover, minimum thickness and the need for and placement of ties, see CD – 04.01.

2. Foundation springing a) is built exactly as described in CD – 01.01. See the recommendations included therein as well.

3. L_1 is the ϕ_1 bar lap length.

4. L_2 is the lap length in the thicker of the ϕ_4 or ϕ_5 reinforcing bars; is the lap length L_3 in the thicker of ϕ_3 or ϕ_5 bars; ϕ_5 is equal to the thicker of the ϕ_3 or ϕ_4 bars.

5. L_4 is the lap length in the thicker of the ϕ_4 or ϕ_5 bars; L_5 is the lap length L_5 in the thicker of the ϕ_5 or ϕ_6 bars; ϕ_6 is equal to the thicker of the ϕ_4 or ϕ_5 bars.

6. See 1.2 and 1.3 for descriptions of how to tie bars and place spacers.

2. STATUTORY LEGISLATION

See EC2 (5).

3. RECOMMENDED ALTERNATIVE CODES

See ACI 318-08 (22), which deals with the subject more extensively than EC2 (5).

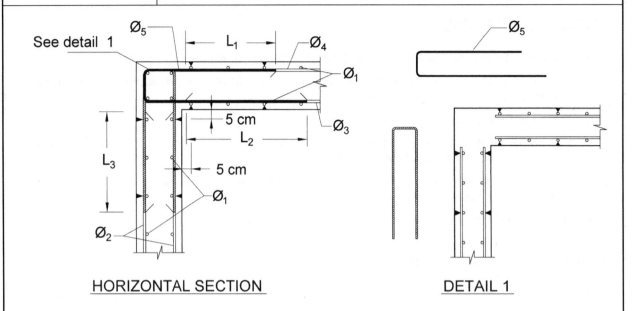

HORIZONTAL SECTION

DETAIL 1

(a) CORNER DETAIL

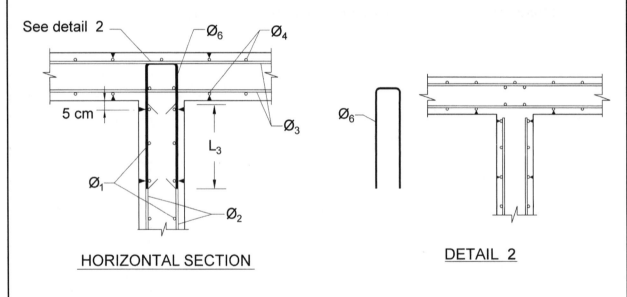

HORIZONTAL SECTION

DETAIL 2

(b) JOINT DETAIL

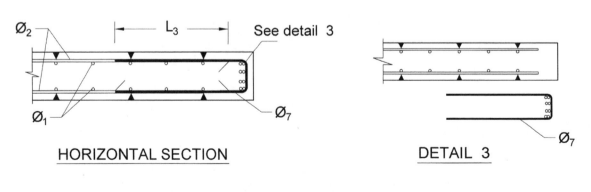

HORIZONTAL SECTION

DETAIL 3

(c) EDGE DETAIL IN SHEAR WALLS

1. RECOMMENDATIONS

1. For cover, minimum thickness and the need for and placement of tie bars, see CD – 04.01.

2. In detail (a), L_1 is the lap length in the thicker of the ϕ_4 or ϕ_5 bars; and L_2 is the lap length in the thicker of the bars. ϕ_3 or ϕ_5. The cross-ties are tied to the horizontal reinforcing bars. ϕ_5 is equal to the thicker of the ϕ_3 or ϕ_4 bars.

3. In details (a), (b) and (c), L_3 is the lap length in the horizontal reinforcement. The hairpins are tied to the intersecting horizontal reinforcing bars.

4. See 1.2 and 1.3 for descriptions of how to tie bars and place spacers.

2. STATUTORY LEGISLATION

See EC2 (5).

3. RECOMMENDED ALTERNATIVE CODES

See ACI 318-08 (22), which deals with the subject more extensively than EC2 (5).

WALLS, SHEAR WALLS AND CORES.
DETAIL OF OPENINGS

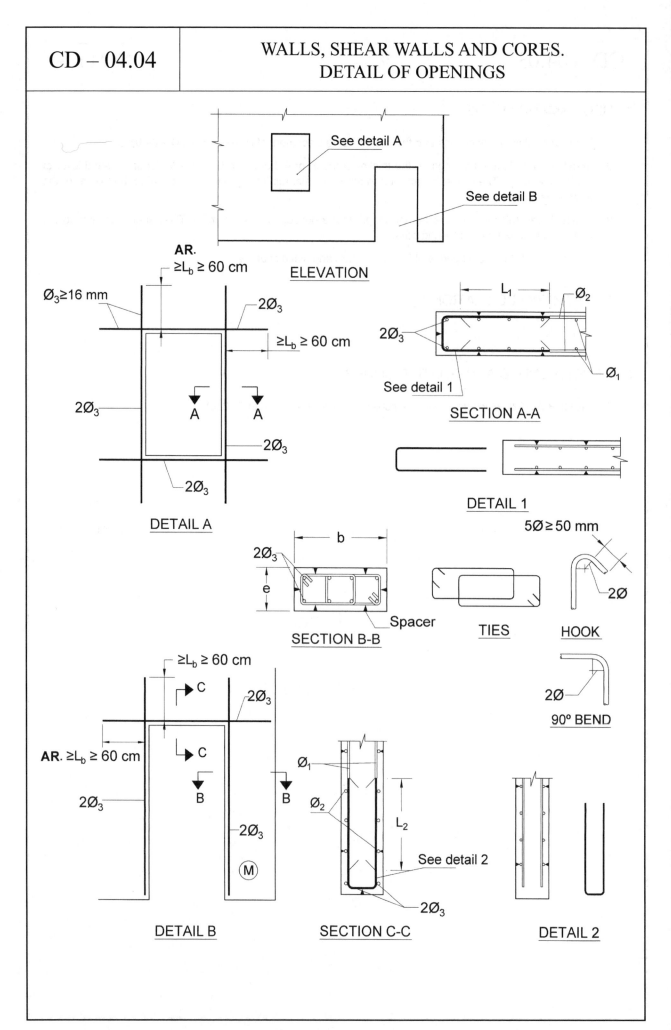

See detail A

See detail B

AR.

$\geq L_b \geq 60$ cm

ELEVATION

$\emptyset_3 \geq 16$ mm

$2\emptyset_3$

$\geq L_b \geq 60$ cm

$2\emptyset_3$

A A

$2\emptyset_3$

$2\emptyset_3$

DETAIL A

L_1

$\emptyset_2$

$2\emptyset_3$

$\emptyset_1$

See detail 1

SECTION A-A

DETAIL 1

$5\emptyset \geq 50$ mm

$2\emptyset$

HOOK

$2\emptyset$

b

$2\emptyset_3$

e

Spacer

SECTION B-B

TIES

$2\emptyset$

90° BEND

$\geq L_b \geq 60$ cm

C

$2\emptyset_3$

AR. $\geq L_b \geq 60$ cm

C

$2\emptyset_3$

B B

$2\emptyset_3$

M

DETAIL B

$\emptyset_1$

$\emptyset_2$

L_2

See detail 2

$2\emptyset_3$

SECTION C-C

DETAIL 2

216

1. RECOMMENDATIONS

1. For cover, minimum thickness and the need for and placement of tie bars, see CD – 04.01.

2. In details A and B, the anchorage length is diameter ϕ_3.

3. L_1 is the lap length in the ϕ_2 horizontal reinforcement. The cross-ties are tied to the vertical reinforcing bars.

4. In detail B, L_2 is the ϕ_1 bar lap length. The cross-ties are tied to the vertical reinforcing bars.

5. Care must be taken in areas such as M in detail B. If $b \leq 4_e$, for all intents and purposes, such areas must be designed as if they were columns. See CD – 03.16.

6. See 1.2 and 1.3 for descriptions of how to tie bars and place spacers.

2. STATUTORY LEGISLATION

See EC2 (5).

3. RECOMMENDED ALTERNATIVE CODES

See ACI 318-08 (22), which deals with the subject more extensively than EC2 (5).

CD – 04.05	WALLS, SHEAR WALLS AND CORES. SPECIAL DETAILS FOR SLIP FORMS (VARIATION 1)

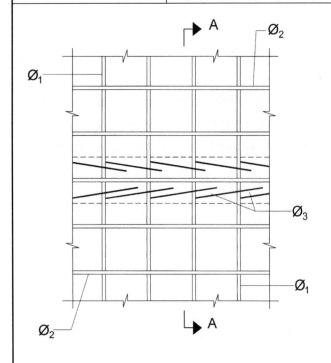

ELEVATION

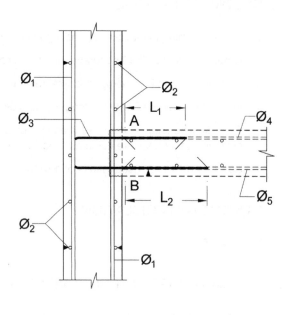

SECTION A-A

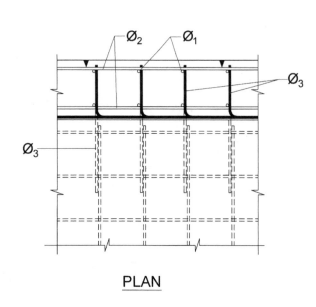

PLAN

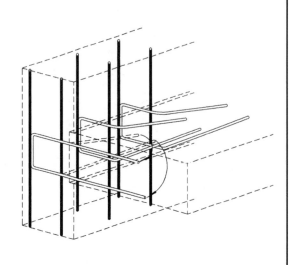

PERSPECTIVE

1. RECOMMENDATIONS

1. L_1 is the lap length in the thicker of the ϕ_3 or ϕ_4 bars.

2. L_2 is the lap length in the thicker of the ϕ_3 or ϕ_5 bars.

3. ϕ_3 is equal to the thicker of the ϕ_4 or ϕ_5 bars.

4. ϕ_2 bars should be in the bent position, securely tied to the grids and in contact with the formwork.

5. After the forms are stripped, contact surface AB is superficially chipped and the hairpins are straightened.

6. The ϕ_2 bars are tied to ϕ_4 and ϕ_5 bars, which must be designed to the same spacing.

7. See 1.2 and 1.3 for descriptions of how to tie bars and place spacers.

2. STATUTORY LEGISLATION

See EC2 (5).

3. RECOMMENDED ALTERNATIVE CODES

See ACI 318-08 (22), which deals with the subject more extensively than EC2 (5).

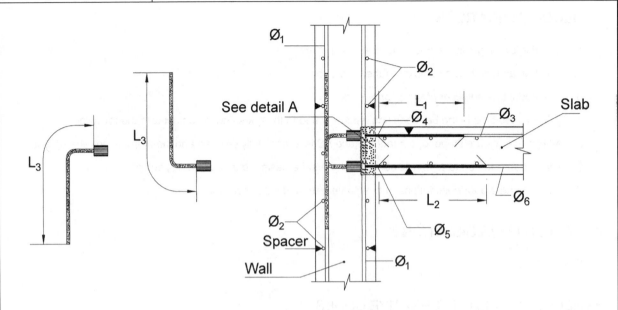

TRANSVERSE CROSS-SECTION

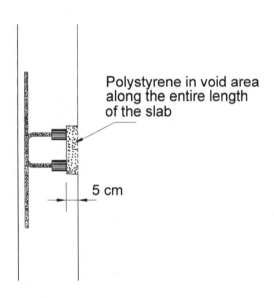

Polystyrene in void area along the entire length of the slab

5 cm

DETAIL A

1. RECOMMENDATIONS

1. L_1 is the ϕ_3 bar lap length.

2. L_2 is the ϕ_6 bar lap length.

3. Depending on the model of the coupler, its threading should be protected against contact with the concrete.

4. L_3 is the anchorage length of the coupler anchorage bars.

5. After the forms are raised, the couplers are exposed and the ϕ_4 and ϕ_5 splice bars are positioned. (Where necessary, the splice bars may have a larger diameter than the slab bars.)

6. The ϕ_4 and ϕ_5 splice bars are tied to the ϕ_3 and ϕ_5 general bars, which should be designed to the same spacing.

7. Before casting the slab, the polystyrene should be removed from the void area. This area should be cleaned and moistened and the concrete should not be poured until its surface dries.

8. See CD – 02.28 for an alternative procedure.

9. See 1.2 and 1.3 for descriptions of how to tie bars and place spacers.

2. STATUTORY LEGISLATION

None in place.

Group 05

Beams and lintels

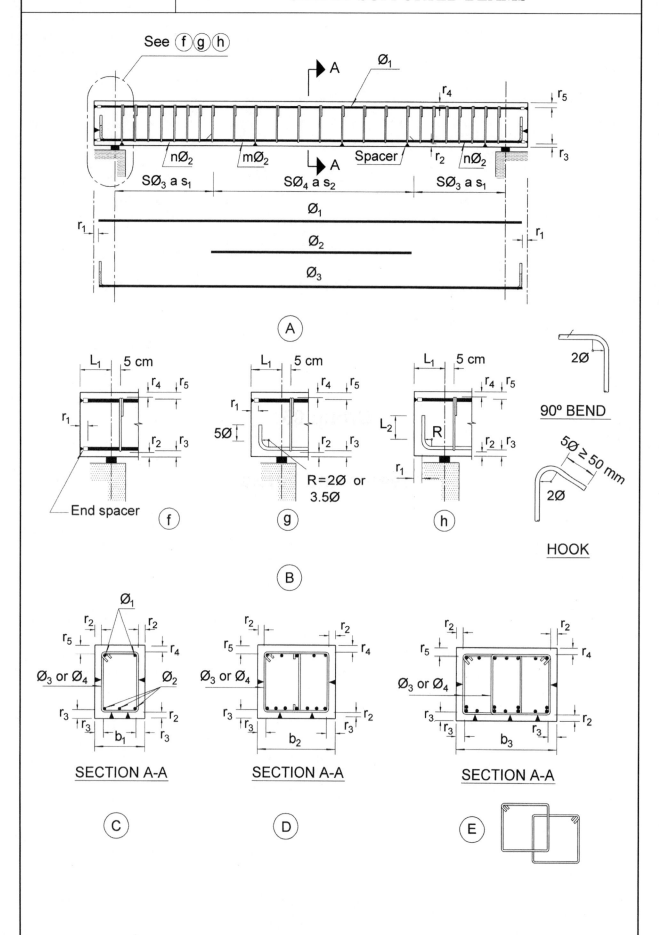

1. RECOMMENDATIONS

1. See CD – 09 for information on support members.

2. $r_1 = 2.5$ cm $\geq \phi_2$.

   $r_2 = 2.5$ cm.

   $r_3 \geq \phi_2$.

   $r_4 = r_2$.

   AR. If the top of the beam is horizontal and exposed to rain or condensed water, $r_4 = 3$ cm.

   $r_5 \geq \phi_1$.

3. See EC2, anchorage lengths, to determine when to use anchor type f, g or h at the end of the ϕ_2 reinforcing bar. (The length is measured from the centreline of the support.)

4. The double tie solution (C) is only valid where $b_1 \leq 65$ cm and ≤ 0.75 d. Where 65 cm $< b_2 \leq 125$ cm, use solution (D). Where 125 cm $< b_3 \leq 185$ cm, use solution (E).

5. Note that the web may need to be reinforced with skin reinforcement.

6. The top of the beam is smoothed with a float or power float.

7. See 1.2 and 1.3 for descriptions of how to tie bars and place spacers.

2. STATUTORY LEGISLATION

See EC2 (5).

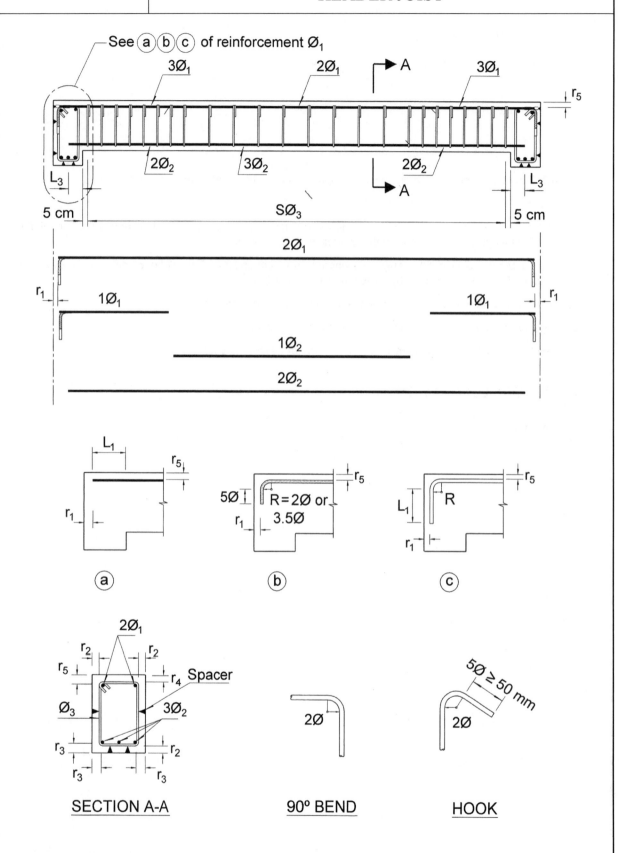

SECTION A-A

90º BEND

HOOK

1. RECOMMENDATIONS

1. $r_1 = 2.5$ cm $\geq \phi_1$.

   $r_2 = 2.5$ cm.

   $r_3 \geq \phi_2$.

   $r_4 = r_2$.

   AR. If the top of the beam is horizontal and exposed to rain or condensed water, $r_4 = 3$ cm.

   $r_5 \geq \phi_1$

2. See EC2, anchorage lengths, to determine when to use anchor type a, b or c at the end of the ϕ_1 reinforcing bar.

3. See EC2 for information on calculating length L_3.

4. Note that the web may need to be reinforced with skin reinforcement.

5. The top of the beam is smoothed with a float or power float.

6. See 1.2 and 1.3 for descriptions of how to tie bars and place spacers.

2. STATUTORY LEGISLATION

See EC2 (5).

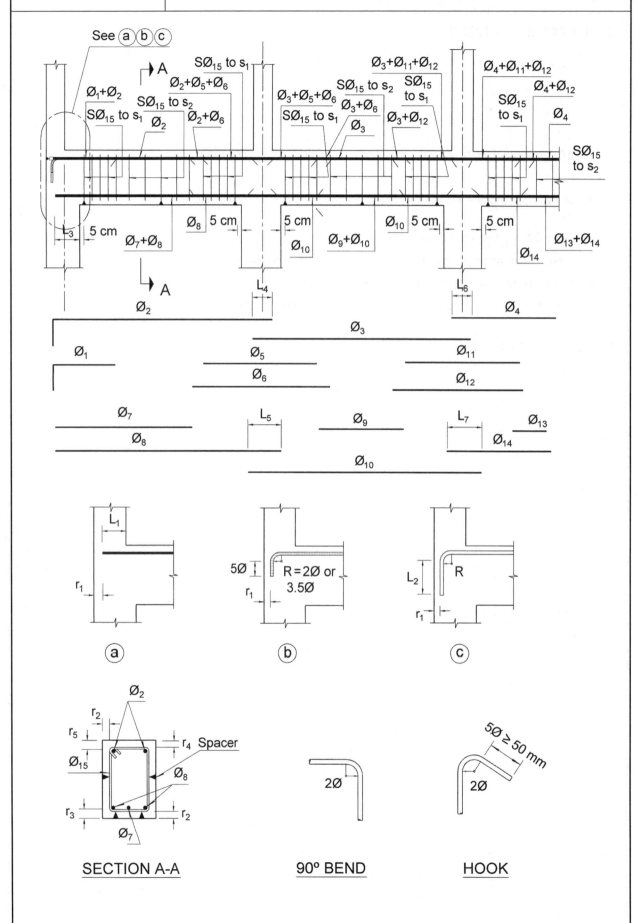

SECTION A-A

90º BEND

HOOK

1. RECOMMENDATIONS

1. $r_1 = 2.5$ cm.

    $r_2 = 2.5$ cm.

    $r_3 \begin{cases} \geq \phi_7 \\ \geq \phi_8 \end{cases}$

    $r_4 = 2.5$ cm.

    If the top of the beam is horizontal and exposed to rain or condensed water, $r_4 = 3$ cm.

    $r_5 \geq \phi_2$.

2. See EC2, anchorage lengths, to determine when to use anchor type a, b or c at the end of the ϕ_1 and ϕ_2 reinforcing bars.

3. See EC2 for information on calculating length L_3.

4. L_4 is the lap length in the thicker of the ϕ_2 or ϕ_3 bars, L_5 is the lap length in the thicker of the ϕ_8 or ϕ_{10} bars L_6 is the lap length in the thicker of ϕ_3 or ϕ_4 bars and L_7 is the lap length in the thicker of the ϕ_{10} or ϕ_{14} bars.

5. Note that the web may need to be reinforced with skin reinforcement.

6. The top of the beam is smoothed with a float or power float.

7. See 1.2 and 1.3 for descriptions of how to tie bars and place spacers.

2. STATUTORY LEGISLATION

See EC2 (5).

BEAMS.
CONTINUOUS LINTELS WITH VARIABLE DEPTH

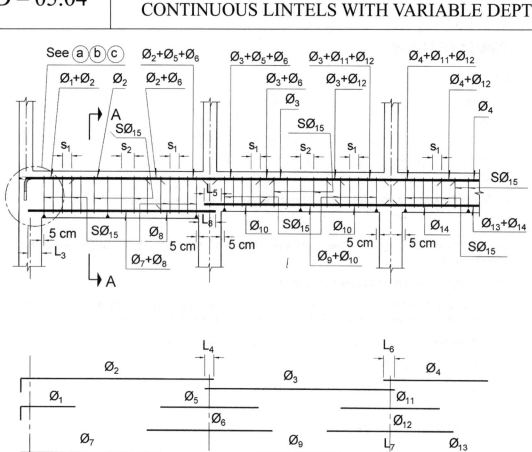

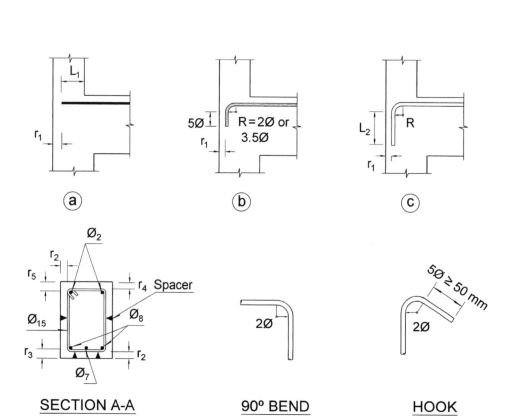

| (a) | (b) | (c) |

SECTION A-A   90° BEND   HOOK

1. RECOMMENDATIONS

1. $r_1 = 2.5$ cm $\begin{cases} \geq \phi_2 \\ \geq \phi_1 \end{cases}$

   $r_2 = 2.5$ cm.

   $r_3 \geq \phi_8$.

   $r_4 = r_2$.

   If the top of the beam is horizontal and exposed to rain or condensed water, $r_4 = 3$ cm.

   $r_5 \geq \phi_2$.

2. See EC2, anchorage lengths, to determine when to use anchor type a, b or c at the end of the ϕ_1 reinforcing bar.

3. See EC2 for the procedure to calculate L_3.

4. L_4 is the lap length in the thicker of the ϕ_2 or ϕ_3 bars, L_6 is the lap length in the thicker of the ϕ_3 or ϕ_4 bars and L_7 is the lap length in the thicker of ϕ_{10} or ϕ_{14} bars.

5. See EC2 for information on calculating anchorage lengths L_5 and L_8.

6. Note that the web may need to be reinforced.

7. The top of the beam is smoothed with a float or power float.

8. See 1.2 and 1.3 for descriptions of how to tie bars and place spacers.

2. STATUTORY LEGISLATION

See EC2 (5).

BEAMS.
STAGGERED LINTELS

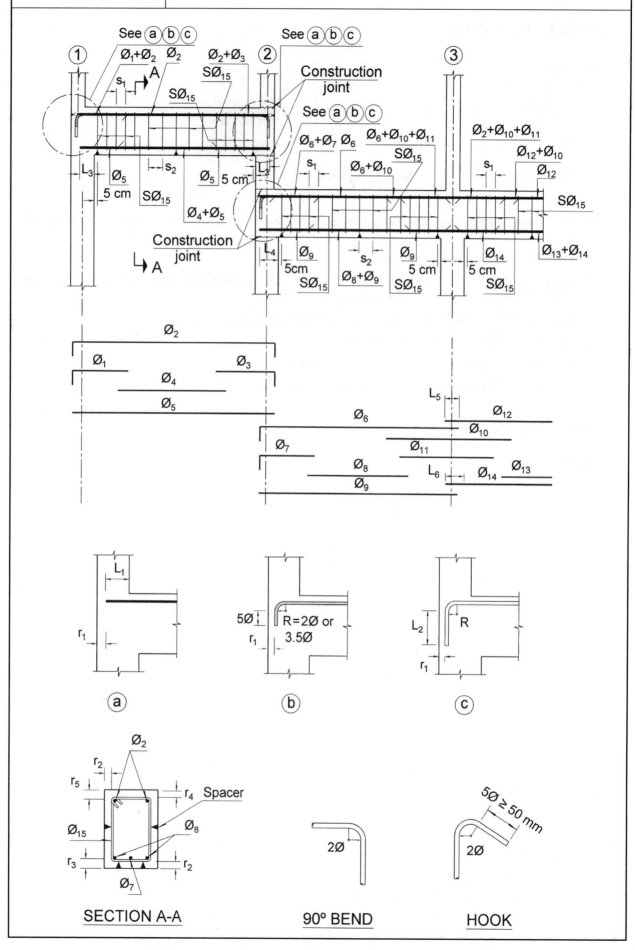

SECTION A-A

90° BEND

HOOK

1. RECOMMENDATIONS

1. See the notes for CD – 05.04. The following refers to the present CD only.

2. See EC2, anchorage lengths, to determine when to use a, b or c at the end of the ϕ_2, ϕ_1, ϕ_3, ϕ_5 and ϕ_7 reinforcing bars.

3. See EC2 for information on calculating L_3 and L_4.

4. Note the construction joints in column 2.

2. STATUTORY LEGISLATION

See EC2 (5).

BEAMS.
STEPPED LINTELS

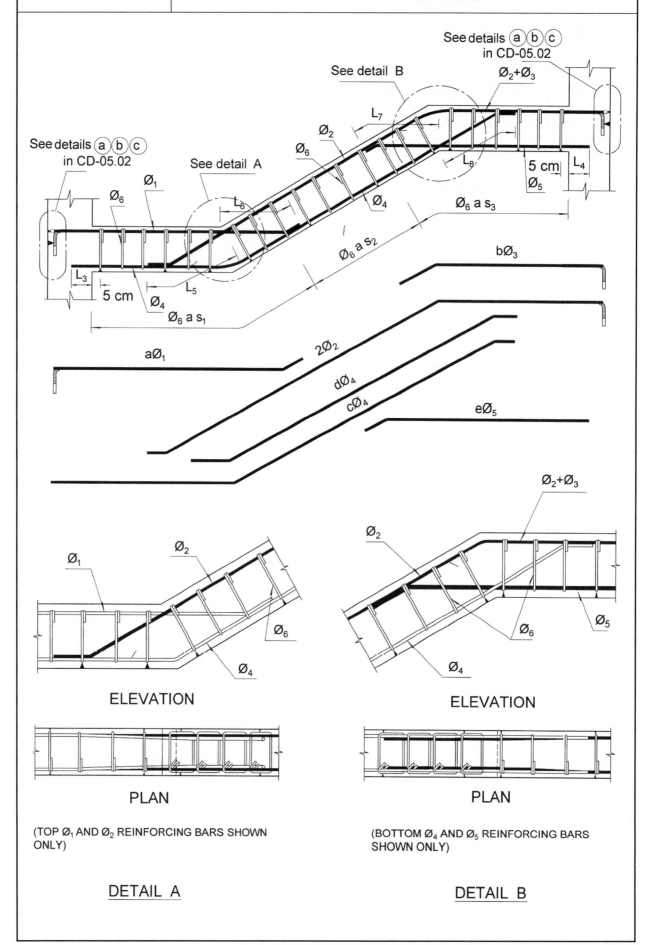

See details (a)(b)(c) in CD-05.02

See detail B

$\varnothing_2 + \varnothing_3$

L_7

$\varnothing_2$

$\varnothing_6$

L_8

5 cm L_4

$\varnothing_5$

See details (a)(b)(c) in CD-05.02

See detail A

$\varnothing_1$

$\varnothing_6$

L_6

$\varnothing_4$

$\varnothing_6$ a s_3

L_3

5 cm

$\varnothing_4$

L_5

$\varnothing_6$ a s_1

$\varnothing_6$ a s_2

$b\varnothing_3$

$a\varnothing_1$

$2\varnothing_2$

$d\varnothing_4$

$c\varnothing_4$

$e\varnothing_5$

$\varnothing_1$

$\varnothing_2$

$\varnothing_6$

$\varnothing_4$

ELEVATION

$\varnothing_2 + \varnothing_3$

$\varnothing_2$

$\varnothing_6$

$\varnothing_5$

$\varnothing_4$

ELEVATION

PLAN

(TOP $\varnothing_1$ AND $\varnothing_2$ REINFORCING BARS SHOWN ONLY)

PLAN

(BOTTOM $\varnothing_4$ AND $\varnothing_5$ REINFORCING BARS SHOWN ONLY)

DETAIL A

DETAIL B

234

1. RECOMMENDATIONS

1. See the notes for CD – 05.04. The following refers to the present CD only.

2. See EC2, anchorage lengths, to determine when to use a, b or c at the end of the ϕ, ϕ_2 and ϕ_3 reinforcing bars.

3. See EC2 for information on calculating L_3 and L_4.

4. L_5, L_6, L_7 and L_8 are the anchorage lengths for the ϕ_2, ϕ_1, ϕ_5 and ϕ_4, bars respectively.

5. Concrete is poured in the beam in the upward direction.

2. STATUTORY LEGISLATION

See EC2 (5).

BEAMS.
EDGE SOFFIT BEAM

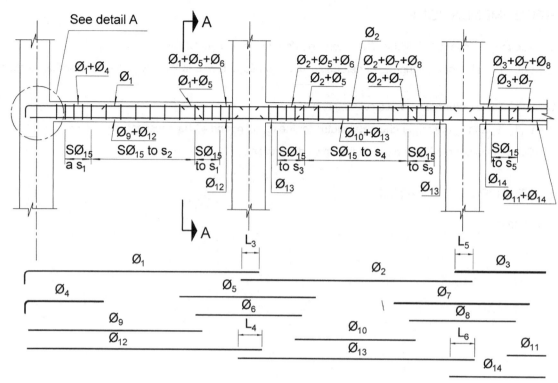

ELEVATION

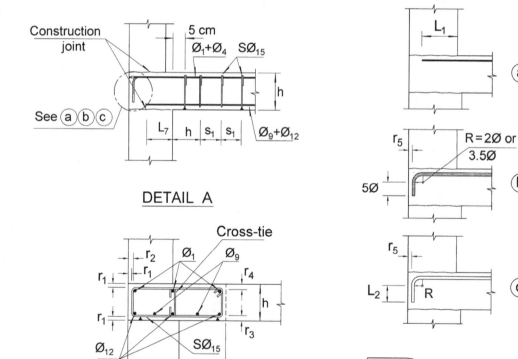

DETAIL A

SECTION A-A

HOOK   90° BEND

STIRRUPS

236

1. RECOMMENDATIONS

1. $r_1 = 2.5$ cm.

$$r_2 \geq \begin{cases} \phi_1 \\ \phi_{12} \end{cases}$$

$$r_3 \geq \begin{cases} \phi_9 \\ \phi_{12} \end{cases}$$

$$r_4 \geq \phi_1.$$

If the top is horizontal and exposed to rain or condensed water, $r_4 = 3$ cm.

$$r_5 = 2.5 \text{ cm} \begin{cases} \geq \phi_1 \\ \geq \phi_4 \end{cases}$$

2. See EC2, anchorage lengths, to determine when to use anchor type a, b or c at the end of the ϕ_1 and ϕ_4 reinforcing bars.

3. See EC2 for information on calculating length L_7.

4. L_3 is the lap length in the thicker of the ϕ_1 or ϕ_2 bars, L_4 is the lap length in the thicker of the ϕ_{12} or ϕ_{13} bars, L_5 is the lap length in the thicker of the ϕ_2 or ϕ_3 bars and L_6 is the lap length in the thicker of the ϕ_{13} or ϕ_{14} bars.

5. The top of the beam is smoothed with a float or power float.

6. See 1.2 and 1.3 for descriptions of how to tie bars and place spacers.

2. STATUTORY LEGISLATION

See EC2 (5).

CD – 05.08	BEAMS. INTERNAL SOFFIT BEAM

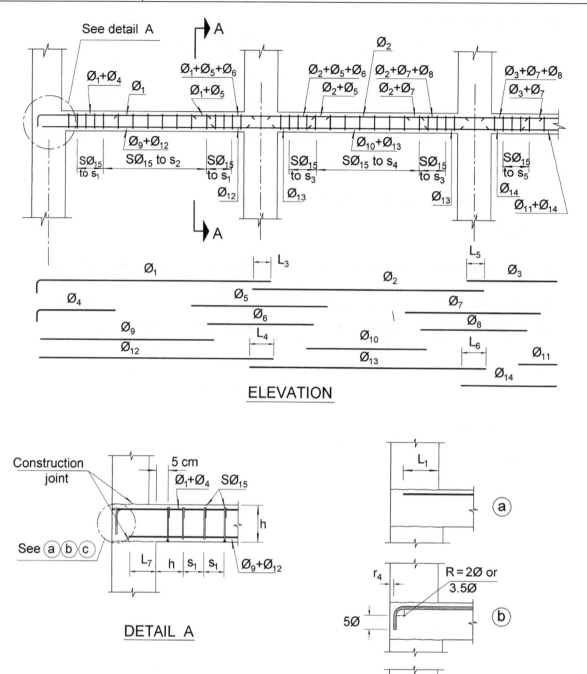

ELEVATION

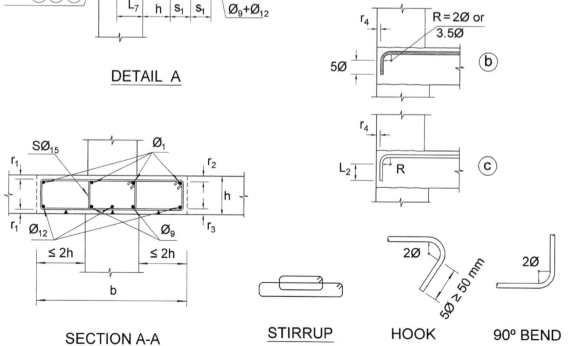

DETAIL A

SECTION A-A

STIRRUP ARRANGEMENT

HOOK

90° BEND

1. RECOMMENDATIONS

1. $r_1 = 2.5$ cm.

   $r_2 \geq \phi_1$.

   If the top is horizontal and exposed to rain or condensed water, $r_1 = 3$ cm.

   $$r_3 \geq \begin{cases} \phi_9 \\ \phi_{12} \end{cases}$$

   $$r_4 = 2.5 \text{ cm} \geq \begin{cases} \phi_1 \\ \phi_4 \end{cases}$$

2. See EC2, anchorage lengths, to determine when to use anchor type a, b or c at the end of the ϕ_1 and ϕ_4 reinforcing bars.

3. See EC2 for information on calculating length L_7.

4. L_3 is the lap length in the thicker of the ϕ_1 or ϕ_2 bars, L_4 is the lap length in the thicker of the ϕ_{12} or ϕ_{13} bars, L_5 is the lap length in the thicker of the ϕ_2 or ϕ_3 bars and L_6 is the lap length in the thicker of the ϕ_{13} or ϕ_{14} bars.

5. The top of the beam is smoothed with a float or power float.

6. See 1.2 and 1.3 for descriptions of how to tie bars and place spacers.

2. STATUTORY LEGISLATION

See EC2 (5).

Construction joint

See details (a)(b)(c) in CD-05.07

5 cm

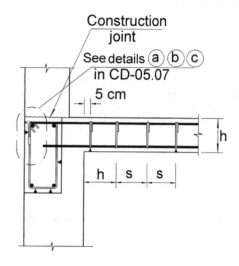

ELEVATION

Construction joint

See details (a)(b)(c) in CD-05.07

5 cm

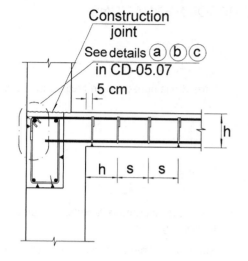

ELEVATION

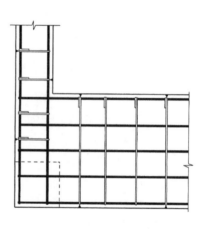

PLAN

EDGE BEAM

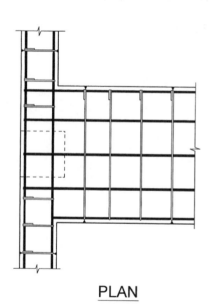

PLAN

INTERNAL SOFFIT BEAM

1. RECOMMENDATIONS

1. See CD – 05.07 and CD – 05.08 for details in general and cover dimensions.

2. In special open stirrups, L_b is the lap length of the bar with the respective diameter.

3. The top of the beam is smoothed with a float or power float.

2. STATUTORY LEGISLATION

See EC2 (5).

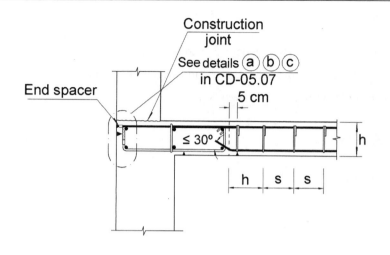

Construction joint

See details (a)(b)(c) in CD-05.07

End spacer

5 cm

≤ 30°

h

h | s | s

ELEVATION

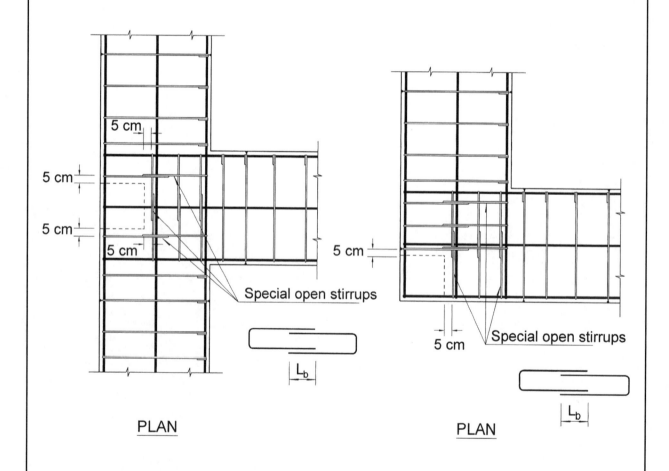

5 cm

5 cm

5 cm

5 cm

Special open stirrups

L_b

PLAN

INTERNAL SOFFIT BEAM

5 cm

5 cm

Special open stirrups

L_b

PLAN

EDGE SOFFIT BEAM

1. RECOMMENDATIONS

1. See CD – 05.07 and CD – 05.08 for details in general and cover dimensions.

2. In special open stirrups, L_b is the lap length of the bar with the respective diameter.

3. The top of the beam is smoothed with a float or power float.

2. STATUTORY LEGISLATION

See EC2 (5).

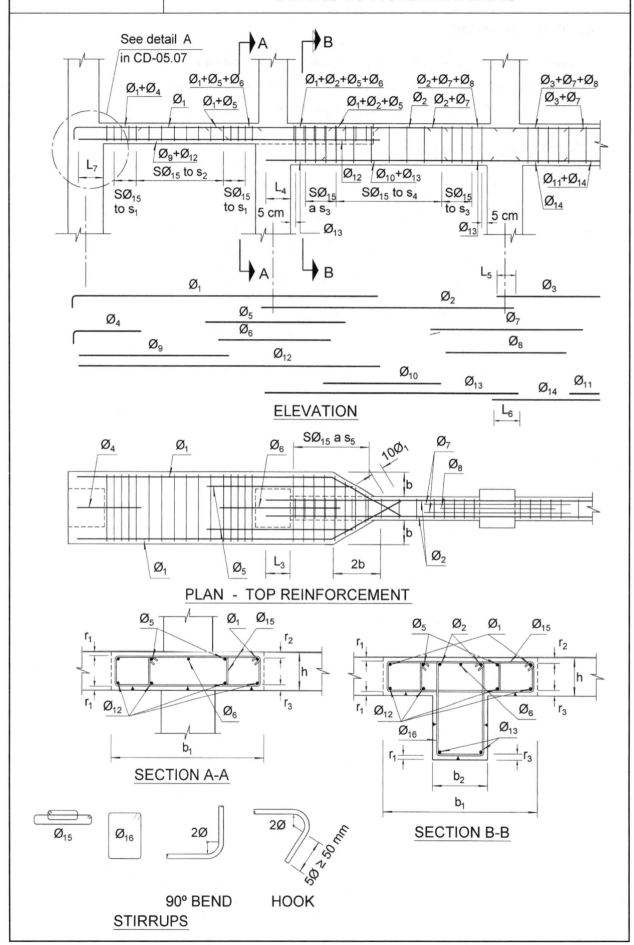

ELEVATION

PLAN - TOP REINFORCEMENT

SECTION A-A

SECTION B-B

90º BEND   HOOK

STIRRUPS

1. RECOMMENDATIONS

1. $r_1 = 2.5$ cm

    $$r_2 \geq \begin{cases} \phi_1 \\ \phi_5 \end{cases}$$

    If the top of the beam is horizontal and exposed to rain or condensed water, $r_2 = 3$ cm.

    $r_3 \geq \phi_{12}$

2. See CD – 05.08 for anchorage and lap lengths.

3. Stirrups of varying dimensions are needed in the transition area.

4. The top of the beam is smoothed with a float or power float.

5. See 1.2 and 1.3 for descriptions of how to tie bars and place spacers.

2. STATUTORY LEGISLATION

See EC2 (5).

BEAMS. TRANSITION FROM EDGE SOFFIT BEAMS TO NORMAL BEAMS

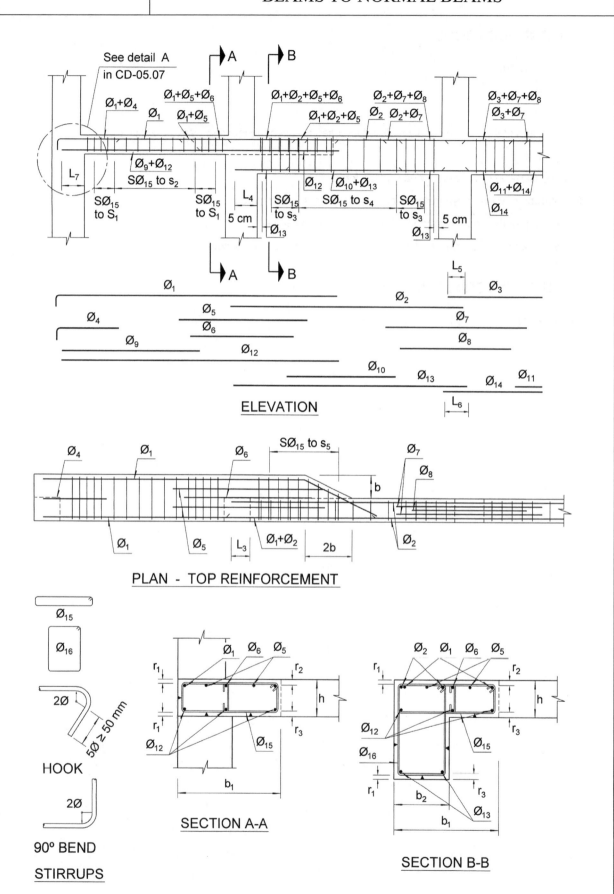

ELEVATION

PLAN - TOP REINFORCEMENT

HOOK

90º BEND

STIRRUPS

SECTION A-A

SECTION B-B

1. RECOMMENDATIONS

1. $r_1 = 2.5$ cm.

    $$r_2 \geq \begin{cases} \phi_1 \\ \phi_2 \\ \phi_5 \end{cases}$$

    If the top of the beam is horizontal and exposed to rain or condensed water, $r_2 = 3$ cm.

    $r_3 \geq \phi_{12}$.

2. See EC2 for anchorage and lap lengths.

3. Stirrups of varying dimensions are needed in the transition area.

4. The top of the beam is smoothed with a float or power float.

5. See 1.2 and 1.3 for descriptions of how to tie bars and place spacers.

2. STATUTORY LEGISLATION

See EC2 (5).

CD – 05.12	BEAMS. JOINT DETAILS

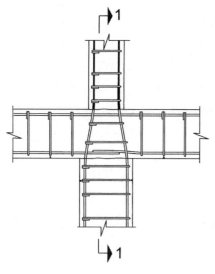

ELEVATION A

ELEVATION B

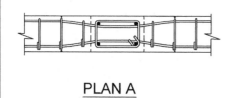

PLAN A

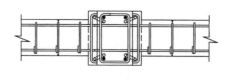

PLAN B

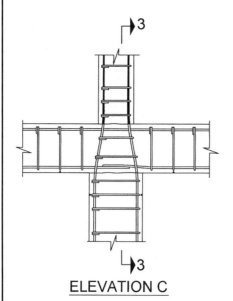

ELEVATION C

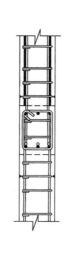  

SECTION A-A   SECTION B-B   SECTION C-C

1. RECOMMENDATIONS

1. See CD – 05.01 to CD – 05.05 for general details.
2. In solution A, the beam reinforcement bars must be shaped.
3. The solutions of choice are B and C, the former in general and the latter for specific cases.
4. See CD – 03.01 to CD – 03.12 for details on reinforcing bar arrangement in nodes.

2. STATUTORY LEGISLATION

See EC2 (5).

3. RECOMMENDED ALTERNATIVE CODES

See ACI 318-08 (22) and ACI (15).

CD – 05.13	BEAMS. INDUSTRIALISED JOINT

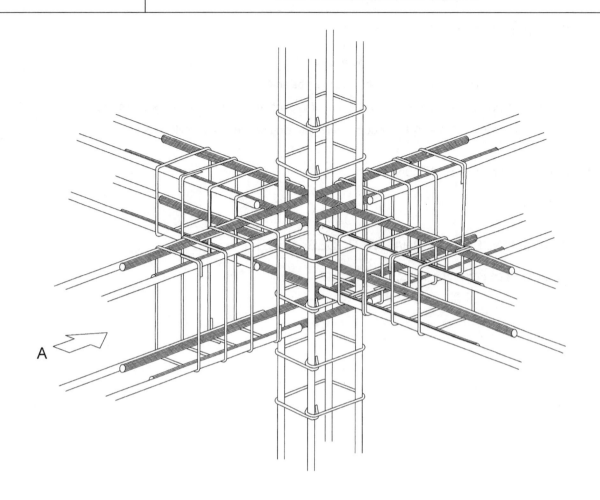

A

PERSPECTIVE

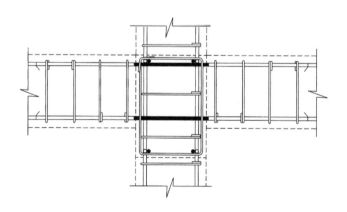

VIEW FROM A

NOTES

1. RECOMMENDATIONS

1. This detail is applicable to highly industrialised systems, in which the 'cages' for each span are prepared in the shop (usually by welding the stirrups to the bars). It is taken from the CEB manual (45).

2. Such systems provide for much speedier reinforcement assembly.

2. SPECIFIC REFERENCES

See CEB (45).

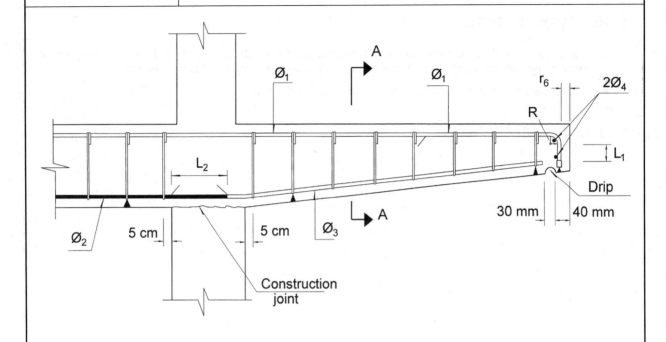

$\varnothing_1$ A $\varnothing_1$ r_6 $2\varnothing_4$ R L_1 Drip

L_2

$\varnothing_2$ 5 cm 5 cm $\varnothing_3$ A 30 mm 40 mm

Construction joint

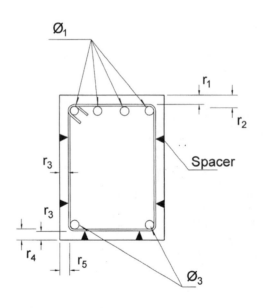

$\varnothing_1$ r_1 r_2 Spacer r_3 r_3 r_4 r_5 $\varnothing_3$

SECTION A-A

1. RECOMMENDATIONS

1. See CD – 05.01 and CD – 05.03 for general considerations.

2. Members exposed to the elements should be fitted with a dripstone.

3. $r_1 = 2.5$ cm.

   If the top is horizontal and exposed to rain or condensed water, $r_1 = 3$ cm.

   $r_2 \geq \phi_1$.

   $r_3 = 2.5$ cm.

   $r_4 \geq \phi_3$

   $r_5 \begin{cases} \geq \phi_1 \\ \geq \phi_3 \end{cases}$

   $r_6 = 2.5$ cm $\geq \phi_1$.

4. If thick bars are used in short cantilevers, length L_1 must be verified.

5. L_2 is the lap length of the thicker of the ϕ_2 or ϕ_3 bars.

6. The top of the beam is smoothed with a float or power float.

2. STATUTORY LEGISLATION

See EC2 (5).

CD – 05.15	BEAMS. ARRANGEMENT OF REINFORCEMENT AT CROSS-SECTION

 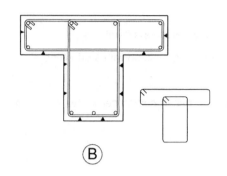 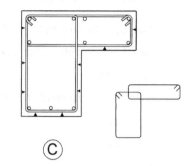

Ⓐ   Ⓑ   Ⓒ

CROSS-SECTIONAL VARIATIONS

Off cuts

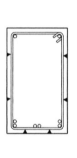  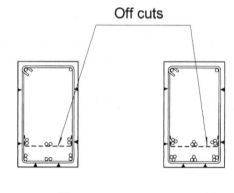

n = 6   n = 9   n = 12   n = 18

USE OF BUNDLES

Off cuts

 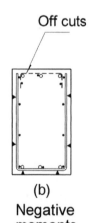

○ - Bars
• - Wires

(a)   (b)
Positive   Negative
moments   moments

USE OF WELDED-WIRE MESH IN PLACE OF STIRRUPS

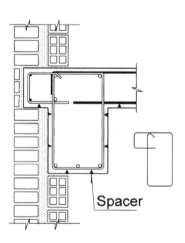

Spacer

BEAM SUPPORTING THE OUTER LAYER OF FACADE ENCLOSURES TO REDUCE THERMAL BRIDGES

1. RECOMMENDATIONS

1. See CD – 05.01 to CD – 05.03 for cover dimensions and associated details.

2. Note that where bundles are used, while the cover dimension is determined in terms of the equivalent diameter, the cover must actually be measured from the outermost bar in the bundle.

2. STATUTORY LEGISLATION

See EC2 (5).

CD – 05.16	BEAMS. CONTRACTION JOINTS

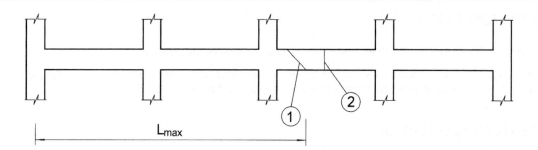

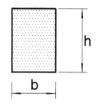

TYPE ①

POSITION: Points with a nil (or very small) bending moment

SLANT: $\approx 45°$

EXECUTION: For b ≤ 30 cm, the natural slope is $\approx 45°$.
For b > 30 cm, form with galvanised welded wire with mesh size under 20 mm or galvanised metal lath, at a 45° angle

FINISH: Natural roughness

TYPE ②

POSITION: Points with nil (or very small) shear

SLANT: Vertical

EXECUTION: Form with galvanised welded wire with mesh size under 20 mm or galvanised metal lath

DISTANCE (L_{max}) BETWEEN JOINTS

TYPE OF CLIMATE	SEASON OF YEAR	
	WARM	COLD
DRY	16 m	20 m
HUMID	20 m	24 m

1. RECOMMENDATIONS

1. **AR.** The suggested times for keeping contraction joints open are listed in the chart below.

Cement content, C	SUMMER	WINTER
$C \leq 250$ kg/m³	3 days	2 days
$250 < C \leq 400$ kg/m³	4 days	3 days
$C > 400$ kg/m³	MUST BE STUDIED	MUST BE STUDIED

Construction joints need not be left open for any specified time.

2. Two joints may be set close to one another (providing they are in positions 1 or 2) while concrete pouring continues. The area left open would subsequently be filled in as per the preceding recommendation. A second option is simply to interrupt concrete pouring during the times specified. (This applies to contraction joints.)

3. Note the specifications for architectural concrete to prevent rough edges from forming (see CD – 05.17).

2. SPECIFIC REFERENCES

See (14), Vol. I, pages 502 to 522.

ARCHITECTURAL CONCRETE VARIATIONS

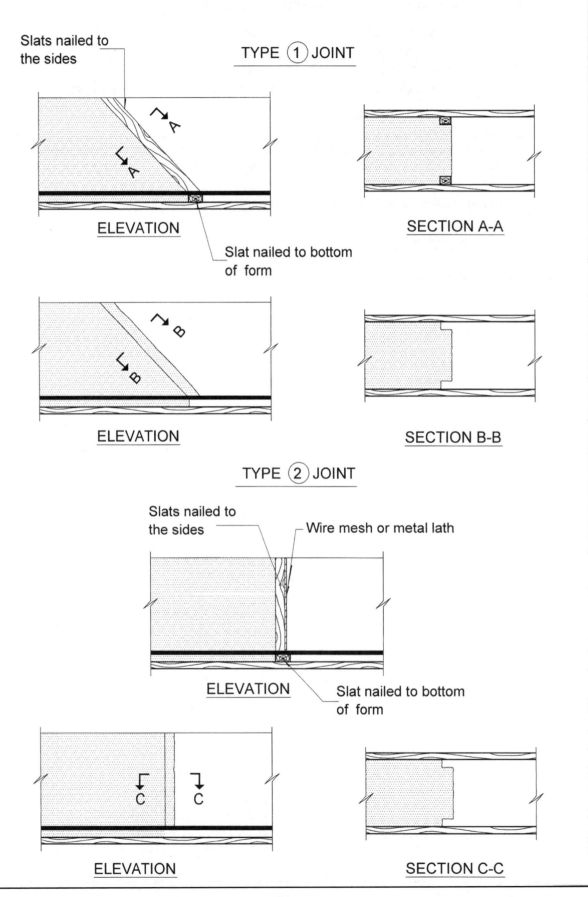

Slats nailed to the sides

TYPE ① JOINT

A

ELEVATION

SECTION A-A

Slat nailed to bottom of form

B

ELEVATION

SECTION B-B

TYPE ② JOINT

Slats nailed to the sides

Wire mesh or metal lath

ELEVATION

Slat nailed to bottom of form

C

ELEVATION

SECTION C-C

1. RECOMMENDATIONS

1. **AR.** The suggested times for keeping contraction joints open are listed in the chart below.

Cement content, C	SUMMER	WINTER
$C \leq 250$ kg/m³	3 days	2 days
$250 < C \leq 400$ kg/m³	4 days	3 days
$C > 400$ kg/m³	MUST BE STUDIED	MUST BE STUDIED

Construction joints need not be left open for any specified time.

2. Two joints may be set close to one another (providing they are in positions 1 or 2 (see CD – 05.16)) while concrete pouring continues. The area left open would subsequently be filled in as per the preceding recommendation. A second option is simply to interrupt concrete pouring during the times specified. (This applies to contraction joints.)

3. Note the specifications for architectural concrete to prevent rough edges from forming.

2. SPECIFIC REFERENCES

See (14), Vol. I, pages 502 to 522.

Group 06

Slabs, ribbed slabs and precast slabs with beam-block and hollow core floor systems

SOLID SLAB

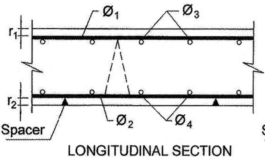

LONGITUDINAL SECTION

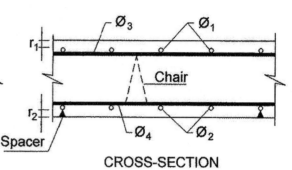

CROSS-SECTION

RIBBED SLAB

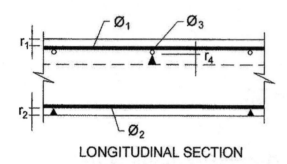

LONGITUDINAL SECTION

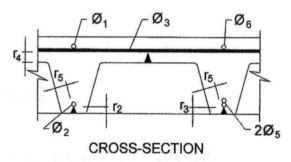

CROSS-SECTION

BEAM-BLOCK FLOOR SYSTEM
(SELF-SUPPORTING JOIST)

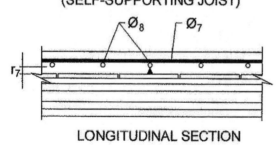

LONGITUDINAL SECTION

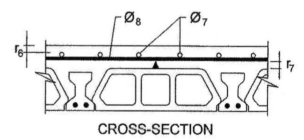

CROSS-SECTION

BEAM-BLOCK FLOOR SYSTEM
(SEMI-SELF-SUPPORTING JOIST)

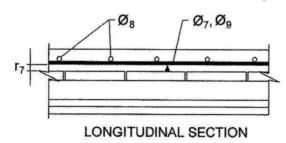

LONGITUDINAL SECTION

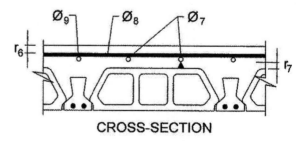

CROSS-SECTION

NOTE: See CD-05.16 for contraction joints

1. RECOMMENDATIONS

1. $r_1 = 2.5$ cm.

   $r_1 \geq \phi_1$.

   If the top is horizontal and exposed to rain or condensed water, $r_1 = 3$ cm.

   $r_2 = 2.5$ cm.

   $r_3 \begin{cases} \geq \phi_2 \\ \geq 1.4\, \phi_5 \end{cases}$

   $r_4 = 2.5$ cm.

   $r_5 = 2.5$ cm.

   $r_6 = 2.5$ cm.

   If the top is horizontal and exposed to rain or condensed water, $r_6 = 3$ cm.

   $r_7 = 2.5$ cm.

2. The top is smoothed with a float or power float.

3. See 1.2 and 1.3 for descriptions of how to tie bars and place spacers.

4. The detail for contraction joints is CD – 05.16. In slabs, the joint is usually as specified in type 1.

2. STATUTORY LEGISLATION

See EC2 (5).

3. SPECIFIC REFERENCES

See (30) and (31).

CD – 06.02	SOLID SLAB. CONNECTION TO BRICK WALL AND CONCRETE BEAMS

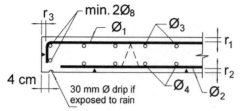

① END REINFORCEMENT IN CANTILEVER

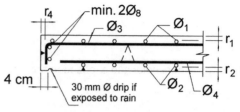

② EDGE REINFORCEMENT ON FREE SIDE

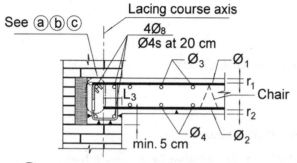

③ SUPPORT ON OUTSIDE WALL

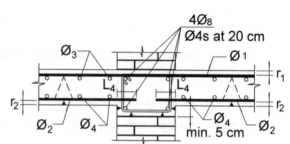

④ SUPPORT ON INSIDE WALL

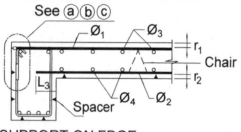

⑤ SUPPORT ON EDGE CONCRETE BEAM

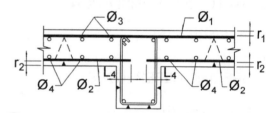

⑥ SUPPORT ON INTERNAL CONCRETE BEAM

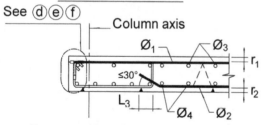

⑦ SUPPORT ON EDGE SOFFIT BEAM

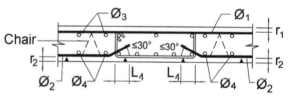

⑧ SUPPORT ON INTERNAL SOFFIT BEAM

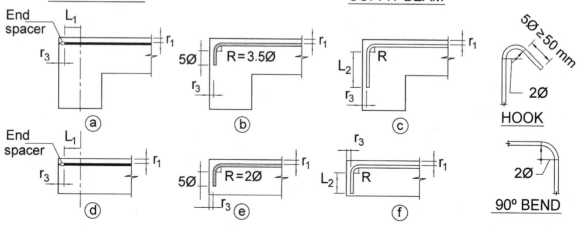

1. RECOMMENDATIONS

1. See CD – 06.01 for the general arrangement.

2. See CD – 05.01 to CD – 05.14 for the concrete cover in beams and lacing courses.

3. $r_1 = 2.5$ cm.

   If the top is horizontal and exposed to rain or condensed water, $r_1 = 3$ cm.

   $r_2 = 2.5$ cm $\geq \phi_2$.

   $r_3 = 2.5$ cm $\geq \phi_1$.

   $r_4 = 2.5$ cm $\geq \phi_3$.

4. See EC2, anchorage lengths, to calculate L_1 and L_2 and to determine when to use a, b, c, d, e or f for anchoring the ϕ_1 bars, and to calculate L_1 and L_2.

5. L_3 is calculated bearing in mind the following.

   The force, T, at the beginning of the anchor (Figure (a)) is in equilibrium with Figure (b).

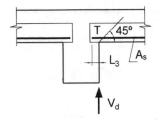

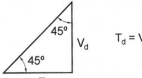

$T_d = V_d$

$$L_3 = \frac{V_d}{A_s f_{yd}} \cdot \ell_b \begin{cases} \geq 10\,\varnothing_2 \\ \geq 10 \text{ cm} \end{cases}$$

Figure (a)   Figure (b)

   where ℓ_b is the anchorage length in the ϕ_2 bar.

6. For negative moments (interior supports), L_4 is calculated as follows.

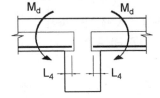

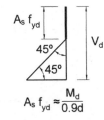

$$L_4 = \frac{V_d - \dfrac{M_d}{0.9d}}{A_s f_{yd}} \cdot \ell_b \begin{cases} \geq 6\,\varnothing_2 \\ \geq 6 \text{ cm} \end{cases}$$

(Shear friction with $\rho = 1$)

$$A_s f_{yd} \approx \frac{M_d}{0.9d}$$

Figure (c)

   (M_d is the design moment at the interior support.)

7. The top is smoothed with a float or power float.

8. See 1.2 and 1.3 for descriptions of how to tie bars and place spacers.

2. STATUTORY LEGISLATION

See EC2 (5).

3. RECOMMENDED ALTERNATIVE CODES

See (31).

4. SPECIFIC REFERENCES

See (14).

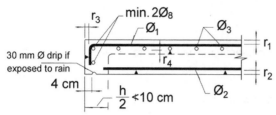

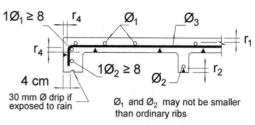

(1) END REINFORCEMENT IN CANTILEVER

(2) EDGE REINFORCEMENT ON FREE SIDE

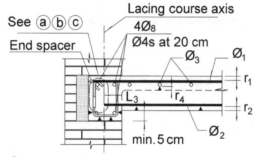

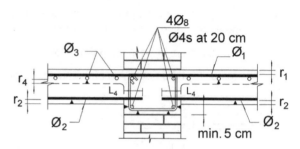

(3) SUPPORT ON OUTSIDE WALL

(4) SUPPORT ON INSIDE WALL

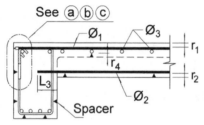

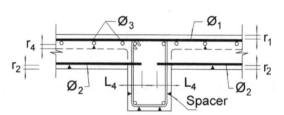

(5) SUPPORT ON EDGE CONCRETE BEAM

(6) SUPPORT ON INTERNAL CONCRETE BEAM

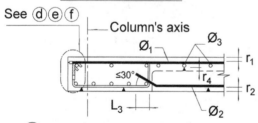

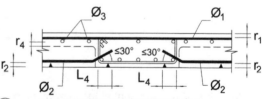

(7) SUPPORT ON EDGE SOFFIT BEAM

(8) SUPPORT ON INTERNAL SOFFIT BEAM

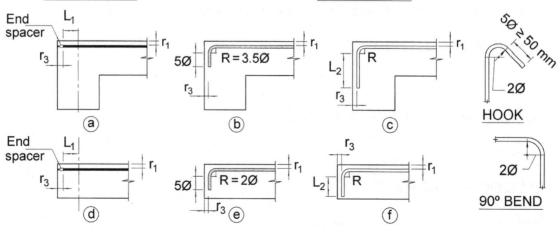

1. RECOMMENDATIONS

1. See CD – 06.01 for the general arrangement.

2. See CD – 05.01 to CD – 05.14 for the concrete cover in beams and lacing courses.

3. $r_1 = 2.5$ cm $\geq \phi_1$.

   If the top is horizontal and exposed to rain or condensed water, $r_1 = 3$ cm.

   $r_2 = 2.5$ cm $\geq \phi_2$.

   $r_3 = 2.5$ cm $\geq \phi_1$.

   $r_4 = 2.5$ cm $\geq \phi_3$.

4. See EC2, anchorage lengths, to calculate L_1 and L_2 and to determine when to use a, b, c, d, e or f for anchoring the ϕ_1 bars.

5. To calculate L_3, see Recommendation 5 in CD – 06.02.

6. To calculate L_4, see Recommendation 6 in CD – 06.02.

7. The top is smoothed with a float or power float.

8. See 1.2 and 1.3 for descriptions of how to tie bars and place spacers.

2. STATUTORY LEGISLATION

See EC2 (5).

3. RECOMMENDED ALTERNATIVE CODES

See (31).

4. SPECIFIC REFERENCES

See (14) pages 487 to 516.

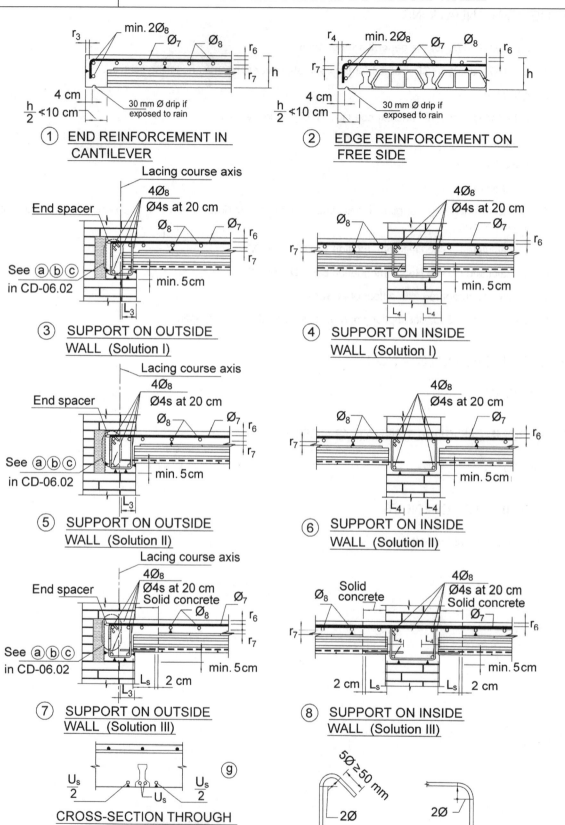

① END REINFORCEMENT IN CANTILEVER

② EDGE REINFORCEMENT ON FREE SIDE

③ SUPPORT ON OUTSIDE WALL (Solution I)

④ SUPPORT ON INSIDE WALL (Solution I)

⑤ SUPPORT ON OUTSIDE WALL (Solution II)

⑥ SUPPORT ON INSIDE WALL (Solution II)

⑦ SUPPORT ON OUTSIDE WALL (Solution III)

⑧ SUPPORT ON INSIDE WALL (Solution III)

CROSS-SECTION THROUGH SOLID CONCRETE AREA (CASES ⑦ AND ⑧)

HOOK

90° BEND

1. RECOMMENDATIONS

1. See CD – 06.01 for the general arrangement.

2. See CD – 05.01 to CD – 05.14 for the concrete cover in beams and lacing courses.

3. $r_3 = 2.5$ cm $\geq \phi_7$.

   $r_4 = 2.5$ cm $\geq \phi_8$.

   $r_6 = 2.5$ cm $\geq \phi_7$.

   If the top is horizontal and exposed to rain or condensed water, $r_6 = 3$ cm.

   $r_7 = 2.5$ cm $\geq \phi_8$.

   L_s is the lap length of the joist reinforcing bar.

4. See Recommendation 4 in CD – 06.02 to calculate L_1 and L_2 and to determine when to use a, b or c for anchoring the ϕ_7 bars.

5. To calculate L_3, see Recommendation 5 in CD – 06.02.

6. To calculate L_4, see Recommendation 6 in CD – 06.02.

7. The top is smoothed with a float or power float.

8. See 1.2 and 1.3 for descriptions of how to tie bars and place spacers.

9. In cases 7 and 8, special attention must be paid to spacer positioning to hold the pieces of anchorage bar in place during concrete pouring (detail (g)). A practical solution is shown in Figure (a) below.

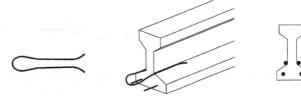

Figure (a)

2. STATUTORY LEGISLATION

See EC2 (5).

3. RECOMMENDED ALTERNATIVE CODES

See (31).

4. SPECIFIC REFERENCES

See (14) pages 487 to 516.

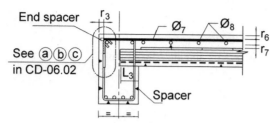

① EDGE CONNECTION WITH REINFORCED CONCRETE BEAM (Solution I)

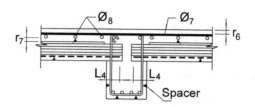

② INTERNAL CONNECTION WITH REINFORCED CONCRETE BEAM (Solution I)

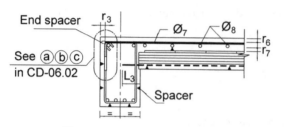

③ EDGE CONNECTION WITH REINFORCED CONCRETE BEAM (Solution II)

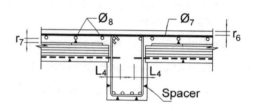

④ INTERNAL CONNECTION WITH REINFORCED CONCRETE BEAM (Solution II)

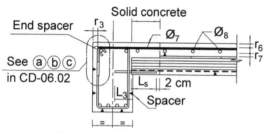

⑤ EDGE CONNECTION WITH REINFORCED CONCRETE BEAM (Solution III)

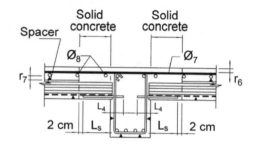

⑥ INTERNAL CONNECTION WITH REINFORCED CONCRETE BEAM (Solution III)

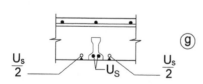

CROSS-SECTION THROUGH SOLID CONCRETE AREA (CASES ⑤ AND ⑥)

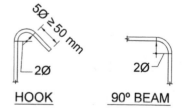

HOOK   90° BEAM

1. RECOMMENDATIONS

1. See CD – 06.01 for the general arrangement.

2. See CD – 05.01 to CD – 05.14 for the concrete cover in beams and tie beams.

3. $r_3 = 2.5$ cm $\geq \phi_7$.

   $r_6 = 2.5$ cm $\geq \phi_7$.

   If the top is horizontal and exposed to rain or condensed water, $r_6 = 3$ cm.

   $r_7 = 2.5$ cm $\geq \phi_8$.

   L_s is the lap length of the joist reinforcing bar.

4. See EC2, anchorage lengths, to calculate L_1 and L_2 and to determine when to use a, b or c for anchoring the ϕ_7 bars.

5. To calculate L_3, see Recommendation 5 in CD – 06.02.

6. To calculate L_4, see Recommendation 6 in CD – 06.02.

7. The top is smoothed with a float or power float.

8. See 1.2 and 1.3 for descriptions of how to tie bars and place spacers.

9. In cases 5 and 6, special attention must be paid to spacer positioning to hold the pieces of anchorage bar in place during concrete pouring (detail (g)). See Recommendation 9 in CD – 06.04 for an alternative solution.

2. STATUTORY LEGISLATION

See EC2 (5).

3. RECOMMENDED ALTERNATIVE CODES

See (31).

4. SPECIFIC REFERENCES

See (14) pages 487 to 516.

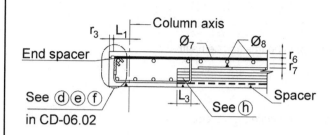

① CONNECTION TO CONCRETE EDGE SOFFIT BEAM (Solution I)

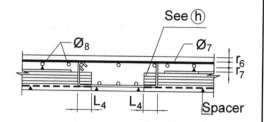

② CONNECTION TO CONCRETE INTERNAL SOFFIT BEAM (Solution I)

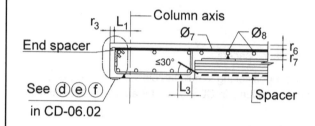

③ CONNECTION TO CONCRETE EDGE SOFFIT BEAM (Solution II)

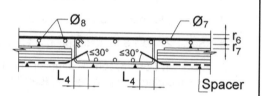

④ CONNECTION TO CONCRETE INTERNAL SOFFIT BEAM (Solution II)

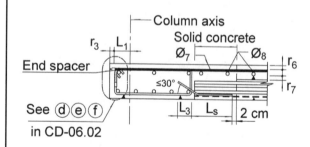

⑤ CONNECTION TO CONCRETE EDGE SOFFIT BEAM (Solution III)

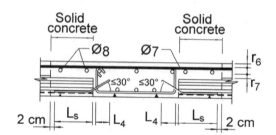

⑥ CONNECTION TO CONCRETE INTERNAL SOFFIT BEAM (Solution III)

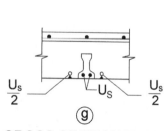

ⓖ CROSS-SECTION THROUGH SOLID CONCRETE AREA (CASES ⑤ AND ⑥)

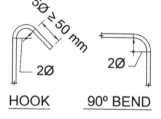

HOOK   90º BEND

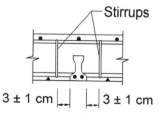

ⓗ (CASES ① AND ②)

1. RECOMMENDATIONS

1. See CD – 06.01 for the general arrangement.

2. See CD – 05.01 to CD – 05.14 for the concrete cover in beams and lacing courses.

3. $r_3 = 2.5$ cm $\geq \phi_7$.

   $r_6 = 2.5$ cm $\geq \phi_7$.

   If the top is horizontal and exposed to rain or condensed water, $r_6 = 3$ cm.

   $r_7 = 2.5$ cm $\geq \phi_8$.

   L_s is the lap length of the joist reinforcing bar.

4. See EC2, anchorage lengths, to calculate L_1 and L_2 and to determine when to use d, e or f for anchoring the ϕ_7 bars.

5. To calculate L_3, see Recommendation 5 in CD – 06.02.

6. To calculate L_4, see Recommendation 6 in CD – 06.02.

7. The top is smoothed with a float or power float.

8. See 1.2 and 1.3 for descriptions of how to tie bars and place spacers.

9. In cases 5 and 6, special attention must be paid to spacer positioning to hold the pieces of anchorage bar in place during concrete pouring (detail (g)). See Recommendation 9 in CD – 06.04 for an alternative solution.

10. Note that in cases 1 and 2, the stirrups in the soffit beam must be positioned so that each joist has one on each side, as shown in detail (h).

2. STATUTORY LEGISLATION

See EC2 (5).

3. RECOMMENDED ALTERNATIVE CODES

See (31).

4. SPECIFIC REFERENCES

See (14) pages 487 to 516.

CD – 06.07	SLABS WITH SELF-SUPPORTING PRESTRESSED CONCRETE JOISTS. CONNECTIONS TO BRICK WALL

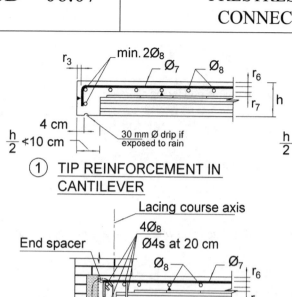

① TIP REINFORCEMENT IN CANTILEVER

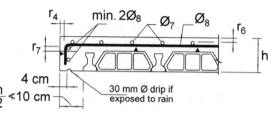

② EDGE REINFORCEMENT ON FREE SIDE

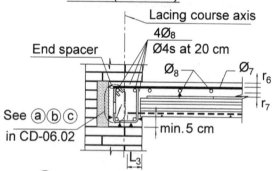

③ SUPPORT ON OUTSIDE WALL (Solution I)

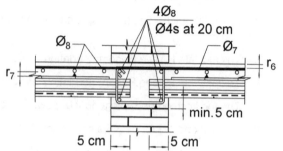

④ SUPPORT ON INSIDE WALL (Solution I)

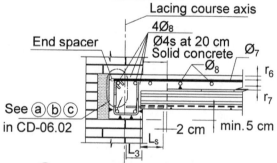

⑤ SUPPORT ON OUTSIDE WALL (Solution II)

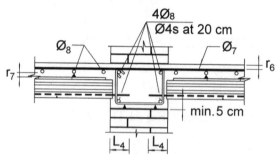

⑥ SUPPORT ON INSIDE WALL (Solution II)

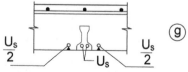

⑦ SUPPORT ON OUTSIDE WALL (Solution III)

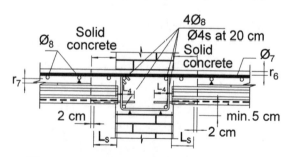

⑧ SUPPORT ON INSIDE WALL (Solution III)

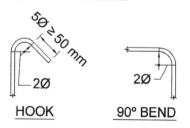

CROSS-SECTION THROUGH SOLID CONCRETE AREA (CASES ⑦ AND ⑧)

HOOK   90° BEND

1. RECOMMENDATIONS

1. See CD – 06.01 for the general arrangement.

2. See CD – 05.01 to CD-05.14 for the concrete cover in beams and lacing courses.

3. $r_3 = 2.5$ cm $\geq \phi_7$.

    $r_4 = 2.5$ cm $\geq \phi_8$.

    $r_6 = 2.5$ cm $\geq \phi_7$.

    If the top is horizontal and exposed to rain or condensed water, $r_6 = 3$ cm.

    $r_7 = 2.5$ cm $\geq \phi_8$.

    L_s is the lap length of the joist reinforcing bar.

4. See EC2, anchorage lengths, to calculate L_1 and L_2 and to determine when to use a, b or c for anchoring the ϕ_7 bars.

5. To calculate L_3, see Recommendation 5 in CD – 06.02.

6. To calculate L_4, see Recommendation 6 in CD – 06.02.

7. The top is smoothed with a float or power float.

8. See 1.2 and 1.3 for descriptions of how to tie bars and place spacers.

9. In cases 7 and 8, special attention must be paid to spacer positioning to hold the pieces of anchorage bar in place during concrete pouring (detail (g)). See Recommendation 9 in CD – 06.04 for an alternative solution.

2. STATUTORY LEGISLATION

See EC2 (5).

3. RECOMMENDED ALTERNATIVE CODES

See (31).

4. SPECIFIC REFERENCES

See (14) pages 487 to 516.

CD – 06.08	SLABS WITH SELF-SUPPORTING PRESTRESSED CONCRETE JOISTS. CONNECTIONS TO REINFORCED CONCRETE BEAMS

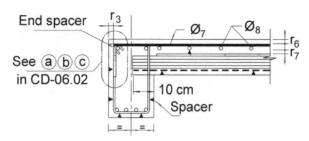

① CONNECTION TO REINFORCED CONCRETE EDGE BEAM (Solution I)

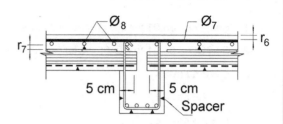

② CONNECTION TO REINFORCED CONCRETE INTERNAL BEAM (Solution I)

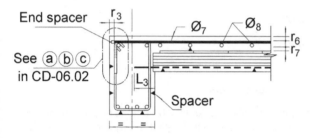

③ CONNECTION TO REINFORCED CONCRETE EDGE BEAM (Solution II)

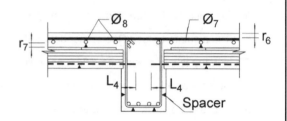

④ CONNECTION TO REINFORCED CONCRETE INTERNAL BEAM (Solution II)

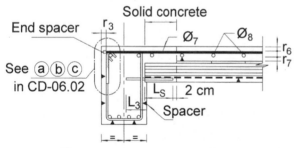

⑤ CONNECTION TO REINFORCED CONCRETE EDGE BEAM (Solution III)

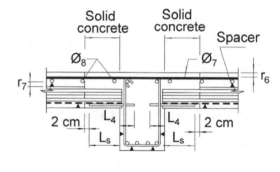

⑥ CONNECTION TO REINFORCED CONCRETE INTERNAL BEAM (Solution III)

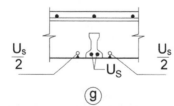

ⓖ CROSS-SECTION THROUGH SOLID CONCRETE AREA (CASES ⑤ AND ⑥)

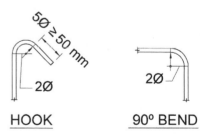

HOOK   90° BEND

1. RECOMMENDATIONS

1. See CD – 06.01 for the general arrangement.

2. See CD – 05.01 to CD – 05.14 for the concrete cover in beams and lacing courses.

3. $r_3 = 2.5$ cm $\geq \phi_7$.

   $r_6 = 2.5$ cm $\geq \phi_7$.

   If the top is horizontal and exposed to rain or condensed water, $r_6 = 3$ cm.

   $r_7 = 2.5$ cm $\geq \phi_8$.

   L_s is the lap length of the joist reinforcing bar.

4. See EC2, anchorage lengths, to calculate L_1 and L_2 and to determine when to use a, b or c for anchoring the ϕ_7 bars.

5. To calculate L_3, see Recommendation 5 in CD – 06.02.

6. To calculate L_4, see Recommendation 6 in CD – 06.02.

7. The top is smoothed with a float or power float.

8. See 1.2 and 1.3 for descriptions of how to tie bars and place spacers.

9. In cases 5 and 6, special attention must be paid to spacer positioning to hold the pieces of anchorage bar in place during concrete pouring (detail (g)). See Recommendation 9 in CD – 06.04 for an alternative solution.

2. STATUTORY LEGISLATION

See EC2 (5).

3. RECOMMENDED ALTERNATIVE CODES

See (31).

4. SPECIFIC REFERENCES

See (14) pages 487 to 516.

CD – 06.09	SLABS WITH SELF-SUPPORTING PRESTRESSED CONCRETE JOISTS. CONNECTIONS TO SOFFIT BEAMS

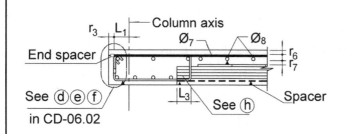

(1) CONNECTION TO CONCRETE EDGE SOFFIT BEAM (Solution I)

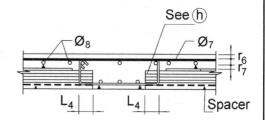

(2) CONNECTION TO CONCRETE INTERNAL SOFFIT BEAM (Solution I)

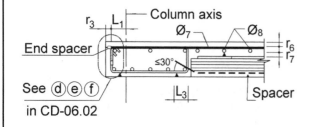

(3) CONNECTION TO CONCRETE EDGE SOFFIT BEAM (Solution II)

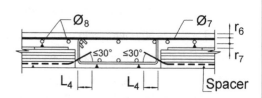

(4) CONNECTION TO CONCRETE INTERNAL SOFFIT BEAM (Solution II)

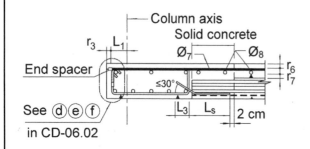

(5) CONNECTION TO CONCRETE EDGE SOFFIT BEAM (Solution III)

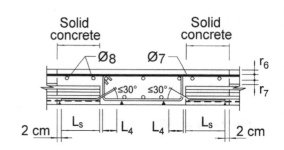

(6) CONNECTION TO CONCRETE INTERNAL SOFFIT BEAM (Solution III)

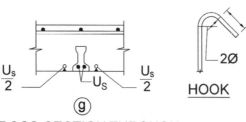

(g)
CROSS-SECTION THROUGH SOLID CONCRETE AREA
(CASES (5) AND (6))

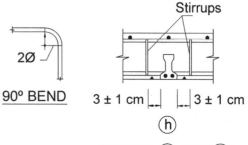

(h)
(CASES (1) AND (2))

1. RECOMMENDATIONS

1. See CD – 06.01 for the general arrangement.

2. See CD – 05.01 to CD – 05.14 for the concrete cover in beams and lacing courses.

3. $r_3 = 2.5$ cm $\geq \phi_7$.

    $r_6 = 2.5$ cm $\geq \phi_7$.

    If the top is horizontal and exposed to rain or condensed water, $r_6 = 3$ cm.

    L_s is the lap length of the joist reinforcing bar.

    $r_7 = 2.5$ cm $\geq \phi_8$.

4. See EC2, anchorage lengths, to calculate L_1 and L_2 and to determine when to use d, e or f for anchoring the ϕ_7 bars.

5. To calculate L_3, see Recommendation 5 in CD – 06.02.

6. To calculate L_4, see Recommendation 6 in CD – 06.02.

7. The top is smoothed with a float or power float.

8. See 1.2 and 1.3 for descriptions of how to tie bars and place spacers.

9. In cases 5 and 6, special attention must be paid to spacer positioning to hold the pieces of anchorage bar in place during concrete pouring (detail (g)). See Recommendation 9 in CD – 06.04 for an alternative solution.

10. Note that in cases 1 and 2, the stirrups in the soffit beam must be positioned so that each joist has one on each side, as shown in detail (h).

2. STATUTORY LEGISLATION

See EC2 (5).

3. RECOMMENDED ALTERNATIVE CODES

See (31).

4. SPECIFIC REFERENCES

See (14) pages 487 to 516.

CD – 06.10	SLABS WITH SEMI-SELF-SUPPORTING REINFORCED CONCRETE JOISTS. CONNECTIONS TO BRICK WALL

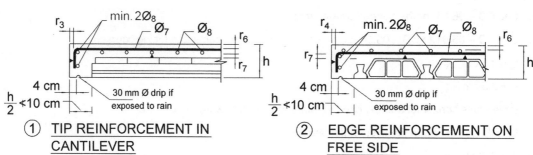

① **TIP REINFORCEMENT IN CANTILEVER**

② **EDGE REINFORCEMENT ON FREE SIDE**

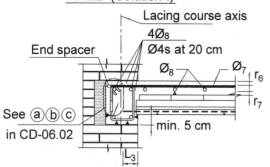

③ **SUPPORT ON OUTSIDE WALL (Solution I)**

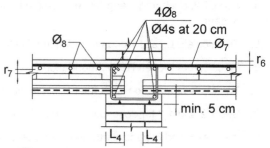

④ **SUPPORT ON INSIDE WALL (Solution I)**

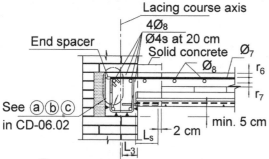

⑤ **SUPPORT ON OUTSIDE WALL (Solution II)**

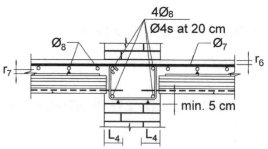

⑥ **SUPPORT ON INSIDE WALL (Solution II)**

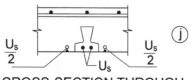

⑦ **SUPPORT ON OUTSIDE WALL (Solution III)**

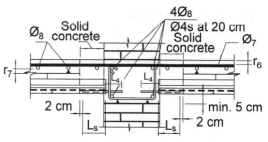

⑧ **SUPPORT ON INSIDE WALL (Solution III)**

CROSS-SECTION THROUGH SOLID CONCRETE AREA (CASES ⑦ AND ⑧)

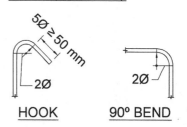

HOOK   90° BEND

1. RECOMMENDATIONS

1. See CD – 06.01 for the general arrangement.

2. See CD – 05.01 to CD – 05.14 for the concrete cover in beams and lacing courses.

3. $r_3 = 2.5$ cm $\geq \phi_7$.

   $r_4 = 2.5$ cm $\geq \phi_8$.

   $r_6 = 2.5$ cm $\geq \phi_7$.

   $r_7 = 2.5$ cm $\geq \phi_8$.

   If the top is horizontal and exposed to rain or condensed water, $r_6 = 3$ cm.

   L_s is the lap length of the joist reinforcing bar.

4. See EC2, anchorage lengths, to calculate L_1 and L_2 and to determine when to use a, b or c for anchoring the ϕ_7 bars.

5. To calculate L_3, see Recommendation 5 in CD – 06.02.

6. To calculate L_4, see Recommendation 6 in CD – 06.02.

7. The top is smoothed with a float or power float.

8. See 1.2 and 1.3 for descriptions of how to tie bars and place spacers.

9. In cases 7 and 8, special attention must be paid to spacer positioning to hold the pieces of anchorage bar in place during concrete pouring (detail (j)). See Recommendation 9 in CD – 06.04 for an alternative solution.

2. STATUTORY LEGISLATION

See EC2 (5).

3. RECOMMENDED ALTERNATIVE CODES

See (31).

4. SPECIFIC REFERENCES

See (14) pages 487 to 516.

CD – 06.11	SLABS WITH SEMI-SELF-SUPPORTING REINFORCED CONCRETE JOISTS. CONNECTIONS TO CONCRETE BEAMS

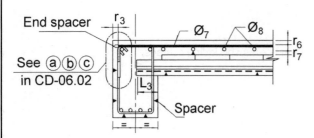

① CONNECTION TO REINFORCED CONCRETE EDGE BEAM (Solution I)

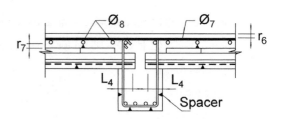

② CONNECTION TO REINFORCED CONCRETE INTERNAL BEAM (Solution I)

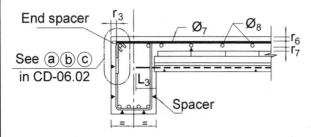

③ CONNECTION TO REINFORCED CONCRETE EDGE BEAM (Solution II)

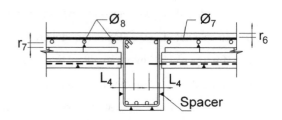

④ CONNECTION TO REINFORCED CONCRETE INTERNAL BEAM (Solution II)

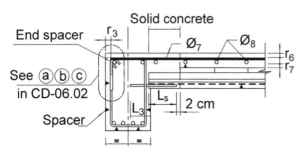

⑤ CONNECTION TO REINFORCED CONCRETE EDGE BEAM (Solution III)

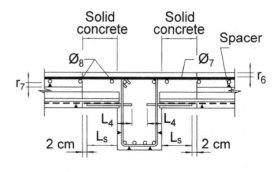

⑥ CONNECTION TO REINFORCED CONCRETE INTERNAL BEAM (Solution III)

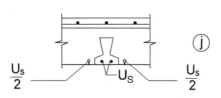

CROSS-SECTION THROUGH SOLID CONCRETE AREA (CASES ⑤ AND ⑥)

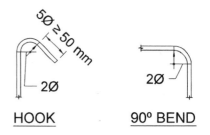

HOOK   90° BEND

1. RECOMMENDATIONS

1. See CD – 06.01 for the general arrangement.

2. See CD – 05.01 to CD – 05.14 for the concrete cover in beams and lacing courses.

3. $r_3 = 2.5$ cm $\geq \phi_7$.

   $r_6 = 2.5$ cm $\geq \phi_7$.

   If the top is horizontal and exposed to rain or condensed water, $r_6 = 3$ cm.

   $r_7 = 2.5$ cm $\geq \phi_8$.

   L_s is the lap length of the joist reinforcing bar.

4. See EC2, anchorage lengths, to calculate L_1 and L_2 and to determine when to use a, b or c for anchoring the ϕ_7 bars.

5. To calculate L_3, see Recommendation 5 in CD – 06.02.

6. To calculate L_4, see Recommendation 6 in CD – 06.02.

7. The top is smoothed with a float or power float.

8. See 1.2 and 1.3 for descriptions of how to tie bars and place spacers.

9. In cases 5 and 6, special attention must be paid to spacer positioning to hold the pieces of anchorage bar in place during concrete pouring (detail (j)). See Recommendation 9 in CD – 06.04 for an alternative solution.

2. STATUTORY LEGISLATION

See EC2.

3. RECOMMENDED ALTERNATIVE CODES

See (31).

4. SPECIFIC REFERENCES

See (14) pages 487 to 516.

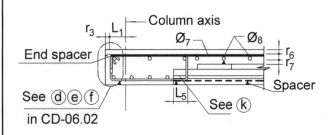

1. CONNECTION TO CONCRETE EDGE SOFFIT BEAM (Solution I)

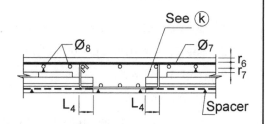

2. CONNECTION TO CONCRETE INTERNAL SOFFIT BEAM (Solution I)

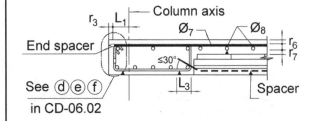

3. CONNECTION TO CONCRETE EDGE SOFFIT BEAM (Solution II)

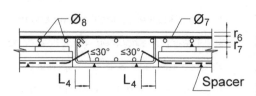

4. CONNECTION TO CONCRETE INTERNAL SOFFIT BEAM (Solution II)

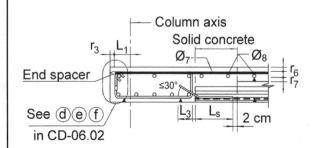

5. CONNECTION TO CONCRETE EDGE SOFFIT BEAM (Solution III)

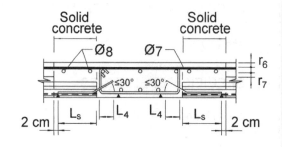

6. CONNECTION TO CONCRETE INTERNAL SOFFIT BEAM (Solution III)

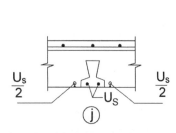

(j) CROSS-SECTION THROUGH SOLID CONCRETE AREA (CASES (5) AND (6))

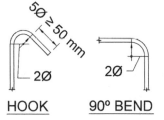

HOOK   90° BEND

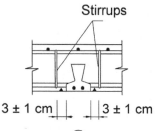

(k) (CASES (1) AND (2))

1. RECOMMENDATIONS

1. See CD – 06.01 for the general arrangement.

2. See CD – 05.01 to CD – 05.14 for the concrete cover in beams and lacing courses.

3. $r_3 = 2.5$ cm $\geq \phi_7$.

    $r_6 = 2.5$ cm $\geq \phi_7$.

    If the top is horizontal and exposed to rain or condensed water, $r_6 = 3$ cm.

    $r_7 = 2.5$ cm $\geq \phi_8$.

    L_s is the lap length of the joist reinforcing bar.

4. See EC2, anchorage lengths, to calculate L_1 and L_2 and to determine when to use d, e or f for anchoring the ϕ_7 bars.

5. To calculate L_3, see Recommendation 5 in CD – 06.02.

6. To calculate L_4, see Recommendation 6 in CD – 06.02.

7. The top is smoothed with a float or power float.

8. See 1.2 and 1.3 for descriptions of how to tie bars and place spacers.

9. In cases 5 and 6, special attention must be paid to spacer positioning to hold the pieces of anchorage bar in place during concrete pouring (detail (j)). See Recommendation 9 in CD – 06.04 for an alternative solution.

10. Note that in cases 1 and 2, the stirrups in the soffit beam must be positioned so that each joist has one on each side, as shown in detail (k).

2. STATUTORY LEGISLATION

See EC2 (5).

3. RECOMMENDED ALTERNATIVE CODES

See (31).

4. SPECIFIC REFERENCES

See (14) pages 487 to 516.

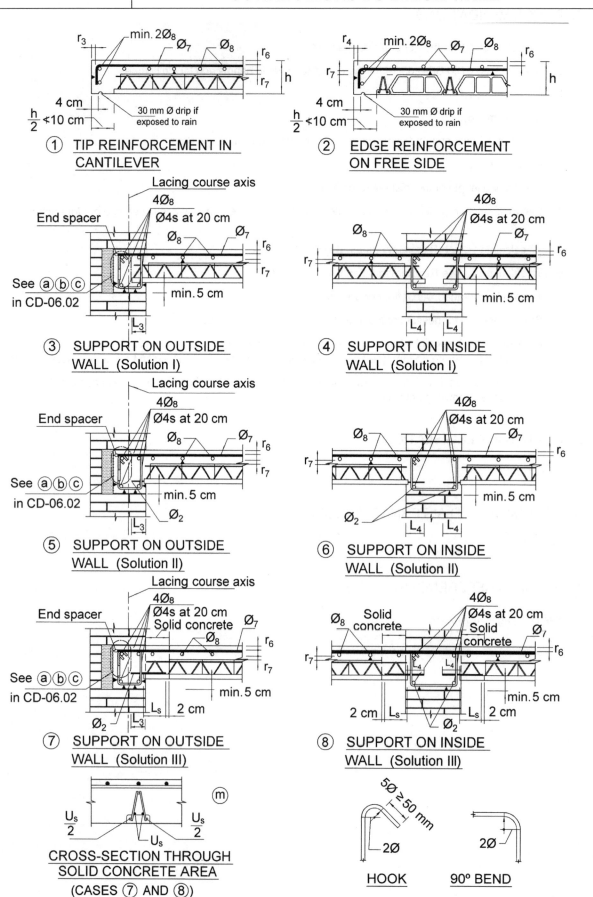

① TIP REINFORCEMENT IN CANTILEVER

② EDGE REINFORCEMENT ON FREE SIDE

③ SUPPORT ON OUTSIDE WALL (Solution I)

④ SUPPORT ON INSIDE WALL (Solution I)

⑤ SUPPORT ON OUTSIDE WALL (Solution II)

⑥ SUPPORT ON INSIDE WALL (Solution II)

⑦ SUPPORT ON OUTSIDE WALL (Solution III)

⑧ SUPPORT ON INSIDE WALL (Solution III)

CROSS-SECTION THROUGH SOLID CONCRETE AREA (CASES ⑦ AND ⑧)

HOOK

90° BEND

1. RECOMMENDATIONS

1. See CD – 06.01 for the general arrangement.

2. See CD – 05.01 to CD – 05.14 for the concrete cover in beams and lacing courses.

3. $r_3 = 2.5$ cm $\geq \phi_7$.

   $r_4 = 2.5$ cm $\geq \phi_8$.

   $r_6 = 2.5$ cm $\geq \phi_7$.

   $r_7 = 2.5$ cm $\geq \phi_8$.

   If the top is horizontal and exposed to rain or condensed water, $r_6 = 3$ cm.

   L_s is the lap length of the joist reinforcing bar.

4. See EC2, anchorage lengths, to calculate L_1 and L_2 and to determine when to use a, b, or c for anchoring the ϕ_7 bars.

5. To calculate L_3, see Recommendation 5 in CD – 06.02.

6. To calculate L_4, see Recommendation 6 in CD – 06.02.

7. The top is smoothed with a float or power float.

8. See 1.2 and 1.3 for descriptions of how to tie bars and place spacers.

9. In cases 7 and 8, special attention must be paid to spacer positioning to hold the pieces of anchorage bar in place during concrete pouring (detail (m)). See Figure (a) below for an alternative solution.

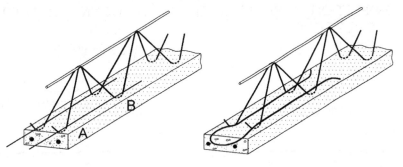

Figure (a)

2. STATUTORY LEGISLATION

See EC2 (5).

3. RECOMMENDED ALTERNATIVE CODES

See (31).

4. SPECIFIC REFERENCES

See (14) pages 487 to 516.

CD – 06.14	SLABS WITH SEMI-SELF-SUPPORTING REINFORCED CONCRETE LATTICE JOISTS. CONNECTIONS TO CONCRETE BEAMS

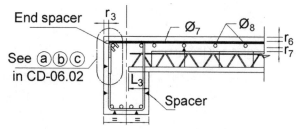

① CONNECTION TO REINFORCED CONCRETE EDGE BEAM
(Solution I)

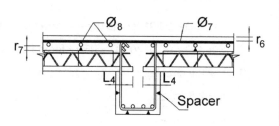

② CONNECTION TO REINFORCED CONCRETE INTERNAL BEAM
(Solution I)

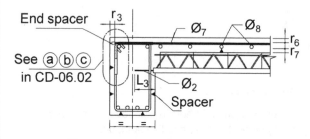

③ CONNECTION TO REINFORCED CONCRETE EDGE BEAM
(Solution II)

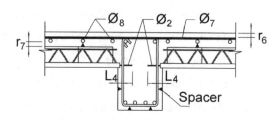

④ CONNECTION TO REINFORCED CONCRETE INTERNAL BEAM
(Solution II)

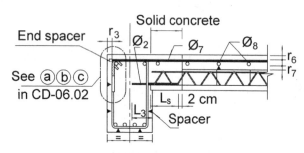

⑤ CONNECTION TO REINFORCED CONCRETE EDGE BEAM
(Solution III)

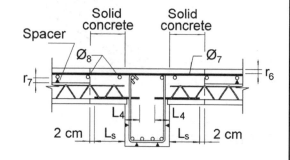

⑥ CONNECTION TO REINFORCED CONCRETE INTERNAL BEAM
(Solution III)

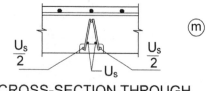

CROSS-SECTION THROUGH SOLID CONCRETE AREA
(CASES ⑤ AND ⑥)

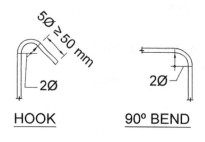

HOOK   90° BEND

1. RECOMMENDATIONS

1. See CD – 06.01 for the general arrangement.

2. See CD – 05.01 to CD – 05.14 for the concrete cover in beams and lacing courses.

3. $r_3 = 2.5$ cm $\geq \phi_7$.

   $r_6 = 2.5$ cm $\geq \phi_7$.

   $r_7 = 2.5$ cm $\geq \phi_8$.

   If the top is horizontal and exposed to rain or condensed water, $r_6 = 3$ cm.

   L_s is the lap length of the joist reinforcing bar.

4. See EC2, anchorage lengths, to calculate L_1 and L_2 and to determine when to use a, b or c for anchoring the ϕ_7 bars.

5. To calculate L_3, see Recommendation 5 in CD – 06.02.

6. To calculate L_4, see Recommendation 6 in CD – 06.02.

7. The top is smoothed with a float or power float.

8. See 1.2 and 1.3 for descriptions of how to tie bars and place spacers.

9. In cases 5 and 6, special attention must be paid to spacer positioning to hold the pieces of anchorage bar in place during concrete pouring (detail (m)). See Recommendation 9 in CD – 06.13 for an alternative solution.

2. STATUTORY LEGISLATION

See EC2 (5).

3. RECOMMENDED ALTERNATIVE CODES

See (31).

4. SPECIFIC REFERENCES

See (14) pages 487 to 516.

CD – 06.15	SLABS WITH SEMI-SELF-SUPPORTING REINFORCED CONCRETE LATTICE JOISTS. CONNECTIONS TO SOFFIT BEAMS

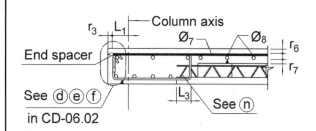

① CONNECTION TO CONCRETE
EDGE SOFFIT BEAM
(Solution I)

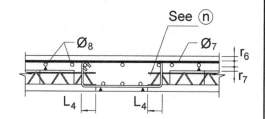

② CONNECTION TO CONCRETE
INTERNAL SOFFIT BEAM
(Solution I)

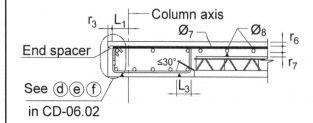

③ CONNECTION TO CONCRETE
EDGE SOFFIT BEAM
(Solution II)

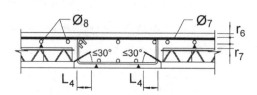

④ CONNECTION TO CONCRETE
INTERNAL SOFFIT BEAM
(Solution II)

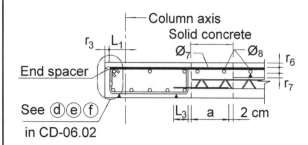

⑤ CONNECTION TO CONCRETE
EDGE SOFFIT BEAM
(Solution III)

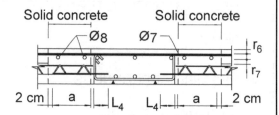

⑥ CONNECTION TO CONCRETE
INTERNAL SOFFIT BEAM
(Solution III)

NOTE:

a is the lap length, but it may be no
shorter than the distance between
two consecutive welds.

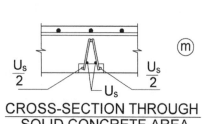

CROSS-SECTION THROUGH
SOLID CONCRETE AREA
(CASES ⑤ AND ⑥)

Stirrups

(CASES ① AND ②)

2Ø

90º BEND

HOOK

1. RECOMMENDATIONS

1. See CD – 06.01 for the general arrangement.
2. See CD – 05.01 to CD – 05.14 for the concrete cover in beams and lacing courses.
3. $r_3 = 2.5$ cm $\geq \phi_7$.

    $r_6 = 2.5$ cm $\geq \phi_7$.

    $r_7 = 2.5$ cm $\geq \phi_8$.

    If the top is horizontal and exposed to rain or condensed water, $r_6 = 3$ cm.

    L_s is the lap length of the joist reinforcing bar.
4. See EC2, anchorage lengths, to calculate L_1 and L_2 and to determine when to use d, e or f for anchoring the ϕ_7 bars.
5. To calculate L_3, see Recommendation 5 in CD – 06.02.
6. To calculate L_4, see Recommendation 6 in CD – 06.02.
7. The top is smoothed with a float or power float.
8. See 1.2 and 1.3 for descriptions of how to tie bars and place spacers.
9. In cases 5 and 6, special attention must be paid to spacer positioning to hold the pieces of anchorage bar in place during concrete pouring (detail (m)). See Recommendation 9 in CD – 06.13 for an alternative solution.
10. Note that in cases 1 and 2, the stirrups in the soffit beam must be positioned so that each joist has one on each side, as shown in detail (n).

2. STATUTORY LEGISLATION

See EC2 (5).

3. RECOMMENDED ALTERNATIVE CODES

See (31).

4. SPECIFIC REFERENCES

See (14) pages 487 to 516.

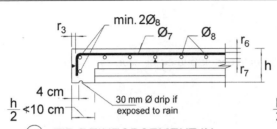

r_3 min. $2\emptyset_8$ $\emptyset_7$ $\emptyset_8$ r_6 r_7 h

4 cm $\frac{h}{2} \leqslant 10$ cm 30 mm $\emptyset$ drip if exposed to rain

① TIP REINFORCEMENT IN CANTILEVER

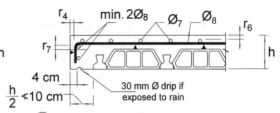

r_4 min. $2\emptyset_8$ $\emptyset_7$ $\emptyset_8$ r_6 r_7 h

4 cm $\frac{h}{2} \leqslant 10$ cm 30 mm $\emptyset$ drip if exposed to rain

② EDGE REINFORCEMENT ON FREE SIDE

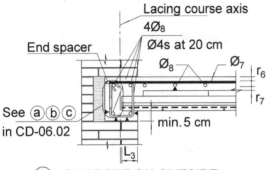

Lacing course axis
$4\emptyset_8$
End spacer $\emptyset$4s at 20 cm
$\emptyset_8$ $\emptyset_7$ r_6 r_7
See ⓐⓑⓒ in CD-06.02 min. 5 cm L_3

③ SUPPORT ON OUTSIDE WALL (Solution I)

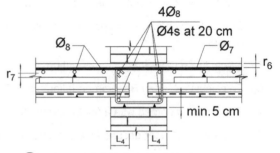

$4\emptyset_8$
$\emptyset$4s at 20 cm
$\emptyset_8$ $\emptyset_7$ r_6 r_7
min. 5 cm L_4 L_4

④ SUPPORT ON INSIDE WALL (Solution I)

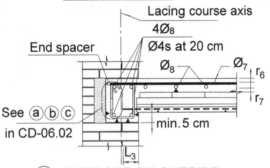

Lacing course axis
$4\emptyset_8$
End spacer $\emptyset$4s at 20 cm
$\emptyset_8$ $\emptyset_7$ r_6 r_7
See ⓐⓑⓒ in CD-06.02 min. 5 cm L_3

⑤ SUPPORT ON OUTSIDE WALL (Solution II)

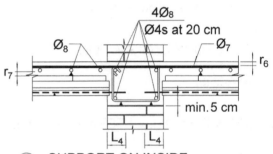

$4\emptyset_8$
$\emptyset$4s at 20 cm
$\emptyset_8$ $\emptyset_7$ r_6 r_7
min. 5 cm L_4 L_4

⑥ SUPPORT ON INSIDE WALL (Solution II)

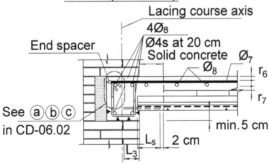

Lacing course axis
$4\emptyset_8$
End spacer $\emptyset$4s at 20 cm
Solid concrete $\emptyset_7$
$\emptyset_8$ r_6 r_7
See ⓐⓑⓒ in CD-06.02 L_s 2 cm L_3 min. 5 cm

⑦ SUPPORT ON OUTSIDE WALL (Solution III)

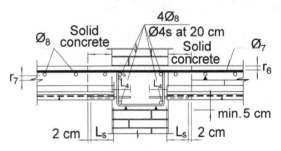

$4\emptyset_8$
Solid concrete $\emptyset$4s at 20 cm
$\emptyset_8$ Solid concrete $\emptyset_7$
r_6 r_7
2 cm L_s L_s 2 cm min. 5 cm

⑧ SUPPORT ON INSIDE WALL (Solution III)

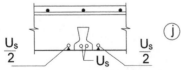

$\frac{U_s}{2}$ $\frac{U_s}{2}$ ⓙ
U_s

CROSS-SECTION THROUGH SOLID CONCRETE AREA
(CASES ⑦ AND ⑧)

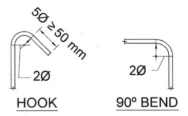

$5\emptyset \geqslant 50$ mm
$2\emptyset$
HOOK

$2\emptyset$
90° BEND

1. RECOMMENDATIONS

1. See CD – 06.01 for the general arrangement.

2. See CD – 05.01 to CD – 05.14 for the concrete cover in beams and lacing courses.

3. $r_3 = 2.5$ cm $\geq \phi_7$.

    $r_6 = 2.5$ cm $\geq \phi_7$.

    $r_7 = 2.5$ cm $\geq \phi_8$.

    If the top is horizontal and exposed to rain or condensed water, $r_6 = 3$ cm.

    L_s is the lap length of the joist reinforcing bar.

4. See EC2, anchorage lengths, to calculate L_1 and L_2 and to determine when to use a, b or c for anchoring the ϕ_7 bars.

5. To calculate L_3, see Recommendation 5 in CD – 06.02.

6. To calculate L_4, see Recommendation 6 in CD – 06.02.

7. The top is smoothed with a float or power float.

8. See 1.2 and 1.3 for descriptions of how to tie bars and place spacers.

9. In cases 7 and 8, special attention must be paid to spacer positioning to hold the pieces of anchorage bar in place during concrete pouring (detail (g)). See Recommendation 9 in CD – 06.04 for an alternative solution.

2. STATUTORY LEGISLATION

See EC2 (5).

3. RECOMMENDED ALTERNATIVE CODES

See (31).

4. SPECIFIC REFERENCES

See (14) pages 487 to 516.

CD – 06.17	SLABS WITH SEMI-SELF-SUPPORTING PRESTRESSED JOISTS. CONNECTIONS TO CONCRETE BEAMS

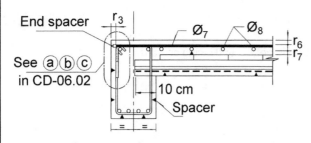

① CONNECTION TO REINFORCED CONCRETE EDGE BEAMS (Solution I)

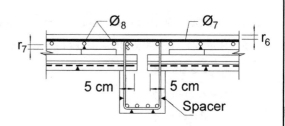

② CONNECTION TO REINFORCED CONCRETE INTERNAL BEAM (Solution I)

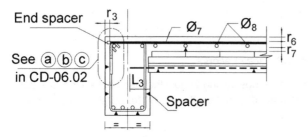

③ CONNECTION TO REINFORCED CONCRETE EDGE BEAMS (Solution II)

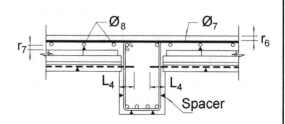

④ CONNECTION TO REINFORCED CONCRETE INTERNAL BEAM (Solution II)

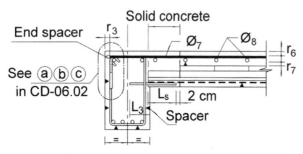

⑤ CONNECTION TO REINFORCED CONCRETE EDGE BEAMS (Solution III)

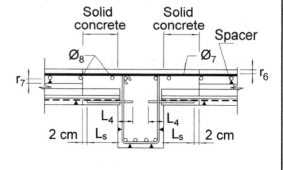

⑥ CONNECTION TO REINFORCED CONCRETE INTERNAL BEAM (Solution III)

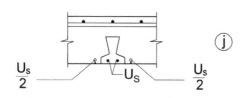

CROSS-SECTION THROUGH SOLID CONCRETE AREA (CASES ⑤ AND ⑥)

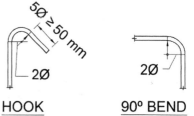

HOOK   90º BEND

1. RECOMMENDATIONS

1. See CD – 06.01 for the general arrangement.

2. See CD – 05.01 to CD – 05.14 for the concrete cover in beams and lacing courses.

3. $r_3 = 2.5$ cm $\geq \phi_7$.

    $r_6 = 2.5$ cm $\geq \phi_7$.

    $r_7 = 2.5$ cm $\geq \phi_8$.

    If the top is horizontal and exposed to rain or condensed water, $r_6 = 3$ cm.

    L_s is the lap length of the joist reinforcing bar.

4. See EC2, anchorage lengths, to calculate L_1 and L_2 and to determine when to use a, b or c for anchoring the ϕ_7 bars.

5. To calculate L_3, see Recommendation 5 in CD – 06.02.

6. To calculate L_4, see Recommendation 6 in CD – 06.02.

7. The top is smoothed with a float or power float.

8. See 1.2 and 1.3 for descriptions of how to tie bars and place spacers.

9. In cases 5 and 6, special attention must be paid to spacer positioning to hold the pieces of anchorage bar in place during concrete pouring (detail (j)). See Recommendation 9 in CD – 06.04 for an alternative solution.

2. STATUTORY LEGISLATION

See EC2 (5).

3. RECOMMENDED ALTERNATIVE CODES

See (31).

4. SPECIFIC REFERENCES

See (14) pages 487 to 516.

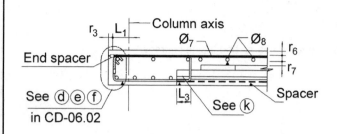

① CONNECTION TO CONCRETE
EDGE SOFFIT BEAM
(Solution I)

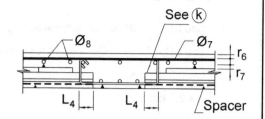

② CONNECTION TO CONCRETE
INTERNAL SOFFIT BEAM
(Solution I)

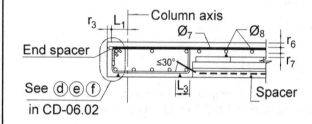

③ CONNECTION TO CONCRETE
EDGE SOFFIT BEAM
(Solution II)

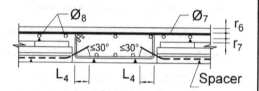

④ CONNECTION TO CONCRETE
INTERNAL SOFFIT BEAM
(Solution II)

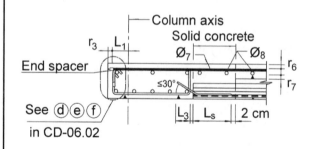

⑤ CONNECTION TO CONCRETE
EDGE SOFFIT BEAM
(Solution III)

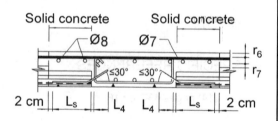

⑥ CONNECTION TO CONCRETE
INTERNAL SOFFIT BEAM
(Solution III)

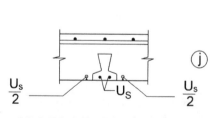

CROSS-SECTION THROUGH
SOLID CONCRETE AREA
(CASES ⑤ AND ⑥)

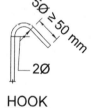

HOOK

90° BEND

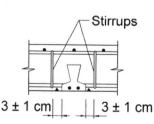

ⓚ
(CASES ① AND ②)

1. RECOMMENDATIONS

1. See CD – 06.01 for the general arrangement.

2. See CD – 05.01 to CD – 05.14 for the concrete cover in beams and lacing courses.

3. $r_3 = 2.5$ cm $\geq \phi_7$.

    $r_6 = 2.5$ cm $\geq \phi_7$.

    $r_7 = 2.5$ cm $\geq \phi_8$.

    If the top is horizontal and exposed to rain or condensed water, $r_6 = 3$ cm.

    L_s is the lap length of the joist reinforcing bar.

4. See EC2, anchorage lengths, to calculate L_1 and L_2 and to determine when to use d, e or f for anchoring the ϕ_7 bars.

5. To calculate L_3, see Recommendation 5 in CD – 06.02.

6. To calculate L_4, see Recommendation 6 in CD – 06.02.

7. The top is smoothed with a float or power float.

8. See 1.2 and 1.3 for descriptions of how to tie bars and place spacers.

9. In cases 5 and 6, special attention must be paid to spacer positioning to hold the pieces of anchorage bar in place during concrete pouring (detail (j)). See Recommendation 9 in CD – 06.04 for an alternative solution.

10. Note that in cases 1 and 2, the stirrups in the soffit beam must be positioned so that each joist has one on each side, as shown in detail (k).

2. STATUTORY LEGISLATION

See EC2 (5).

3. RECOMMENDED ALTERNATIVE CODES

See (31).

4. SPECIFIC REFERENCES

See (14) pages 487 to 516.

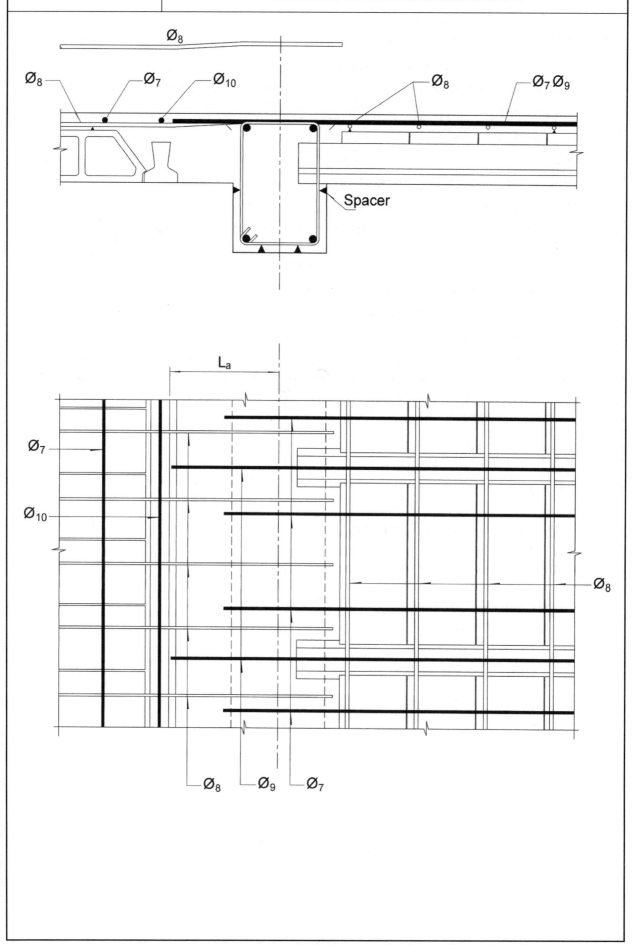

CD – 06.19   PRECAST BEAM AND BLOCK FLOOR SYSTEMS. CHANGE IN BEAM DIRECTION

1. RECOMMENDATIONS

1. The detail shows a floor slab with semi-self-supporting joists. The arrangement is analogous to other types of joist.

2. See CD – 06.01 for the general arrangement.

3. See CD – 05.01 to CD – 05.14 for the concrete cover in beams and lacing courses.

4. See CD – 06.01 for the concrete cover.

5. Note the negative moment ϕ_9 reinforcing bar on the right of the floor slab. It should be anchored in the top slab on the left side of the floor slab for a length L_a equal to its anchorage length.

6. The top is smoothed with a float or power float.

2. STATUTORY LEGISLATION

See EC2 (5).

3. RECOMMENDED ALTERNATIVE CODES

See (31).

4. SPECIFIC REFERENCES

See (14) pages 487 to 516.

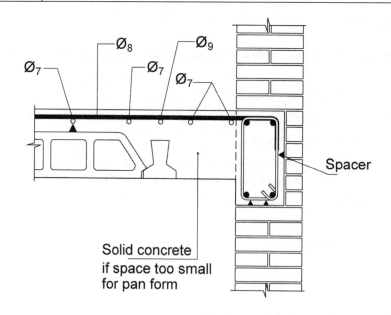

$\emptyset_7$ $\emptyset_8$ $\emptyset_9$ $\emptyset_7$ $\emptyset_7$

Spacer

Solid concrete
if space too small
for pan form

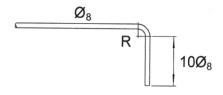

$\emptyset_8$
R
$10\emptyset_8$

300

1. RECOMMENDATIONS

1. The detail shows a floor slab with semi-self-supporting joists. The arrangement is analogous to other types of joist.

2. See CD – 06.01 for the general arrangement.

3. See CD – 05.01 to CD – 05.14 for the concrete cover in beams and tie beams.

4. See CD – 06.01 for the concrete cover.

5. Particular care must be taken when placing the ϕ_8 reinforcing bars and their anchorage in the beam or tie beam. This is essential to prevent cracks from appearing parallel to the joists on the inner side of the floor slab.

6. The top is smoothed with a float or power float.

2. STATUTORY LEGISLATION

See EC2 (5).

3. RECOMMENDED ALTERNATIVE CODES

See (31).

4. SPECIFIC REFERENCES

See (14) pages 487 to 516.

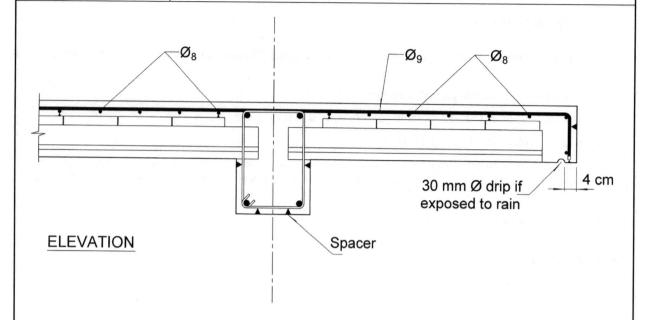

$\emptyset_8$ $\emptyset_9$ $\emptyset_8$

30 mm Ø drip if
exposed to rain

4 cm

ELEVATION

Spacer

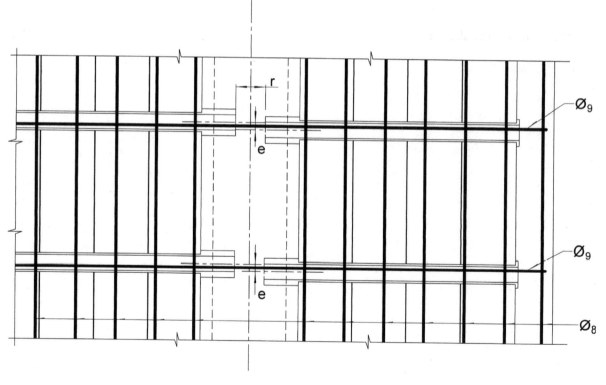

r

$\emptyset_9$

e

$\emptyset_9$

e

$\emptyset_8$

PLAN

1. RECOMMENDATIONS

1. The detail shows a floor slab with semi-self-supporting joists. The arrangement is analogous to other types of joist.

2. See CD – 06.01 for the general arrangement.

3. See CD – 05.01 to CD – 05.14 for the concrete cover in beams and lacing courses.

4. See CD – 06.01 for the concrete cover.

5. The top is smoothed with a float or power float.

6. **AR.** The maximum value for deviation e is e ≤ r.

2. STATUTORY LEGISLATION

See EC2 (5).

3. RECOMMENDED ALTERNATIVE CODES

See (31).

4. SPECIFIC REFERENCES

See (14) pages 487 to 516.

CD – 06.22	PRECAST BEAM AND BLOCK FLOOR SYSTEMS. CANTILEVER WITHOUT EXTENDED JOISTS

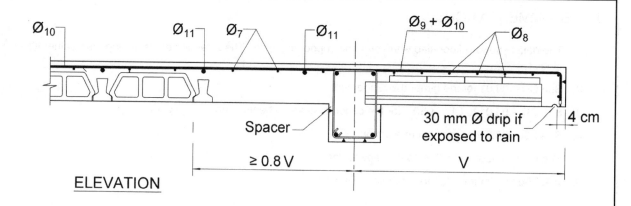

ELEVATION

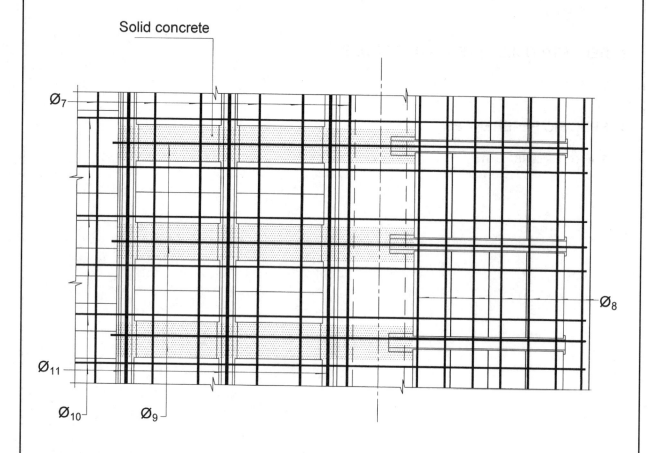

PLAN

1. RECOMMENDATIONS

1. The detail shows a floor slab with semi-self-supporting joists. The arrangement is analogous to other types of joist.

2. See CD – 06.01 for the general arrangement.

3. See CD – 05.01 to CD – 05.14 for the concrete cover in beams and lacing courses.

4. See CD – 06.01 for the concrete cover.

5. In the solid concrete area on the left, the blocks should be sealed off to protect them from the flowing concrete. Care should be taken to maintain the concrete within the specified width.

6. The top is smoothed with a float or power float.

2. STATUTORY LEGISLATION

See EC2 (5).

3. RECOMMENDED ALTERNATIVE CODES

See (31).

4. SPECIFIC REFERENCES

See (14) pages 487 to 516.

BEAM AND BLOCK FLOOR SYSTEMS.
OPENINGS

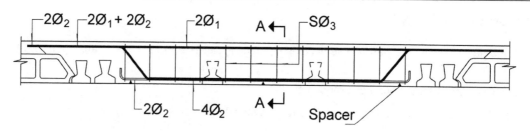

$2\emptyset_2$ $2\emptyset_1 + 2\emptyset_2$ $2\emptyset_1$ A $S\emptyset_3$

$2\emptyset_2$ $4\emptyset_2$ A   Spacer

ELEVATION

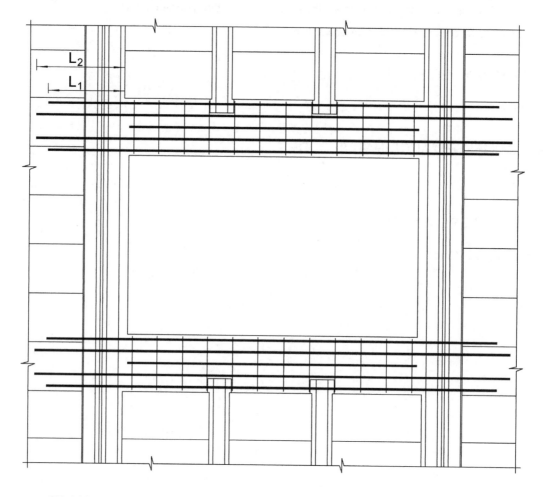

L_2
L_1

PLAN

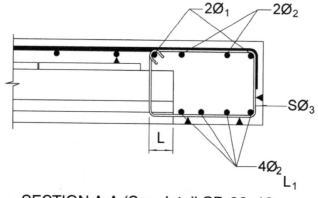

$2\emptyset_1$ $2\emptyset_2$

$S\emptyset_3$

L

$4\emptyset_2$
L_1

SECTION A-A (See detail CD-06. 12
for design values of L_3, L_5)

306

1. RECOMMENDATIONS

1. The detail shows a floor slab with semi-self-supporting joists. The arrangement is analogous to other types of joist.

2. See CD – 06.01 for the general arrangement.

3. See CD – 05.01 to CD – 05.14 for the concrete cover in beams and lacing courses.

4. See CD – 06.01 for the concrete cover.

5. L_1 is the anchorage length in the ϕ_1 bar and L_2 is the anchorage length in the ϕ_2 bar.

6. The top is smoothed with a float or power float.

2. STATUTORY LEGISLATION

See EC2 (5).

3. RECOMMENDED ALTERNATIVE CODES

See (31).

4. SPECIFIC REFERENCES

See (14) pages 487 to 516.

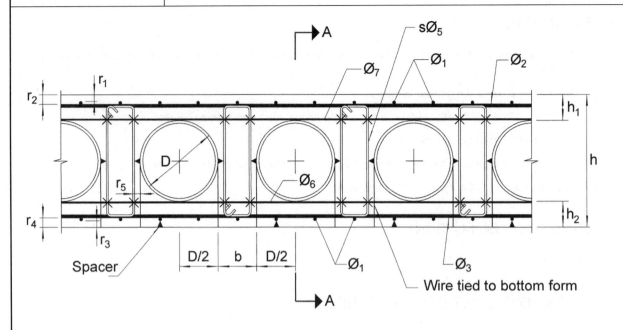

CROSS-SECTION

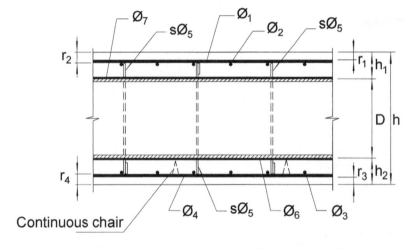

LONGITUDINAL SECTION A-A THROUGH TUBE CROWN

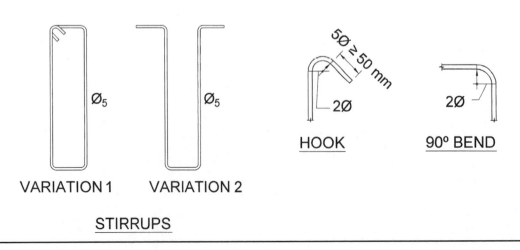

VARIATION 1   VARIATION 2

STIRRUPS

1. RECOMMENDATIONS

1. This solution is worthwhile only for bridge decks, floor slabs with very long spans and similar.

2. $r_1 = 2.5$ cm $\geq \phi_1$.

    If the top is horizontal and exposed to rain or condensed water, $r_1 = 3$ cm.

    This is likewise applicable for bridge decks if the top is not waterproofed.

    $r_2 \geq \phi_2$.

    $r_3 = 2.5$ cm $\geq \phi_1$.

    $r_4 \geq \phi_3$.

    The ϕ_6 and ϕ_7 bars are tied at all bar–stirrup intersections.

3. If the ϕ_2 and ϕ_3 reinforcing bar spacing is coordinated, the stirrups may be tied to them with no need to wrap around reinforcing bars.

4. If the deck is prestressed, the variation 2 stirrup should be used to allow space for positioning the ducts.

5. Particular attention should be paid to the possible flotation of the cardboard tubes used to lighten the structure. One possible solution is to use a ϕ_7 bar as in arrangement 1, winding wires around the tubes and tying them to the bottom forms.

6. The tubes rest on spacers or continuous chairs set underneath them, depending on the case, and on the ϕ_6 bar.

2. STATUTORY LEGISLATION

See EC2 (5).

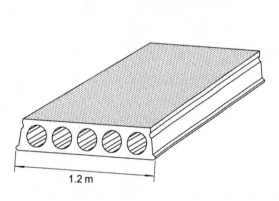

① GENERAL VIEW

1.2 m

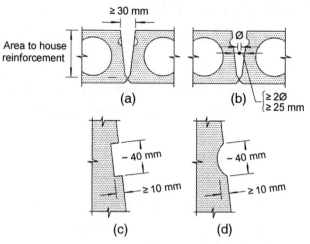

② TYPE OF JOINTS

≥ 30 mm

Area to house reinforcement

(a)

(b) { ≥ 2Ø
{ ≥ 25 mm

~ 40 mm

≥ 10 mm

(c)

~ 40 mm

≥ 10 mm

(d)

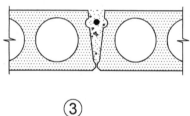

③

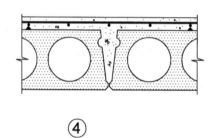

④

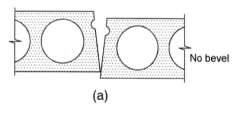

(a)

No bevel

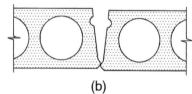

(b)

⑤ BOTTOM BEVELLED EDGE FOR HOLLOW CORES WITH EXPOSED UNDERSIDE

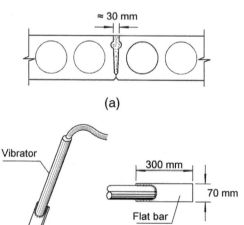

≈ 30 mm

(a)

Vibrator

300 mm

70 mm

Flat bar

Flat bar

(b)

⑥ JOINT VIBRATION

(Not necessary if self-compacting mortar is used)

1. RECOMMENDATIONS

1. The width of a standard hollow core panel is usually 1.20 m (4 ft in the US). Other widths can be made to order, however.

2. The cores are normally circular. In large depths they may be oval, with the long axis in the vertical direction.

3. The joints are standardised in European Standard prEN 1168-1 (32).

4. Hollow cores can be used for normal spans and live loads with no concrete topping. For long spans or large loads, concrete topping is needed.

5. Note detail 5. If the hollow core is exposed to view, the bevelled edge is always essential, but particularly if the members are prestressed.

6. Today self-compacting mortar is recommended to fill in the joints. Otherwise, the concrete would have to be vibrated and since normal vibrators cannot fit in the joint, a plate would have to be attached to the tip.

2. STATUTORY LEGISLATION

See EN 1168-1 (32).

3. SPECIFIC REFERENCES

See (33).

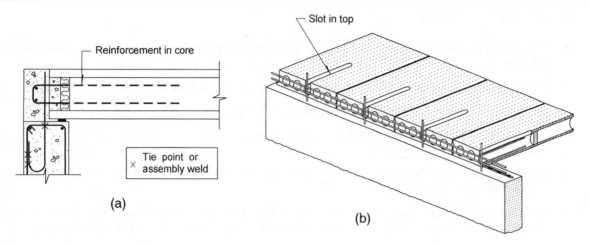

Reinforcement in core

Slot in top

Tie point or assembly weld

(a)

(b)

① SUPPORT ON BEAM OR WALL OVER RUBBER OR NEOPRENE STRIP

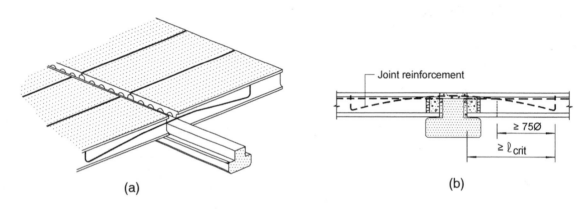

Joint reinforcement

$\geq 75\emptyset$

$\geq \ell_{crit}$

(a)

(b)

② CONTINUOUS SUPPORT ON RUBBER OR NEOPRENE STRIP

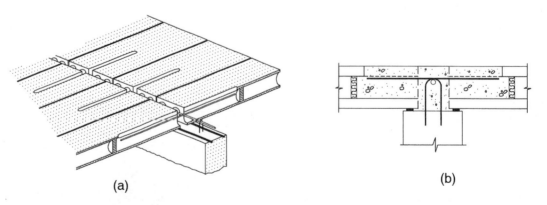

(a)

(b)

③ REINFORCEMENT HOUSED IN CONCRETE-FILLED CORES, BY CUTTING THROUGH THE TOP SLAB

1. RECOMMENDATIONS

1. If plastic stoppers are used to close off the cores, they are fitted to the end of the core, whereas if polystyrene stoppers are used, they are inserted a certain distance into the core. The second solution yields better connections.

2. The slots in the top slab are made with special machinery.

2. STATUTORY LEGISLATION

See EN 1168-1 (32).

3. SPECIFIC REFERENCES

See (33).

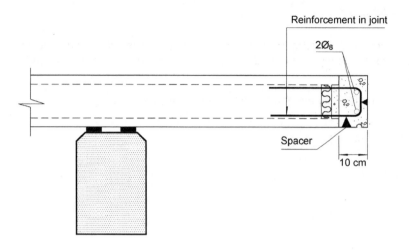

Reinforcement in joint

2Ø8

Spacer

10 cm

① TIE BEAM IN CANTILEVER TIP

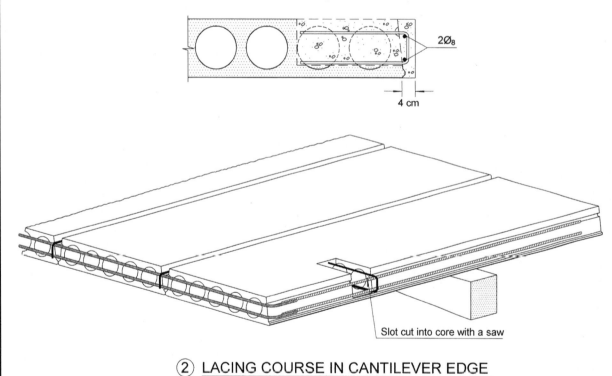

2Ø8

4 cm

Slot cut into core with a saw

② LACING COURSE IN CANTILEVER EDGE

1. RECOMMENDATIONS

1. Tip and edge lacing courses are needed for many reasons, but especially to secure the railings.
2. The slot in solution 2 is cut with a circular saw.

2. STATUTORY LEGISLATION

See EN 1168-1 (32).

3. SPECIFIC REFERENCES

See (33).

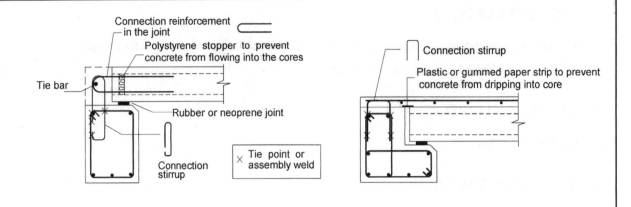

(a)   (b)

①   SIMPLY SUPPORTED CONNECTION ON EDGE

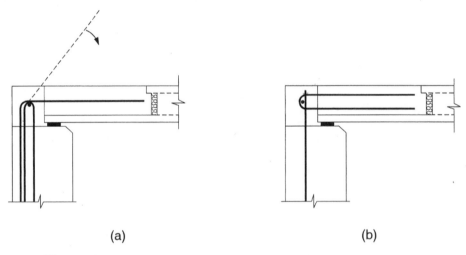

(a)   (b)

②   SIMPLY SUPPORTED CONNECTION ON EDGE
WITH SLOT IN TOP PLATE

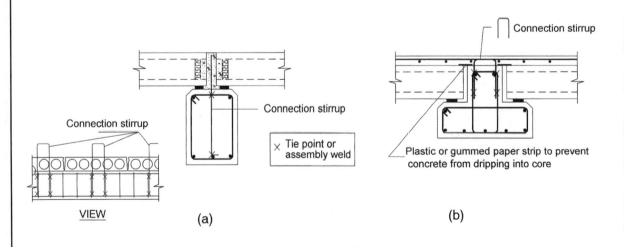

(a)   (b)

③   SIMPLY SUPPORTED CONNECTION IN INTERMEDIATE
SUPPORTS

1. RECOMMENDATIONS

1. Hollow cores should never rest directly on beams or vertical panels. A rubber strip or neoprene pad must always be placed in between the two members. Otherwise, transverse bending stress will appear in the hollow core that may cause longitudinal cracking.

2. STATUTORY LEGISLATION

See EN 1168-1 (32).

3. SPECIFIC REFERENCES

See (33).

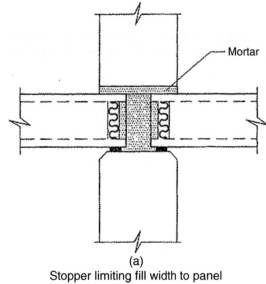

(a)
Stopper limiting fill width to panel

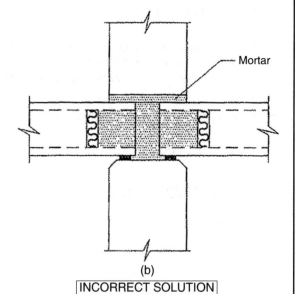

(b)
INCORRECT SOLUTION
When the fill is included in the cores, cracks tend to form due to the discontinuity generated

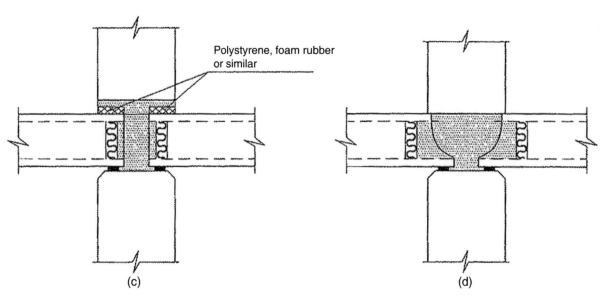

(c)

(d)

SIMPLY SUPPORTED INTERMEDIATE CONNECTIONS IN BUILDINGS
WITH PANELS

NOTES

1. RECOMMENDATIONS

1. Particular attention should be paid to solution b, which is complex.

2. STATUTORY LEGISLATION

See EN 1168-1 (32).

3. SPECIFIC REFERENCES

See (33).

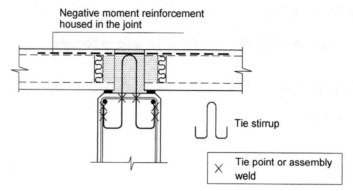

Negative moment reinforcement
housed in the joint

Tie stirrup

× Tie point or assembly
weld

① REINFORCEMENT HOUSED IN JOINTS

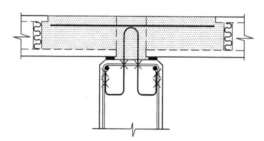

② REINFORCEMENT HOUSED IN CONCRETE-FILLED CORES

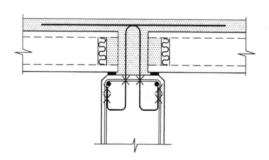

③ REINFORCEMENT HOUSED IN TOP SLAB

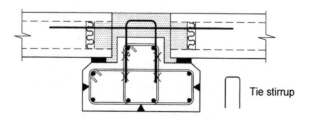

Tie stirrup

④ SUPPORT ON BEAM WITH REINFORCEMENT IN JOINTS

| CD – 06.30 | NOTES |

1. RECOMMENDATIONS

1. Solution 2 calls for cutting through the top slab in the respective cores.

2. STATUTORY LEGISLATION

See EN 1168-1 (32).

3. SPECIFIC REFERENCES

See (33).

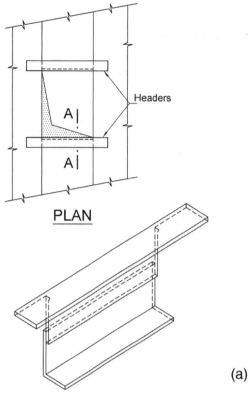

Headers

PLAN

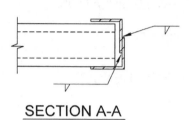

SECTION A-A

(a)

① DETAIL OF STEEL HEADER

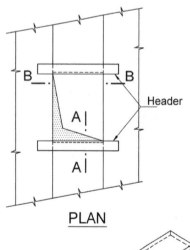

Header

PLAN

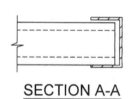

SECTION A-A

SECTION B-B

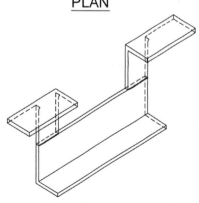

(b)

② DETAIL OF STEEL HEADER

CD – 06.31	NOTES

1. RECOMMENDATIONS

1. A steel beam is the most practical solution. See (32) for the distribution of the point support reactions.

2. STATUTORY LEGISLATION

See EN 1168-1 (32).

3. SPECIFIC REFERENCES

See (33).

Group 07

Flat slabs

FLAT SLABS.
TOP REINFORCEMENT

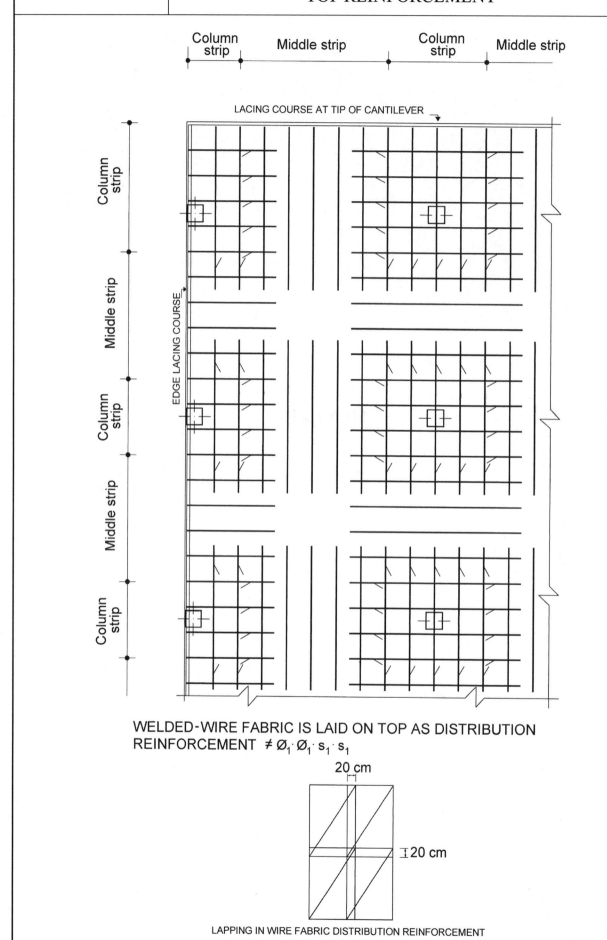

Column strip | Middle strip | Column strip | Middle strip

LACING COURSE AT TIP OF CANTILEVER

EDGE LACING COURSE

Column strip · Middle strip · Column strip · Middle strip · Column strip

WELDED-WIRE FABRIC IS LAID ON TOP AS DISTRIBUTION
REINFORCEMENT $\neq \varnothing_1 \cdot \varnothing_1 \cdot s_1 \cdot s_1$

20 cm

20 cm

LAPPING IN WIRE FABRIC DISTRIBUTION REINFORCEMENT

1. RECOMMENDATIONS

1. See CD – 07.04 for concrete cover and other details.

2. The top reinforcement runs in the direction of the larger span, on top of and in contact with the perpendicular reinforcement.

3. See CD – 07.05, CD – 07.06, CD – 07.07 and CD – 07.08 for possible punching shear reinforcement, which is not shown in this solution.

4. The general arrangement is valid for slabs with pockets and solid slabs, although in that case the parallel reinforcing bars should be spaced at no more than 25 cm.

5. The top is smoothed with floats or a power float.

6. See 1.2 and 1.3 for descriptions of how to tie bars and place spacers.

2. STATUTORY LEGISLATION

See EC2 (5) (although the code contains scant information on the subject).

3. RECOMMENDED ALTERNATIVE CODES

See ACI 318-08 (22).

4. SPECIFIC REFERENCES

See (30).

FLAT SLABS.
BOTTOM REINFORCEMENT

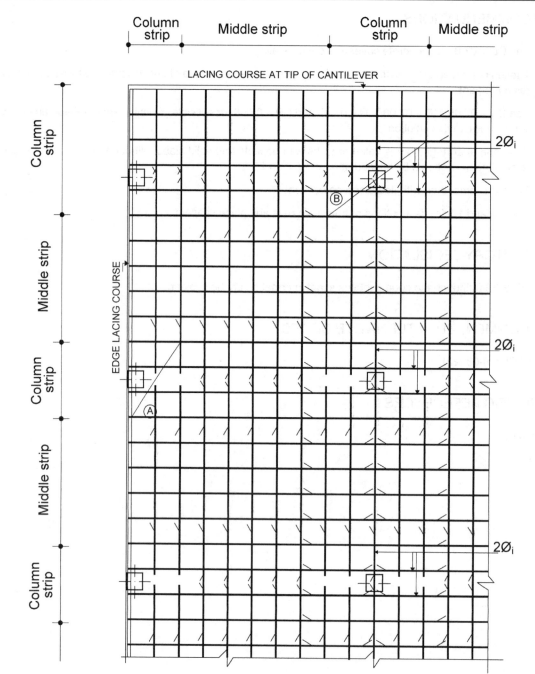

Column strip | Middle strip | Column strip | Middle strip

LACING COURSE AT TIP OF CANTILEVER

Column strip

Middle strip

EDGE LACING COURSE

Column strip

Middle strip

Column strip

$2\emptyset_i$

$2\emptyset_i$

$2\emptyset_i$

B

A

THE ENTIRE BOTTOM OF ABACUS SHOULD BE REINFORCED WITH WELDED-WIRE FABRIC

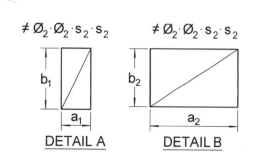

$\neq \emptyset_2 \cdot \emptyset_2 \cdot s_2 \cdot s_2$

b_1

a_1

DETAIL A

$\neq \emptyset_2 \cdot \emptyset_2 \cdot s_2 \cdot s_2$

b_2

a_2

DETAIL B

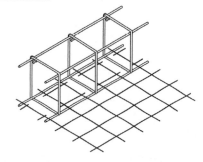

DETAIL OF LAPPING BETWEEN WELDED-WIRE FABRIC ON BOTTOM AND EDGE TIE BEAM

CD – 07.02	NOTES

1. RECOMMENDATIONS

1. See CD – 07.04 for concrete cover and other details.

2. The top reinforcement runs in the direction of the larger span, on top of and in contact with the perpendicular reinforcement.

3. See CD – 07.05, CD – 07.06, CD – 07.07 and CD – 07.08 for possible punching shear reinforcement, which is not shown in this solution.

4. The general arrangement is valid for slabs with pockets and solid slabs, although in that case the parallel reinforcing bars should be spaced at no more than 25 cm.

5. **AR.** The two ϕ_1 reinforcing bars (integrity reinforcement) overlap by their lap length in the area over the column (see (22)).

6. See 1.2 and 1.3 for descriptions of how to tie bars and place spacers.

2. STATUTORY LEGISLATION

See EC2 (5) (although the code contains scant information on the subject).

3. RECOMMENDED ALTERNATIVE CODES

See ACI 318-08 (22).

4. SPECIFIC REFERENCES

See (30).

(A) COLUMN STRIP

See detail A

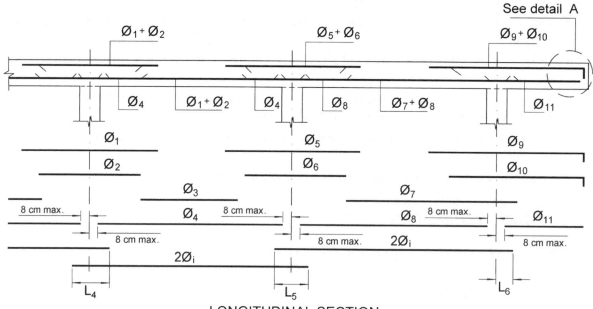

LONGITUDINAL SECTION

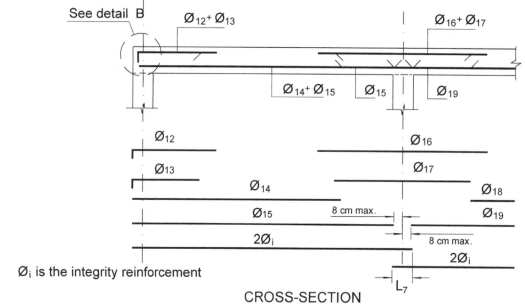

$\emptyset_i$ is the integrity reinforcement

CROSS-SECTION

See Ⓐ Ⓑ Ⓒ   See Ⓐ Ⓑ Ⓒ

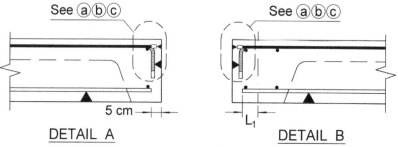

DETAIL A   DETAIL B

(THE DETAIL FOR THE EDGE LACING COURSE SHOWN IN CD-07.04)

1. RECOMMENDATIONS

1. See CD – 07.04 for concrete cover and other details.

2. The top reinforcement runs in the direction of the larger span, on top of and in contact with the perpendicular reinforcement.

3. See CD – 07.05, CD – 07.06, CD – 07.07 and CD – 07.08 for possible punching shear reinforcement, which is not shown in this detail.

4. The general arrangement is valid for slabs with pockets and solid slabs, although in that case the parallel reinforcing bars should be spaced at no more than 25 cm.

5. See 1.6 to determine when to use anchor type a, b or c in details A and B.

6. See EC2 (5) for the procedure to calculate L_1 (detail B).

7. The top is smoothed with floats or a power float.

8. See 1.2 and 1.3 for descriptions of how to tie bars and place spacers.

2. STATUTORY LEGISLATION

See EC2 (5) (although the code contains scant information on the subject).

3. RECOMMENDED ALTERNATIVE CODES

See ACI 318-08 (22).

4. SPECIFIC REFERENCES

See (30).

(B) MIDDLE STRIP

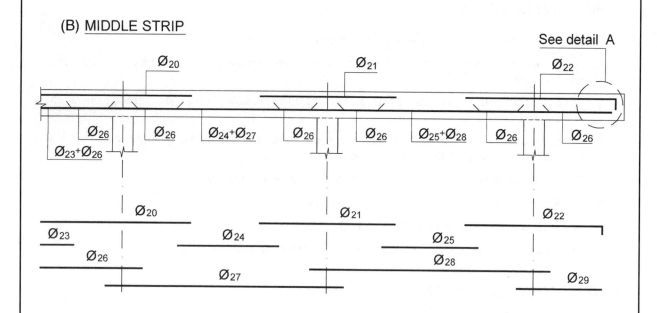

LONGITUDINAL SECTION

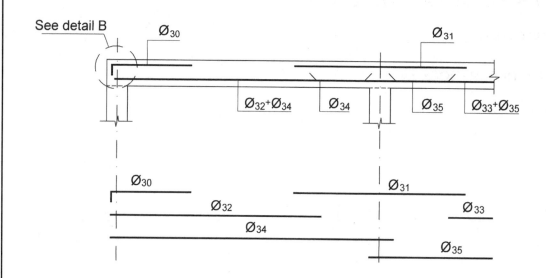

CROSS-SECTION

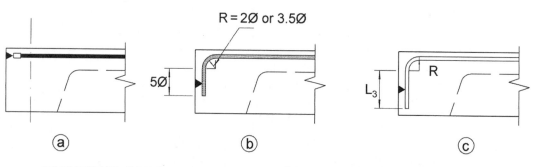

(THE DETAIL FOR THE EDGE LACING COURSE SHOWN IN CD-07.04)

1. RECOMMENDATIONS

1. See CD – 07.04 for concrete cover and other details.

2. The top reinforcement runs in the direction of the larger span, on top of and in contact with the perpendicular reinforcement.

3. See CD – 07.05, CD – 07.06, CD – 07.07 and CD – 07.08 for possible punching shear reinforcement, which is not shown in this detail.

4. The general arrangement is valid for slabs with pockets and solid slabs, although in that case the parallel reinforcing bars should be spaced at no more than 25 cm.

5. See 1.6 to determine when to use anchor type a, b or c in details A and B.

6. See EC2 for the procedure to calculate L_1 (detail B).

7. The top is smoothed with floats or a power float.

8. See 1.2 and 1.3 for descriptions of how to tie bars and place spacers.

2. STATUTORY LEGISLATION

See EC2 (5) (although the code contains scant information on the subject).

3. RECOMMENDED ALTERNATIVE CODES

See ACI 318-08 (22).

4. SPECIFIC REFERENCES

See (30).

(A) CROSS-SECTIONAL DRAWING OF REINFORCEMENT ARRANGEMENT
(INCLUDING WELDED-WIRE FABRIC)

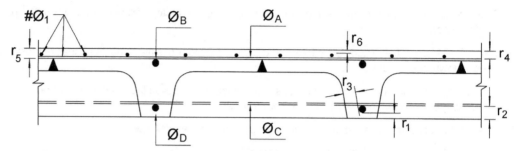

(NB: The perpendicular $\emptyset_A$ bars rest on the welded-wire fabric. The parallel bars run underneath.)

(B) WELDED-WIRE FABRIC IN AREAS NOT CALLED FOR IN THE DESIGN

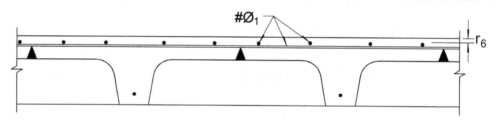

(See note in A re the position of #$\emptyset_1$)

(C) EDGE LACING COURSE (See detail B in CD-07.03)   STIRRUPS

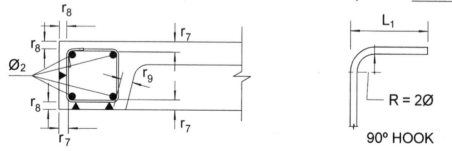

(D) LACING COURSE AT TIP OF CANTILEVER (See detail A in CD-07.03)

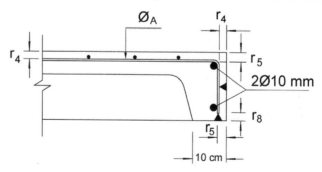

1. RECOMMENDATIONS

1. $r_1 = 2.5 \text{ cm} \geq \phi_D$.

   $r_2 = \phi_C$.

   r_3 must be large enough to accommodate the concrete; in any case $\geq \phi_D$ and never less than 2.5 cm.

   $r_4 = 2.5 \text{ cm} \geq \phi_A$.

   $r_5 = 2.5 \text{ cm} \geq \phi_B$.

   $r_6 = 2.5 \text{ cm} \geq \phi_1$.

   $r_7 \geq \phi_2$.

   $r_8 = 2.5 \text{ cm} \geq \phi_2$.

   $r_9 \geq \phi_2$.

   The maximum aggregate size should be chosen carefully.

2. L_1 should be sized so that the stirrup closure complies with the lap length.

3. If the top is horizontal and exposed to rain or condensed water, concrete covers r_6 and r_8 should be 3 cm.

4. See Recommendation 2 in CD – 07.01 and CD – 07.02.

2. STATUTORY LEGISLATION

See EC2 (5) (although the code contains scant information on the subject).

3. RECOMMENDED ALTERNATIVE CODES

See ACI 318-08 (22).

4. SPECIFIC REFERENCES

See (30).

CD – 07.05	FLAT SLABS. PUNCHING SHEAR REINFORCEMENT (VARIATION 1) (1 of 2)

INTERNAL COLUMN

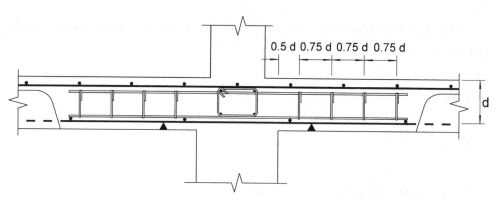

ELEVATION

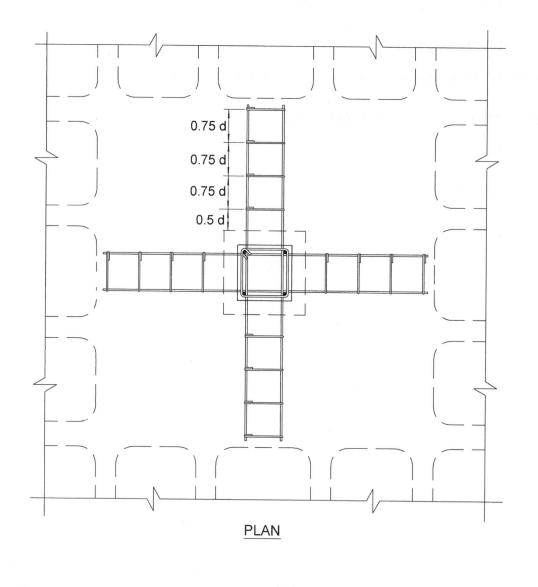

PLAN

1. RECOMMENDATIONS

1. The symbol d is the effective depth in the respective direction.

2. See CD – 07.03 and CD – 07.04 for concrete cover and other details.

3. See 1.2 and 1.3 for descriptions of how to tie bars and place spacers.

2. STATUTORY LEGISLATION

See EC2 (5).

3. RECOMMENDED ALTERNATIVE CODES

See ACI 318-08 (22).

4. SPECIFIC REFERENCES

See (30).

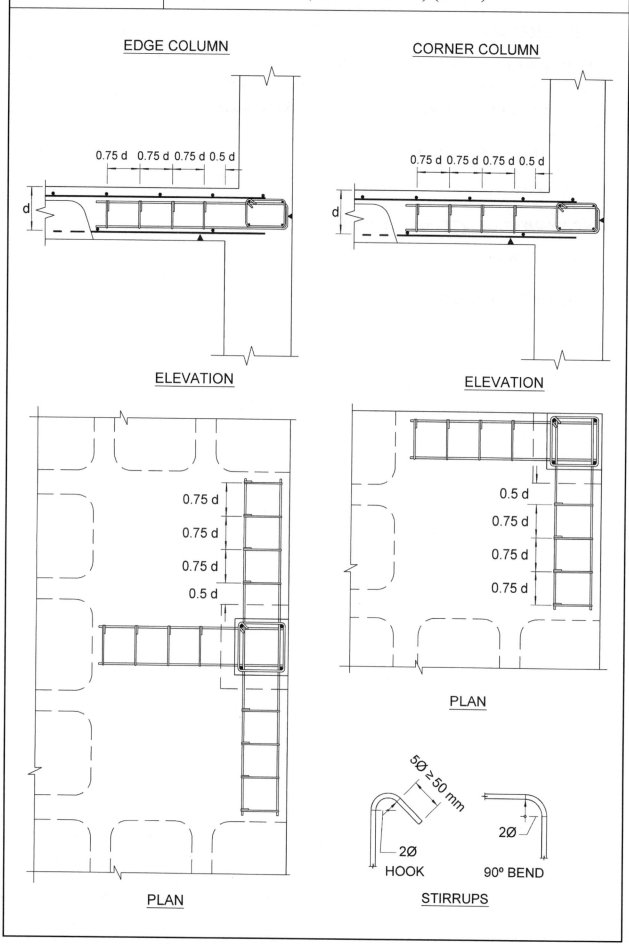

EDGE COLUMN

0.75 d   0.75 d   0.75 d   0.5 d

d

ELEVATION

CORNER COLUMN

0.75 d   0.75 d   0.75 d   0.5 d

d

ELEVATION

0.75 d
0.75 d
0.75 d
0.5 d

PLAN

0.5 d
0.75 d
0.75 d
0.75 d

PLAN

5Ø ≥ 50 mm

2Ø
HOOK

2Ø
90° BEND

STIRRUPS

1. RECOMMENDATIONS

1. The symbol d is the effective depth in the respective direction.

2. See CD – 07.03 and CD – 07.04 for concrete cover and other details.

3. See 1.2 and 1.3 for descriptions of how to tie bars and place spacers.

2. STATUTORY LEGISLATION

See EC2 (5).

3. RECOMMENDED ALTERNATIVE CODES

See ACI 318-08 (22).

4. SPECIFIC REFERENCES

See (30).

CD – 07.06	FLAT SLABS. PUNCHING SHEAR REINFORCEMENT (VARIATION 2) (1 of 2)

INTERNAL COLUMN

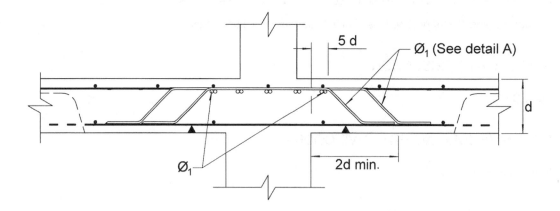

ELEVATION

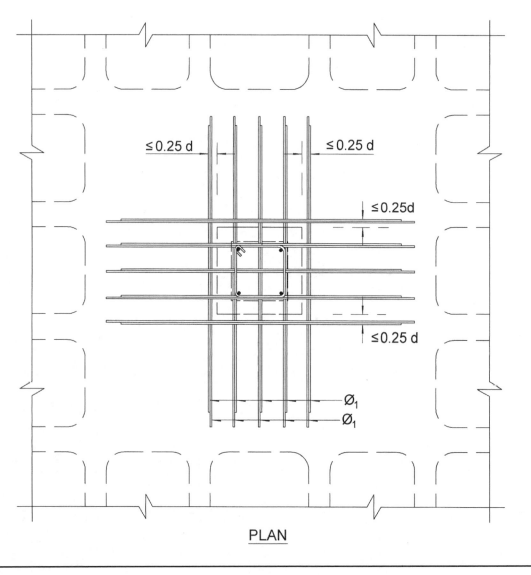

PLAN

1. RECOMMENDATIONS

1. The symbol d is the effective depth in the respective direction.

2. See CD – 07.03 and CD – 07.04 for concrete cover and other details.

3. L_1 is the ϕ_1 bar anchorage length.

4. See 1.2 and 1.3 for descriptions of how to tie bars and place spacers.

2. STATUTORY LEGISLATION

See EC2 (5).

3. RECOMMENDED ALTERNATIVE CODES

See ACI 318-08 (22).

4. SPECIFIC REFERENCES

See (30).

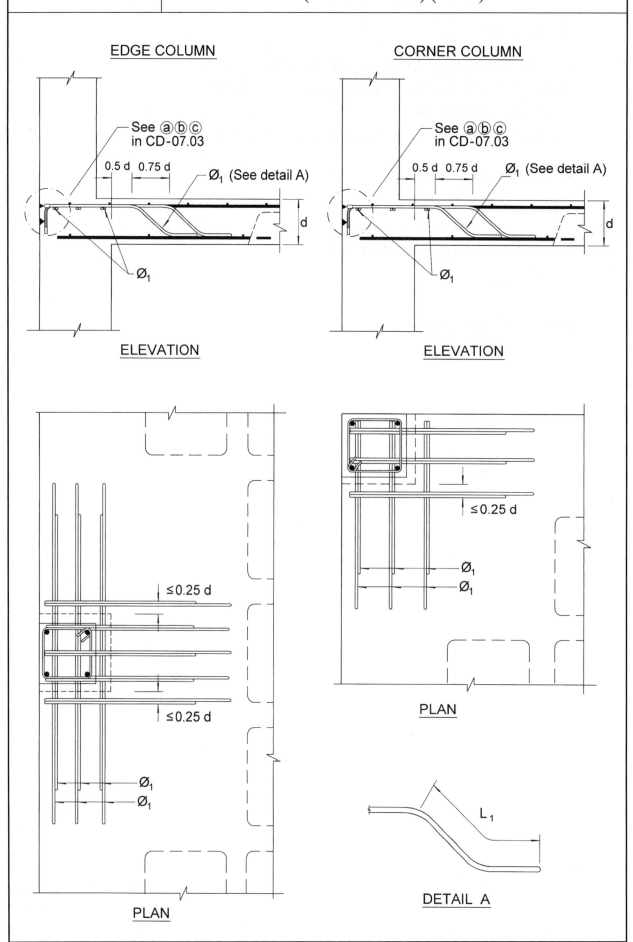

EDGE COLUMN

See ⓐⓑⓒ in CD-07.03

0.5 d   0.75 d $\varnothing_1$ (See detail A)

d

$\varnothing_1$

ELEVATION

CORNER COLUMN

See ⓐⓑⓒ in CD-07.03

0.5 d   0.75 d $\varnothing_1$ (See detail A)

d

$\varnothing_1$

ELEVATION

≤ 0.25 d

$\varnothing_1$

$\varnothing_1$

PLAN

≤ 0.25 d

≤ 0.25 d

$\varnothing_1$

$\varnothing_1$

PLAN

L_1

DETAIL A

1. RECOMMENDATIONS

1. The symbol d is the effective depth in the respective direction.
2. See CD – 07.03 and CD – 07.04 for concrete cover and other details.
3. L_1 is the ϕ_1 bar anchorage length.
4. See 1.2 and 1.3 for descriptions of how to tie bars and place spacers.

2. STATUTORY LEGISLATION

See EC2 (5).

3. RECOMMENDED ALTERNATIVE CODES

See ACI 318-08 (22).

4. SPECIFIC REFERENCES

See (30).

A) <u>INTERNAL COLUMN</u>

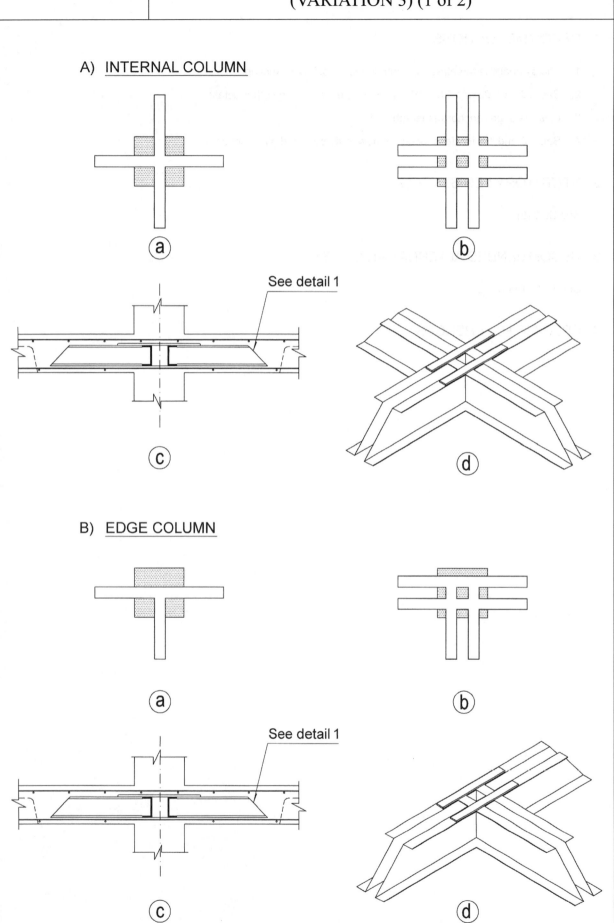

See detail 1

ⓐ

ⓑ

ⓒ

ⓓ

B) <u>EDGE COLUMN</u>

See detail 1

ⓐ

ⓑ

ⓒ

ⓓ

1. RECOMMENDATIONS

1. See CD – 07.03 and CD – 07.04 for concrete cover and other details.

2. $r_1 = 2.5 \text{ cm} \geq \phi_B.$ $\quad r_3 \geq \phi_C.$

   $r_2 \geq \phi_A.$ $\quad\quad\quad\quad r_4 = 2.5 \text{ cm} \geq \phi_D.$

3. Particular attention should be paid to vibration near and underneath the profiles.

2. STATUTORY LEGISLATION

This type of solution is not envisaged in EC2 (5).

3. RECOMMENDED ALTERNATIVE CODES

See ACI 318-08 (22).

4. SPECIFIC REFERENCES

See (30).

C) <u>CORNER COLUMN</u>

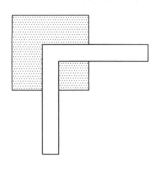

(a)

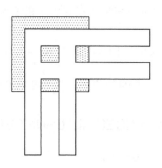

(b)

See ⓐⓑⓒ
in CD-07.03

See detail 1

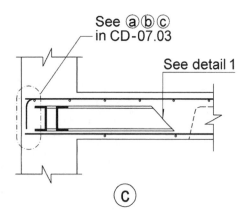

(c)

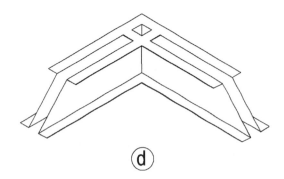

(d)

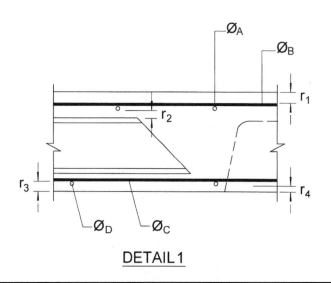

$\emptyset_A$

$\emptyset_B$

r_1

r_2

r_3

r_4

$\emptyset_D$

$\emptyset_C$

<u>DETAIL 1</u>

1. RECOMMENDATIONS

1. See CD – 07.03 and CD – 07.04 for concrete cover and other details.

2. $r_1 = 2.5$ cm $\geq \phi_B$. $r_3 \geq \phi_C$.

   $r_2 \geq \phi_A$. $r_4 = 2.5$ cm $\geq \phi_D$.

3. Particular attention should be paid to vibration near and underneath the profiles.

2. STATUTORY LEGISLATION

This type of solution is not envisaged in EC2 (5).

3. RECOMMENDED ALTERNATIVE CODES

See ACI 318-08 (22).

4. SPECIFIC REFERENCES

See (30).

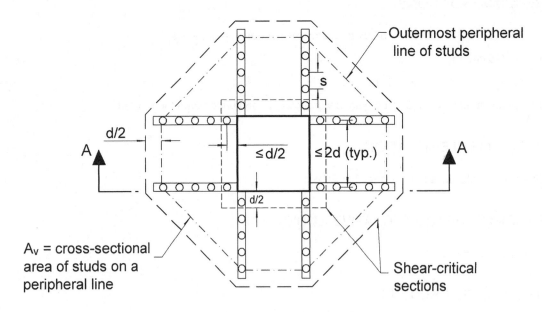

Outermost peripheral line of studs

s

d/2

$\leq d/2$

$\leq 2d$ (typ.)

d/2

A_v = cross-sectional area of studs on a peripheral line

Shear-critical sections

INTERNAL COLUMN

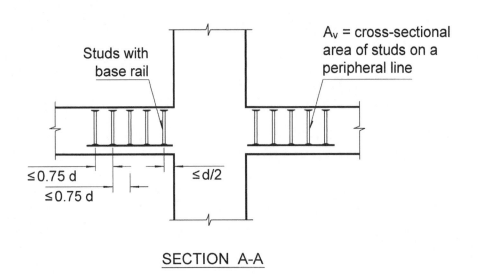

Studs with base rail

A_v = cross-sectional area of studs on a peripheral line

$\leq 0.75\,d$

$\leq 0.75\,d$

$\leq d/2$

SECTION A-A

1. RECOMMENDATIONS

None.

2. STATUTORY LEGISLATION

See EC2 (5).

3. RECOMMENDED ALTERNATIVE CODES

See ACI 318-08 (22).

4. SPECIFIC REFERENCES

See (30).

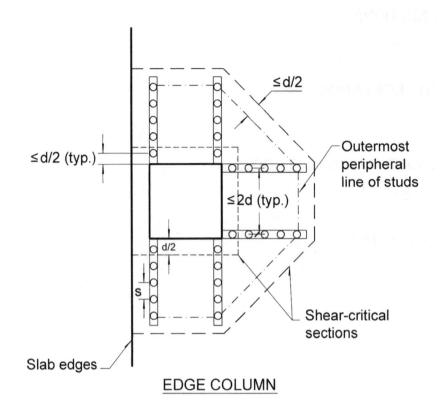

EDGE COLUMN

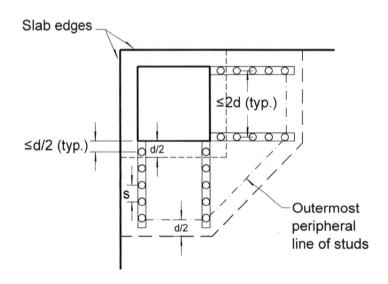

CORNER COLUMN

1. RECOMMENDATIONS

None.

2. STATUTORY LEGISLATION

See EC2 (5).

3. RECOMMENDED ALTERNATIVE CODES

See ACI 318-08 (22).

4. SPECIFIC REFERENCES

See (30).

FLAT SLABS.
DROP PANELS

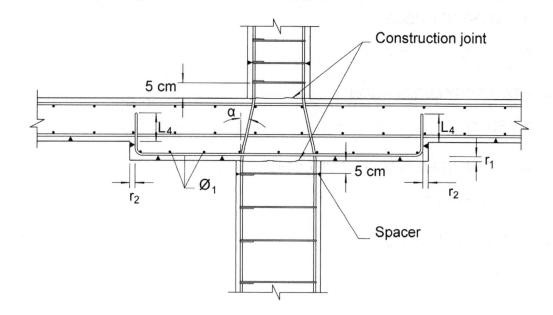

Construction joint

5 cm

L_4

α

$\varnothing_1$

r_2

5 cm

L_4

r_1

r_2

Spacer

INTERNAL COLUMN

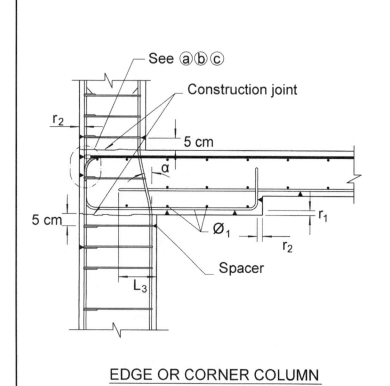

See ⓐⓑⓒ

Construction joint

r_2

5 cm

α

5 cm

$\varnothing_1$

r_1

r_2

Spacer

L_3

EDGE OR CORNER COLUMN

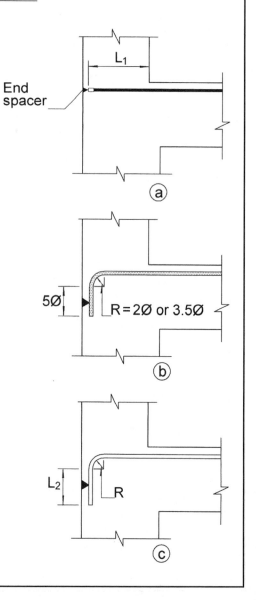

L_1

End
spacer

ⓐ

$5\varnothing$

$R = 2\varnothing$ or $3.5\varnothing$

ⓑ

L_2

R

ⓒ

1. RECOMMENDATIONS

1. $L_4 = 10\,\phi_1$.

2. $r_1 = 2.5$ cm $\geq \phi_1$.

   $r_2 = 2.5$ cm $\geq \phi_1$.

3. For column transition details, depending on whether $\tan\alpha$ is smaller or greater than 1/6, see CD – 03.04 to CD – 03.14.

4. While column ties are not needed in the node or drop areas of inner columns, they are necessary in edge and corner columns.

5. See EC2, anchorage lengths, to calculate lengths L_1 and L_2 in details a, b and c.

6. See Recommendation 5 in CD – 06.02 for the procedure to calculate L_3.

7. See CD – 07.03 and CD – 07.04 for concrete cover and other details.

8. See 1.2 and 1.3 for descriptions of how to tie bars and place spacers.

2. STATUTORY LEGISLATION

See EC2 (5).

3. RECOMMENDED ALTERNATIVE CODES

See ACI 318-08 (22).

4. SPECIFIC REFERENCES

See (30).

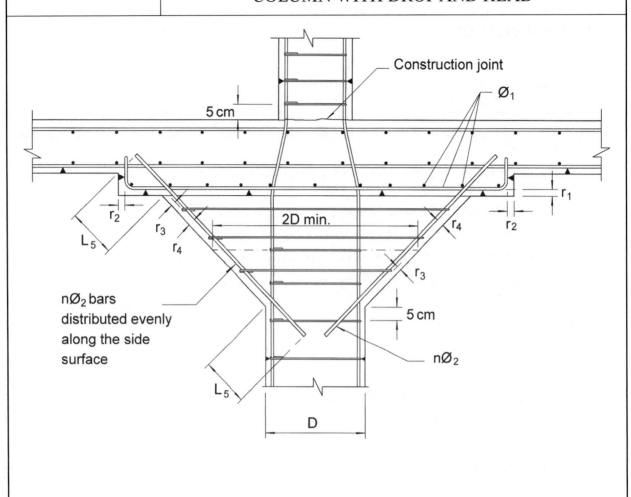

Construction joint

$\varnothing_1$

5 cm

r_1

r_2

r_3

r_4

2D min.

r_4

r_2

L_5

r_3

$n\varnothing_2$ bars
distributed evenly
along the side
surface

5 cm

$n\varnothing_2$

L_5

D

1. RECOMMENDATIONS

1. See the Recommendations for CD – 07.09.

2. $r_3 = 2.5$ cm.

3. $r_4 \geq \phi_2$.

4. Length L_5 is the ϕ_2 bar anchorage length.

5. Ties are not needed in the upper part of the head, the drop or the slab.

6. See CD – 07.03 and CD – 07.04 for concrete cover and other details.

7. See 1.2 and 1.3 for descriptions of how to tie bars and place spacers.

2. STATUTORY LEGISLATION

See EC2 (5).

3. RECOMMENDED ALTERNATIVE CODES

See ACI 318-08 (22).

4. SPECIFIC REFERENCES

See (30).

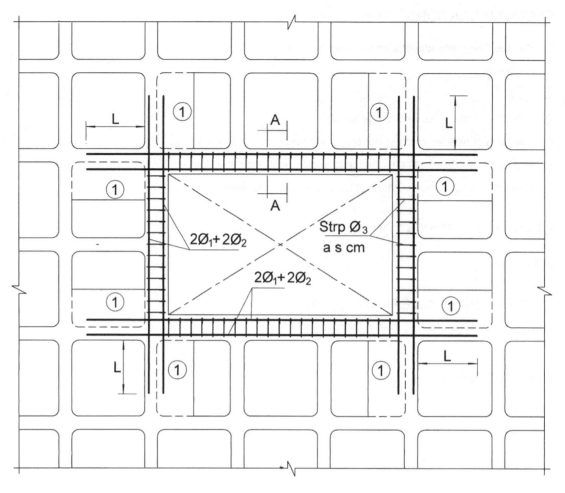

PLAN

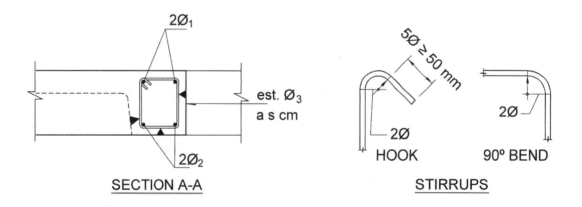

SECTION A-A

STIRRUPS

HOOK

90° BEND

1. RECOMMENDATIONS

1. See CD – 07.04 for concrete cover and general details.

2. Width b may be larger than the normal rib width in the edge area. If necessary around the supports, fill area 1 totally or partially with concrete.

3. Length L is the length of the ϕ_1 (top) and ϕ_2 (bottom) reinforcing bars.

2. STATUTORY LEGISLATION

See EC2 (5).

3. RECOMMENDED ALTERNATIVE CODES

See ACI 318-08 (22).

4. SPECIFIC REFERENCES

See (30).

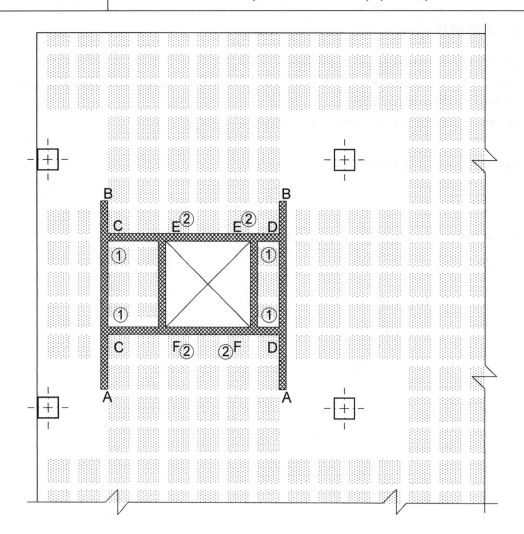

PLAN

1. RECOMMENDATIONS

1. See CD – 07.04 for concrete cover and general details.

2. The header width in the support areas may be greater than the rib width. In this case, areas 1 or 2 must be totally or partially filled in with concrete.

3. Lengths L_1, L_3 and L_5 are the anchorage lengths of the ϕ_1, ϕ_3 and ϕ_5 bars, respectively.

4. To calculate lengths L_3, L_4 and L_6, see Recommendations 5 and 6 in CD – 06.02.

2. STATUTORY LEGISLATION

See EC2 (5).

3. RECOMMENDED ALTERNATIVE CODES

See ACI 318-08 (22).

4. SPECIFIC REFERENCES

See (30).

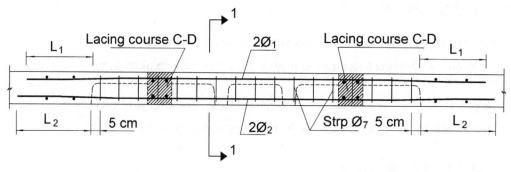

LACING COURSE A-B

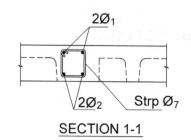

SECTION 1-1

LACING COURSE C-D

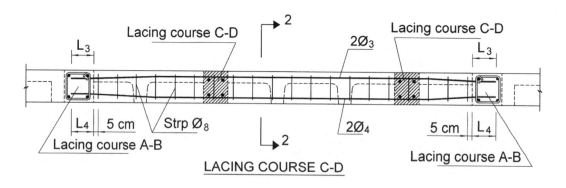

SECTION 2-2

STIRRUPS

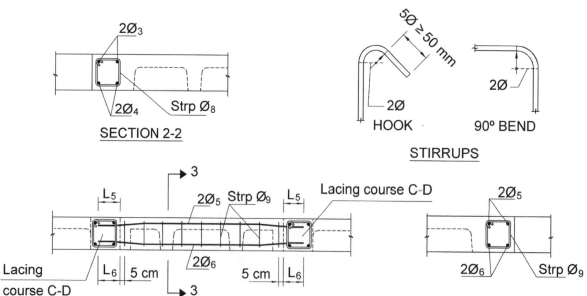

LACING COURSE E-F

SECTION 3-3

1. RECOMMENDATIONS

1. See CD – 07.04 for concrete cover and general details.

2. The header width in the support areas may be greater than the rib width. In this case, areas 1 or 2 must be totally or partially filled in with concrete.

3. Lengths L_1, L_3 and L_5 are the anchorage lengths of the ϕ_1, ϕ_3 and ϕ_5 bars, respectively.

4. To calculate lengths L_3, L_4 and L_6, see Recommendations 5 and 6 in CD – 06.02.

2. STATUTORY LEGISLATION

See EC2 (5).

3. RECOMMENDED ALTERNATIVE CODES

See ACI 318-08 (22).

4. SPECIFIC REFERENCES

See (30).

Group 08

Stairs

A) TREAD AND RISER DIMENSIONS

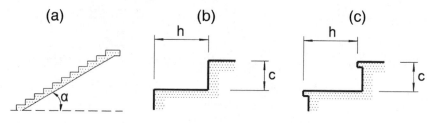

(a) (b) (c)

$$\tan \alpha = c/h$$

Most common expressions relating tread and riser c and h

① (Blondel) $h + 2c = k$ $\begin{cases} k = 59 \text{ cm (residential buildings)} \\ k = 66 \text{ cm (public buildings)} \end{cases}$

② (Neufert) $2c + h = 61 \text{ to } 64 \text{ cm}$
or $h - c = 12 \text{ cm}$

B) LANDINGS

ⓐ

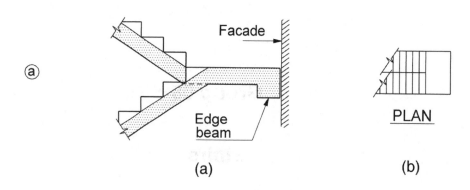

(a) (b)

ⓑ

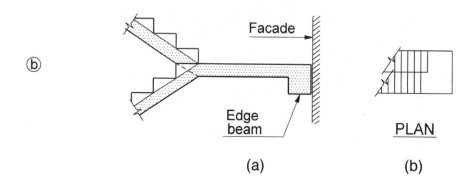

(a) (b)

ⓒ

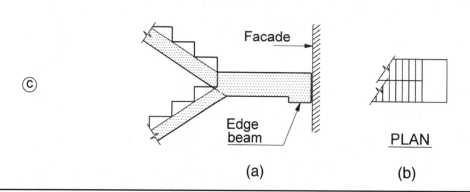

(a) (b)

1. RECOMMENDATIONS

1. Construction details vary depending on the landing solution chosen, particularly in the landing itself.

2. STATUTORY LEGISLATION

Some countries have nationwide and others merely municipal codes.

3. SPECIFIC REFERENCES

See Chapter 62 in (30).

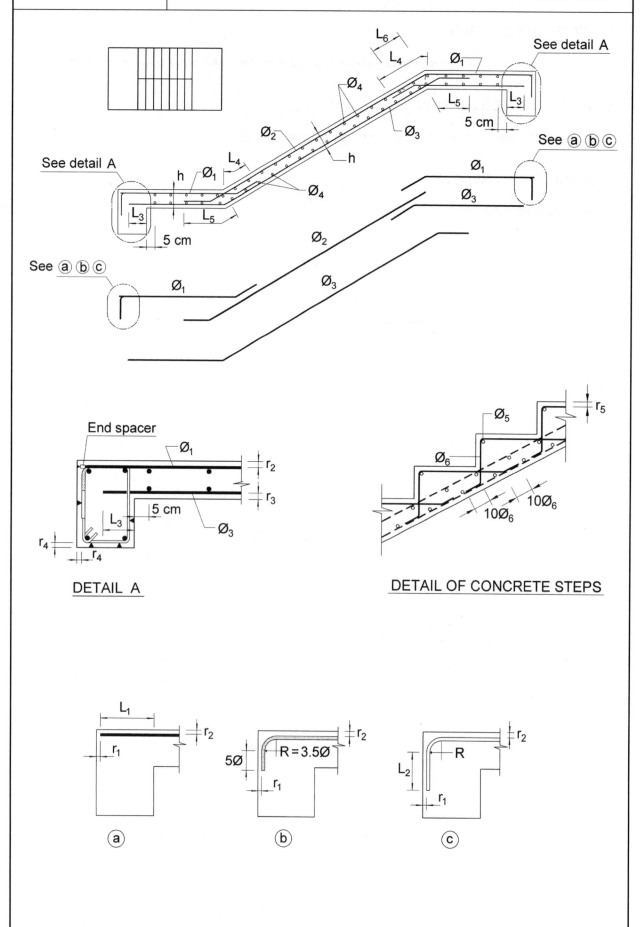

DETAIL A

DETAIL OF CONCRETE STEPS

a

b

c

1. RECOMMENDATIONS

1. $r_1 = 2.5$ cm $\geq \phi_1$.

   $r_2 = 2.5$ cm $\geq \phi_1$.

   $r_3 = 2.5$ cm $\geq \phi_3$.

   $r_4 = r_2$.

   $r_5 = r_2$.

2. See EC2, anchorage lengths, to determine when to use anchor type a, b or c at the end of the ϕ_1 reinforcing bar.

3. See Recommendation 5 in CD – 06.02 for the procedure to calculate L_3.

4. Where the steps are to be made of concrete, the ϕ_6, ϕ_2 and ϕ_3 reinforcing bars should be spaced equally for attachment of the ϕ_6 and ϕ_5 bars during concrete pouring.

5. Concrete is laid from the bottom up in stairs.

6. The top of the stairs is smoothed with a float or power float, unless the design calls for some special treatment.

7. See 1.2 and 1.3 for descriptions of how to tie bars and place spacers.

2. STATUTORY LEGISLATION

See EC2 (5).

3. SPECIFIC REFERENCES

See (30), Chapter 62.

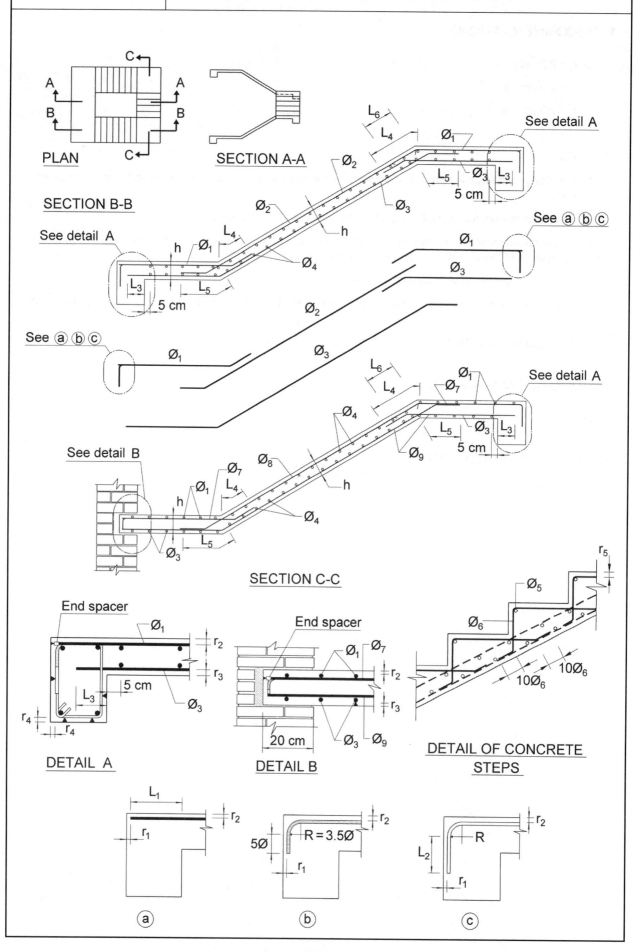

| CD – 08.03 | STAIRS. REINFORCEMENT ARRANGEMENT FOR STAIRS WITH THREE FLIGHTS |

PLAN

SECTION A-A

SECTION B-B

See detail A

See ⓐ ⓑ ⓒ

See detail A

See ⓐ ⓑ ⓒ

See detail B

SECTION C-C

End spacer

DETAIL A

End spacer

DETAIL B

DETAIL OF CONCRETE STEPS

ⓐ

ⓑ

ⓒ

1. RECOMMENDATIONS

1. See CD – 08.02.

2. STATUTORY LEGISLATION

See CD – 08.01.

3. SPECIFIC REFERENCES

See Chapter 62 in (30).

CD – 08.04	STAIRS. FOUNDATION FOR STARTING FLIGHT

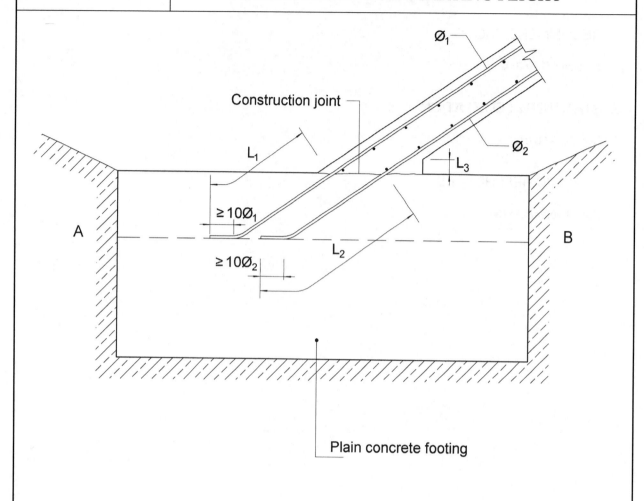

1. RECOMMENDATIONS

1. See CD – 08.02.

2. L_1 is the lap length in the ϕ_1 bar and L_2 is the lap length in the ϕ_2 bar.

3. Plane AB is a construction joint plane on which the rebar rests.

4. Dimension L_3 varies depending on the forms used for the stairflight, but should not be less than 10 cm.

2. STATUTORY LEGISLATION

See CD – 08.01.

3. SPECIFIC REFERENCES

See Chapter 62 in (30).

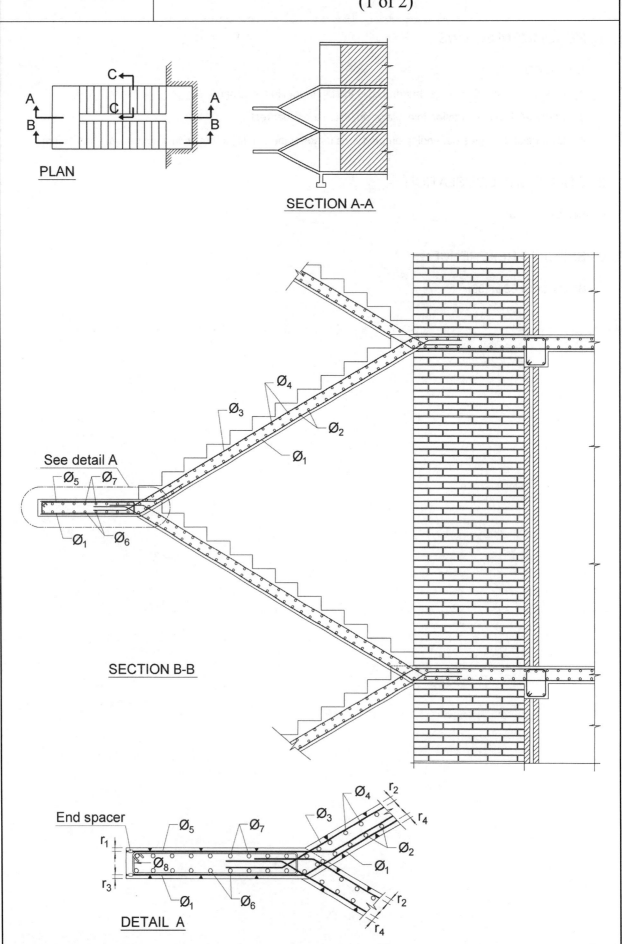

C

A

B

C

A

B

PLAN

SECTION A-A

$\varnothing_4$

$\varnothing_3$

$\varnothing_2$

$\varnothing_1$

See detail A

$\varnothing_5$ $\varnothing_7$

$\varnothing_1$ $\varnothing_6$

SECTION B-B

End spacer

$\varnothing_5$ $\varnothing_7$ $\varnothing_3$ $\varnothing_4$ r_2

r_1 r_4

$\varnothing_8$ $\varnothing_2$

$\varnothing_1$

r_3

$\varnothing_1$ $\varnothing_6$ r_2

DETAIL A r_4

1. RECOMMENDATIONS

1. $r_1 = 2.5$ cm $\geq \phi_5$.

   $r_2 = 2.5$ cm $\geq \phi_3$.

   $r_3 = 2.5$ cm $\geq \phi_1$.

   $r_4 = 2.5$ cm $\geq \phi_1$.

2. If the cantilevered landings are exposed, $r_1 = 3$ cm.

3. Concrete is laid in the stairflight of stairs from the bottom up.

4. The top of the stairs is smoothed with a float or power float.

5. See 1.2 and 1.3 for descriptions of how to tie bars and place spacers.

2. STATUTORY LEGISLATION

See EC2 (5).

3. SPECIFIC REFERENCES

See Chapter 62 in (30).

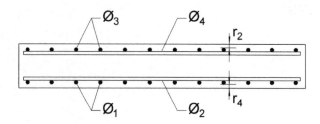

SECTION C-C

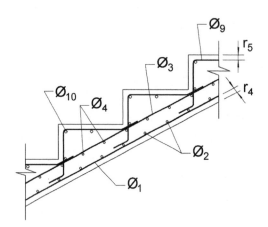

DETAIL OF CONCRETE STEPS

1. RECOMMENDATIONS

1. $r_1 = 2.5$ cm $\geq \phi_5$.

   $r_2 = 2.5$ cm $\geq \phi_3$.

   $r_3 = 2.5$ cm $\geq \phi_1$.

   $r_4 = 2.5$ cm $\geq \phi_1$.

2. If the cantilevered landings are exposed, $r_1 = 3$ cm.

3. Concrete is laid in the stairflight from the bottom up.

4. The top of the stairs is smoothed with a float or power float.

5. See 1.2 and 1.3 for descriptions of how to tie bars and place spacers.

2. STATUTORY LEGISLATION

See EC2 (5).

3. SPECIFIC REFERENCES

See Chapter 62 in (30).

Group 09

Bearings

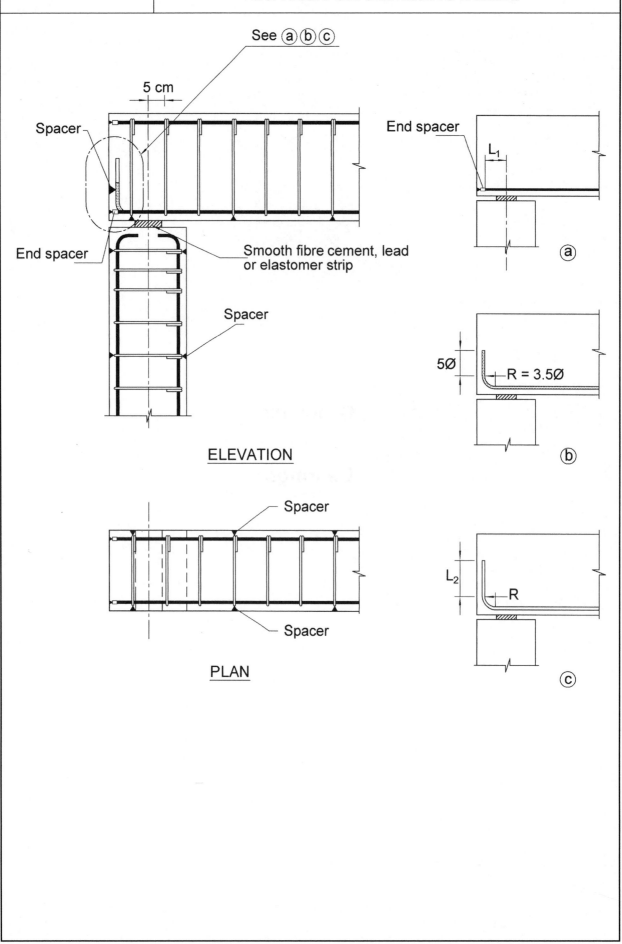

CD – 09.01	BEARINGS. DEVICE FOR CENTRING LOADS

See ⓐⓑⓒ

5 cm

Spacer

End spacer

Smooth fibre cement, lead
or elastomer strip

Spacer

ELEVATION

Spacer

Spacer

PLAN

End spacer

L_1

ⓐ

$5\varnothing$ — $R = 3.5\varnothing$

ⓑ

L_2

R

ⓒ

CD – 09.01	NOTES

1. RECOMMENDATIONS

1. See CD – 05.01 for concrete cover and general details.

2. See 1.5 to determine when to use a, b or c for anchorage of the bottom reinforcement.

3. The bearing strip can be held in position by setting it in a layer of plaster on the top of the column. The plaster, which serves as a bed for the formwork, is subsequently removed. Alternatively, an appropriate glue may be used.

2. STATUTORY LEGISLATION

See EC2 (5).

3. SPECIFIC REFERENCES

See (30).

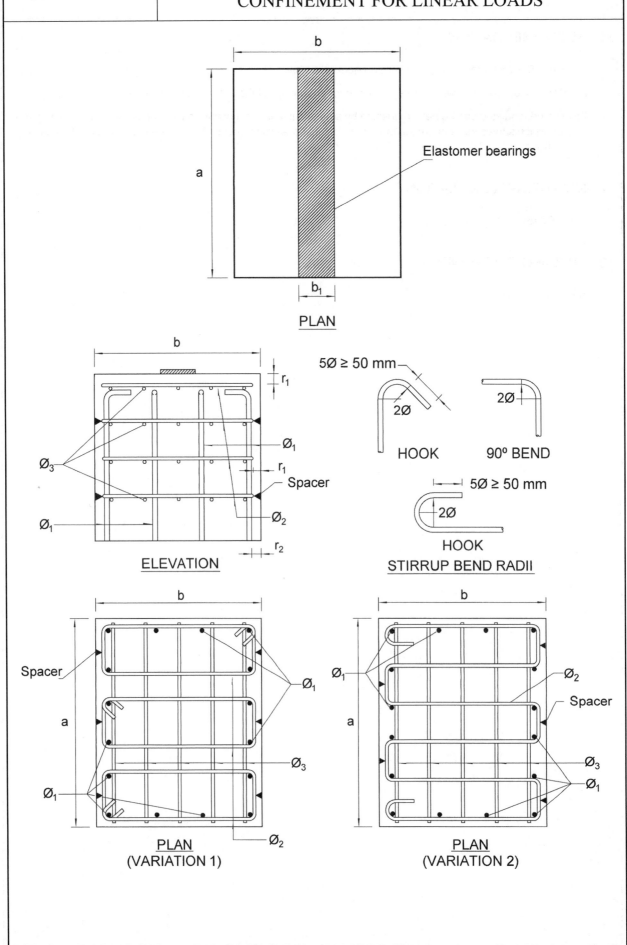

CD – 09.02	BEARINGS. CONFINEMENT FOR LINEAR LOADS

b

a

Elastomer bearings

b_1

PLAN

b

r_1

$\emptyset_3$

$\emptyset_1$

r_1

Spacer

$\emptyset_2$

$\emptyset_1$

r_2

ELEVATION

$5\emptyset \geq 50$ mm

$2\emptyset$

HOOK

$2\emptyset$

90º BEND

$5\emptyset \geq 50$ mm

$2\emptyset$

HOOK

STIRRUP BEND RADII

b

Spacer

a

$\emptyset_1$

$\emptyset_3$

$\emptyset_1$

$\emptyset_2$

PLAN
(VARIATION 1)

b

$\emptyset_1$

$\emptyset_2$

Spacer

a

$\emptyset_3$

$\emptyset_1$

PLAN
(VARIATION 2)

1. RECOMMENDATIONS

1. $r_1 = 2.5$ cm.

   $r_2 \geq \phi_1$.

2. The bearing pad can be held in position by setting it in a bed of plaster on the top of the column. The plaster, which serves as a bed for the formwork, is subsequently removed. Alternatively, an appropriate glue may be used.

2. STATUTORY LEGISLATION

See EC2.

3. RECOMMENDED ALTERNATIVE CODES

See EN 1337-1 (34), EN 1337-3 (35).

4. SPECIFIC REFERENCES

See (30).

CD – 09.03	BEARINGS. ELASTOMER BEARINGS

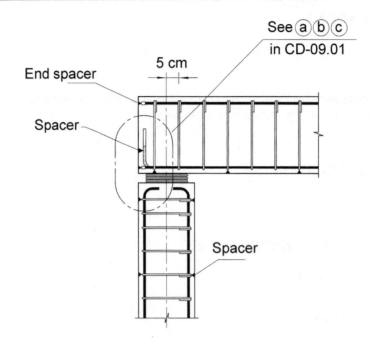

ELEVATION
VARIATION 1

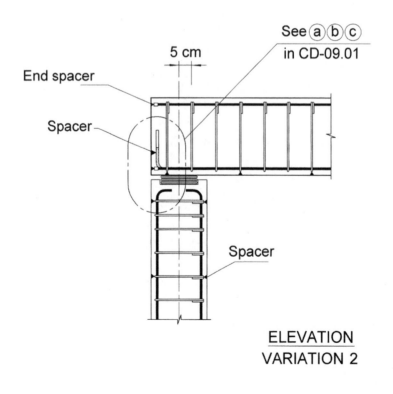

ELEVATION
VARIATION 2

1. RECOMMENDATIONS

1. To hoop the bearing, the solution set out in CD – 09.01 may suffice, or the arrangement in CD – 09.02 may be required.

2. The purpose of the second solution is to prevent the bearing from 'creeping' in the event of longitudinal variations in the lintel due to thermal effects or creep.

2. STATUTORY LEGISLATION

See EN 1337-1 (34) and EN 1337-3 (35).

3. SPECIFIC REFERENCES

See (30).

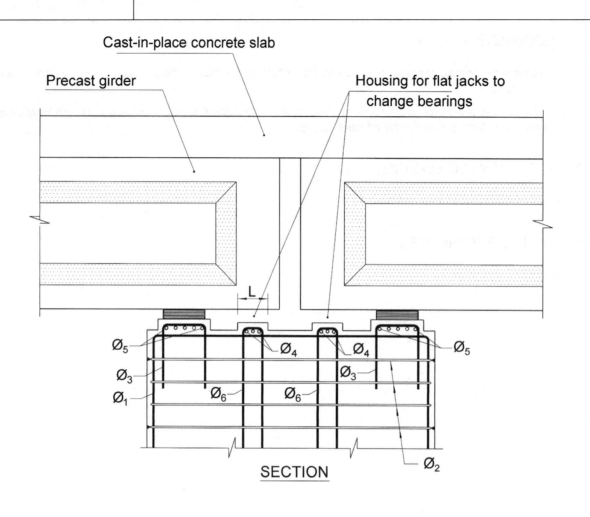

Cast-in-place concrete slab

Precast girder

Housing for flat jacks to change bearings

SECTION

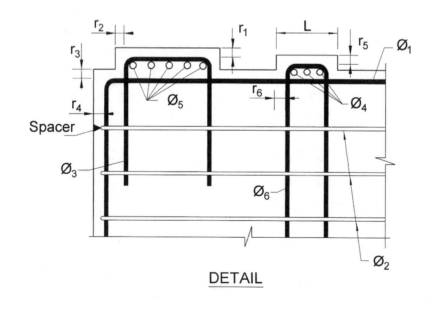

Spacer

DETAIL

1. RECOMMENDATIONS

1. Provision for housing flat jacks may often be advisable to change bearings in precast girder bridges.

2. Dimension L should be established in accordance with the flat jacks to be used.

3. $r_1 = 2.5$ cm. $\qquad r_2 = 2.5$ cm $\geq \phi_6$. $\qquad r_3 = 2.5$ cm $\geq \phi_1$.

   $r_4 = 2.5$ cm. $\qquad r_5 = 2.5$ cm $\geq \phi_4$. $\qquad r_6 = 2.5$ cm $\geq \phi_4$.

2. STATUTORY LEGISLATION

See EN 1337-1 (34) and 1337-3 (35).

3. SPECIFIC REFERENCES

See (30).

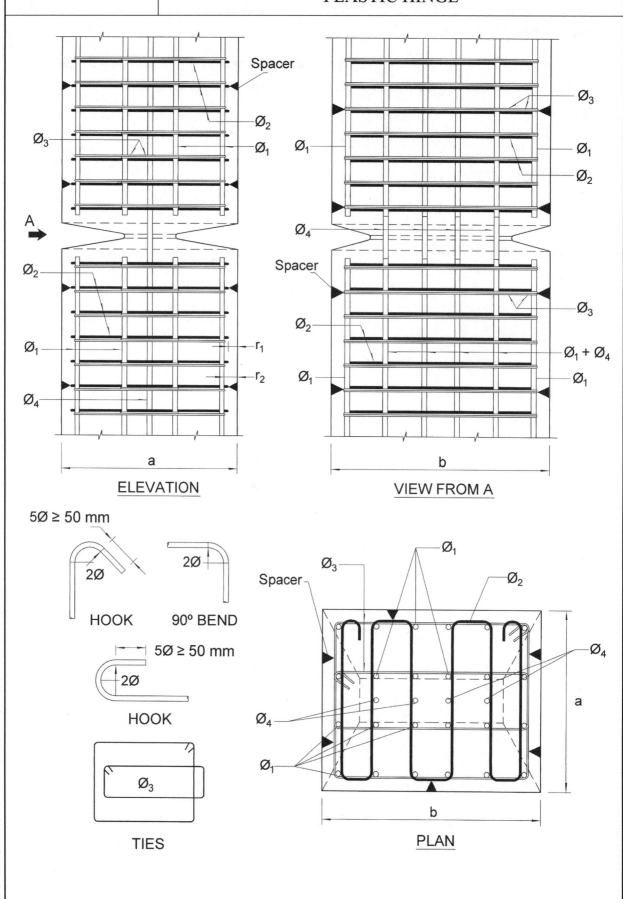

CD – 09.05

BEARINGS.
PLASTIC HINGE

Spacer
$\varnothing_2$
$\varnothing_1$
$\varnothing_3$
A
$\varnothing_2$
$\varnothing_1$
r_1
r_2
$\varnothing_4$
a

ELEVATION

$\varnothing_3$
$\varnothing_1$
$\varnothing_2$
$\varnothing_4$
Spacer
$\varnothing_2$
$\varnothing_1$
$\varnothing_3$
$\varnothing_1 + \varnothing_4$
$\varnothing_1$
b

VIEW FROM A

$5\varnothing \geq 50$ mm
$2\varnothing$
HOOK

$2\varnothing$
$90°$ BEND

$5\varnothing \geq 50$ mm
$2\varnothing$
HOOK

$\varnothing_3$
TIES

$\varnothing_3$
Spacer
$\varnothing_1$
$\varnothing_2$
$\varnothing_4$
$\varnothing_4$
$\varnothing_1$
a
b

PLAN

386

1. RECOMMENDATIONS

1. $r_1 = 2.5$ cm.

2. $r_2 \geq \phi_1$.

3. The ϕ_4 bars may not be necessary, depending on the case.

2. STATUTORY LEGISLATION

See EC2 (5).

3. SPECIFIC REFERENCES

See (30), (39), (37).

Group 10

Brackets and dapped-end beams

BRACKETS
(VARIATION 1)

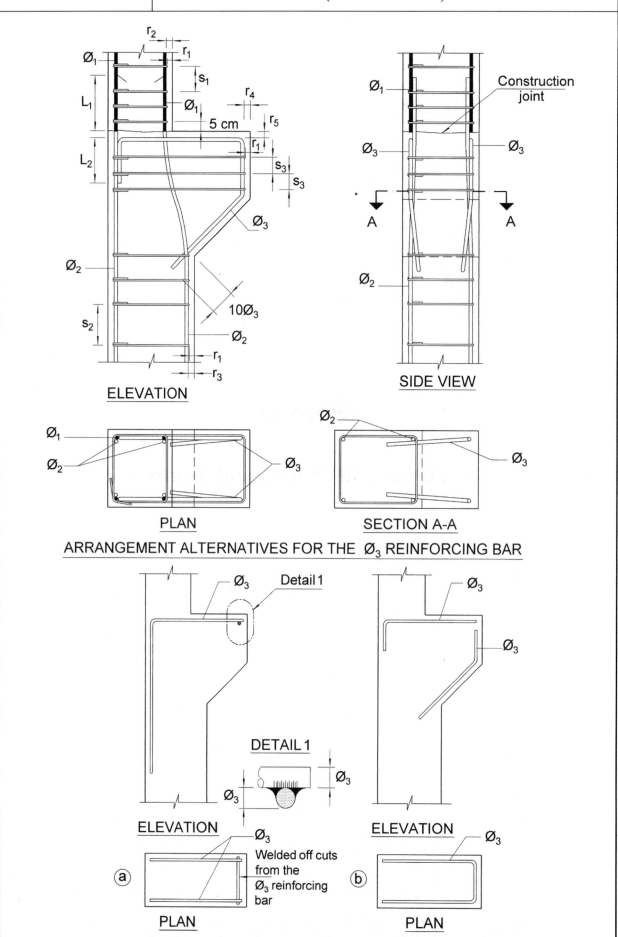

ELEVATION

SIDE VIEW

Construction joint

PLAN

SECTION A-A

ARRANGEMENT ALTERNATIVES FOR THE $\varnothing_3$ REINFORCING BAR

Detail 1

DETAIL 1

ELEVATION

ELEVATION

Welded off cuts from the $\varnothing_3$ reinforcing bar

(a)

(b)

PLAN

PLAN

1. RECOMMENDATIONS

1. $r_1 = 2.5$ cm.

2. $r_2 \geq \phi_1$.

3. $r_3 \geq \phi_2$.

4. $r_4 \geq \phi_3$.

5. $r_5 \geq \phi_3$. If the top of the bracket is exposed to rain or condensed water, $r_5 = 3$ cm.

6. L_1 is the ϕ_1 bar lap length.

7. L_2 can be calculated as per EC2, anchorage lengths.

8. The top of the bracket is smoothed with a float or power float.

9. See 1.4 for welding details.

10. See 1.2 and 1.3 for descriptions of how to tie bars and place spacers.

2. STATUTORY LEGISLATION

See EC2 (5).

3. RECOMMENDED ALTERNATIVE CODES

See ACI 318-08 (22).

4. SPECIFIC REFERENCES

See (38) and (39).

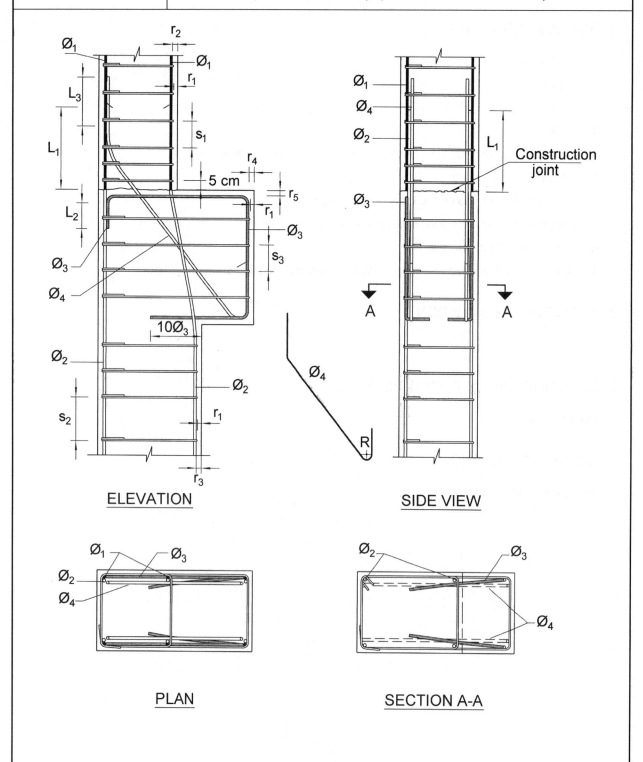

ELEVATION

SIDE VIEW

PLAN

SECTION A-A

CD – 10.02		NOTES

1. RECOMMENDATIONS

1. For cover specifications, see CD – 10.01.

2. L_1 is the ϕ_1 bar lap length.

3. L_2 can be calculated as per EC2, anchorage lengths.

4. L_3 is the lap length in the larger of the reinforcing bars, ϕ_1 or ϕ_4.

5. The top of the bracket is smoothed with a float or power float.

6. See 1.2 and 1.3 for descriptions of how to tie bars and place spacers.

2. STATUTORY LEGISLATION

See EC2 (5).

3. RECOMMENDED ALTERNATIVE CODES

See ACI 318-08 (22).

4. SPECIFIC REFERENCES

See (38) and (39).

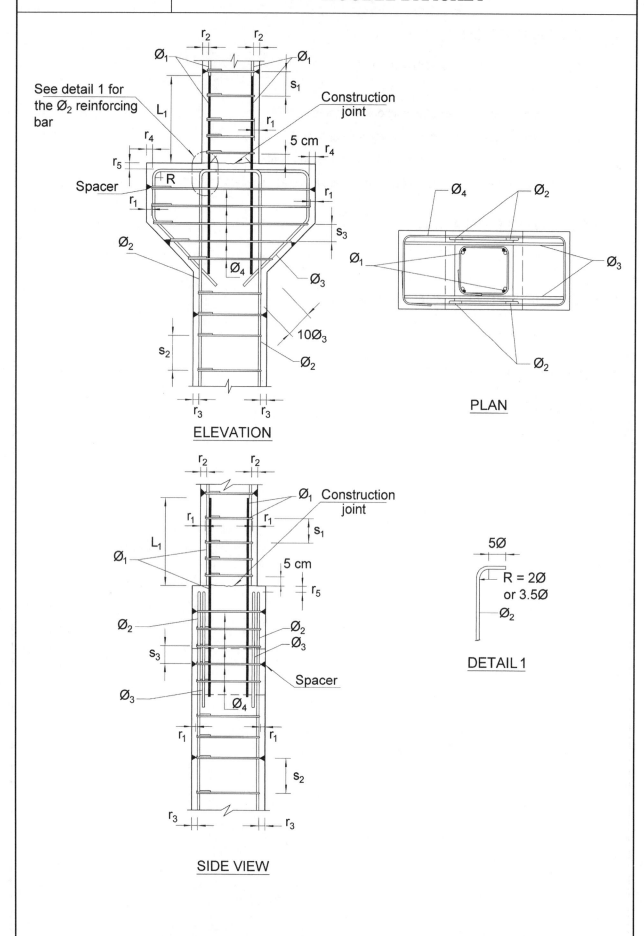

CD – 10.03

BRACKETS.
DOUBLE BRACKET

See detail 1 for
the $\emptyset_2$ reinforcing
bar

Construction
joint

5 cm

Spacer

$10\emptyset_3$

ELEVATION

PLAN

Construction
joint

5 cm

Spacer

SIDE VIEW

$5\emptyset$

$R = 2\emptyset$
or $3.5\emptyset$

$\emptyset_2$

DETAIL 1

394

1. RECOMMENDATIONS

1. $r_1 \geq 2.5$ cm.

2. $r_2 \geq \phi_1$.

3. $r_3 \geq \phi_2$.

4. $r_4 \geq \phi_3$.

5. $r_5 \geq \phi_3$. If the top of the bracket is exposed to rain or condensed water, $r_5 = 3$ cm.

6. L_1 is the ϕ_1 bar lap length.

7. The top of the bracket is smoothed with a float or power float.

8. See 1.2 and 1.3 for descriptions of how to tie bars and place spacers.

2. STATUTORY LEGISLATION

See EC2 (5).

3. RECOMMENDED ALTERNATIVE CODES

See ACI 318-08 (22).

4. SPECIFIC REFERENCES

See (38) and (39).

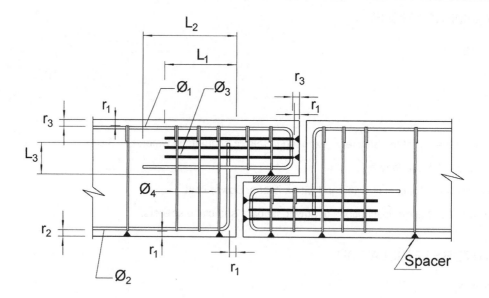

ELEVATION

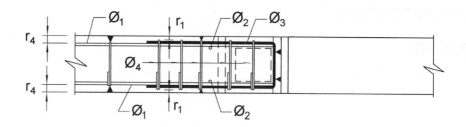

PLAN

1. RECOMMENDATIONS

1. $r_1 = 2.5$ cm.

    $r_2 \geq \phi_2$.

    $r_3 \geq \phi_1$.

    $r_4 \geq \begin{cases} \phi_1 \\ \phi_2 \end{cases}$

2. Care should be taken with regard to inward deviation in the vertical zones of the ϕ_2 bar.

3. L_1 is the ϕ_3 bar lap length.

4. L_2 is the ϕ_1 bar lap length.

5. L_3 is the ϕ_2 bar lap length.

6. For bearings, see CD – 09.01 to CD – 09-04.

7. See 1.2 and 1.3 for descriptions of how to tie bars and place spacers.

2. STATUTORY LEGISLATION

See EC2 (5).

3. SPECIFIC REFERENCES

See (39).

Group 11

Ground slabs and galleries

GROUND SLABS.
TYPICAL SECTION

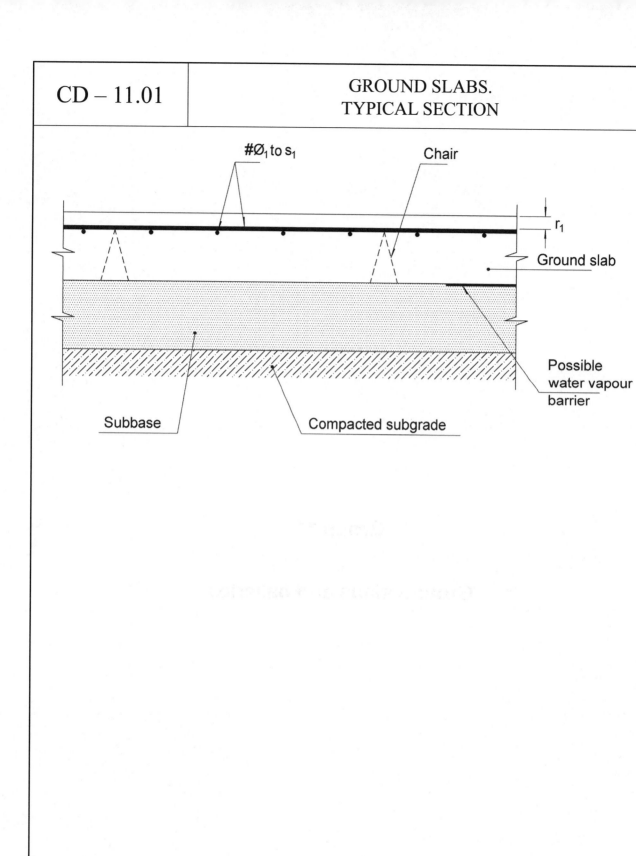

1. RECOMMENDATIONS

1. Welded-wire mesh is normally used for ϕ_1 to s_1 grids (# ϕ_1 to s_1).

2. Inter-mesh lapping need be no greater than 20 cm in any direction.

3. $r_1 = 5$ cm. (This is to ensure that the gravel will not roll over the transverse wires as the slab is finished with a vibrating screed.)

4. A water vapour barrier will be needed if moisture-sensitive materials are to be stored on the slab.

5. The percentage of silt in the subbase should ensure a smooth post-compaction surface.

6. The equipment needed for the surface finish must be specified.

7. See 1.2 and 1.3 for descriptions of how to tie bars and place spacers.

2. STATUTORY LEGISLATION

None in place.

3. RECOMMENDED ALTERNATIVE CODES

See ACI (40).

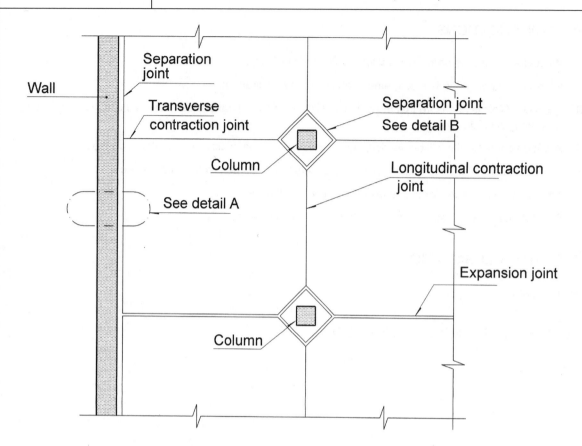

GENERAL LAYOUT. PLAN VIEW

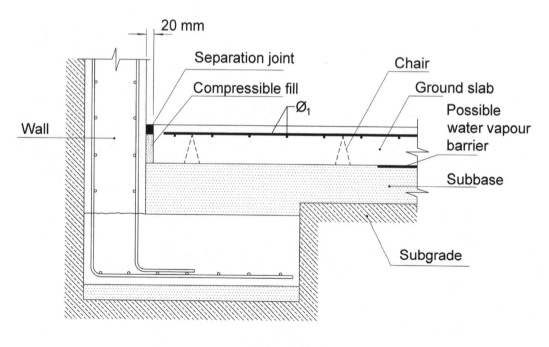

DETAIL A
WALL SEPARATION JOINT

1. RECOMMENDATIONS

1. Welded-wire mesh is normally used for ϕ_1 to s_1 grids (# ϕ_1 to s_1).

2. Inter-mesh lapping need be no greater than 20 cm in any direction.

3. A water vapour barrier will be needed if moisture-sensitive materials are to be stored on the slab.

4. The percentage of silt in the subbase should ensure a smooth post-compaction surface.

5. The equipment needed for the surface finish must be specified.

6. The maximum distance between contraction joints is 7 m.

7. The maximum distance between expansion joints ranges from 20 m to 40 m, depending on the climate.

8. A subbase should be laid between the slab and the top of the footing to avoid direct contact that would cause the slab to crack.

9. See 1.2 and 1.3 for descriptions of how to tie bars and place spacers.

2. STATUTORY LEGISLATION

None in place.

3. RECOMMENDED ALTERNATIVE CODES

See ACI (40).

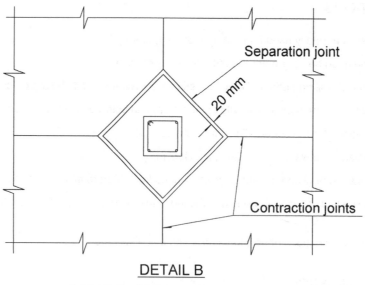

DETAIL B
COLUMN SEPARATION JOINT

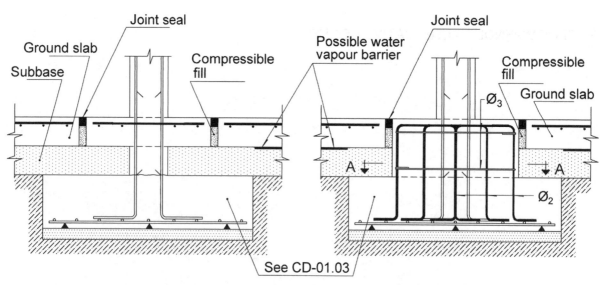

ELEVATION (VARIATION 1)

ELEVATION (VARIATION 2)

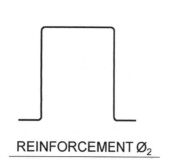

REINFORCEMENT $\varnothing_2$

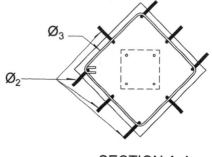

SECTION A-A

1. RECOMMENDATIONS

1. Welded-wire mesh is normally used for ϕ_1 to s_1 grids (# ϕ_1 to s_1).

2. Inter-mesh lapping need be no greater than 20 cm in any direction.

3. A water vapour barrier will be needed if moisture-sensitive materials are to be stored on the floor.

4. The percentage of silt in the subbase should ensure a smooth post-compaction surface.

5. The equipment needed for the surface finish must be specified.

6. The maximum distance between contraction joints is 7 m.

7. The maximum distance between expansion joints ranges from 20 m to 40 m, depending on the climate.

8. A subbase should be laid between the slab and the top of the footing to avoid direct contact that would cause the slab to crack.

9. See 1.2 and 1.3 for descriptions of how to tie bars and place spacers.

2. STATUTORY LEGISLATION

None in place.

3. RECOMMENDED ALTERNATIVE CODES

See ACI (40).

CD – 11.03	GROUND SLABS. CONTRACTION JOINTS

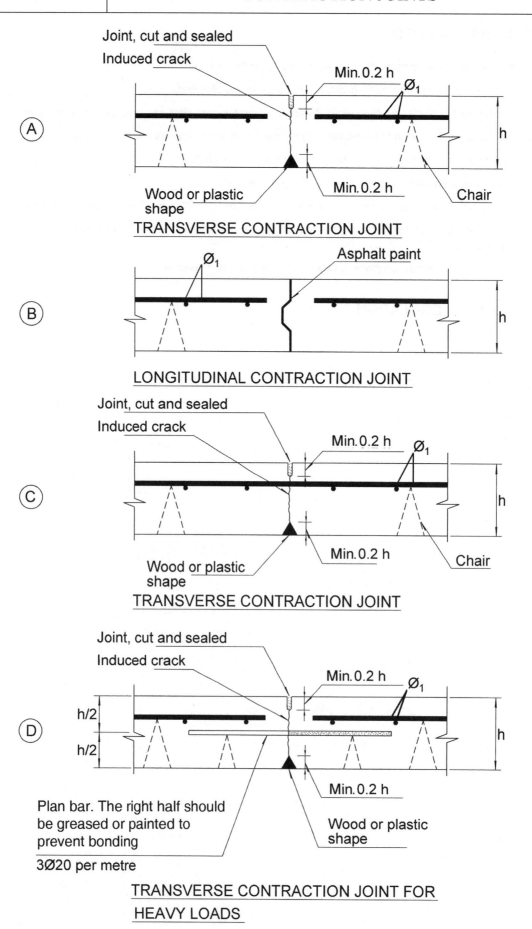

Joint, cut and sealed
Induced crack
Min. 0.2 h
$\emptyset_1$
h
Wood or plastic shape
Min. 0.2 h
Chair

TRANSVERSE CONTRACTION JOINT

A

$\emptyset_1$
Asphalt paint
h

LONGITUDINAL CONTRACTION JOINT

B

Joint, cut and sealed
Induced crack
Min. 0.2 h
$\emptyset_1$
h
Wood or plastic shape
Min. 0.2 h
Chair

TRANSVERSE CONTRACTION JOINT

C

Joint, cut and sealed
Induced crack
Min. 0.2 h
$\emptyset_1$
h/2
h/2
h
Plan bar. The right half should be greased or painted to prevent bonding
3$\emptyset$20 per metre
Min. 0.2 h
Wood or plastic shape

TRANSVERSE CONTRACTION JOINT FOR HEAVY LOADS

D

1. RECOMMENDATIONS

1. Welded-wire mesh is normally used for ϕ_1 to s_1 grids (# ϕ_1 to s_1).

2. Inter-mesh lapping need be no greater than 20 cm in any direction.

3. A water vapour barrier will be needed if moisture-sensitive materials are to be stored on the floor.

4. The percentage of silt in the subbase should ensure a smooth post-compaction surface.

5. The equipment needed for the surface finish must be specified.

6. The maximum distance between contraction joints is 7 m.

7. The maximum distance between expansion joints ranges from 20 m to 40 m, depending on the climate.

8. Variations A and B are not recommended where traffic is heavy.

9. See 1.2 and 1.3 for descriptions of how to tie bars and place spacers.

2. STATUTORY LEGISLATION

None in place.

3. RECOMMENDED ALTERNATIVE CODES

See ACI (40).

GROUND SLABS.
EXPANSION JOINTS

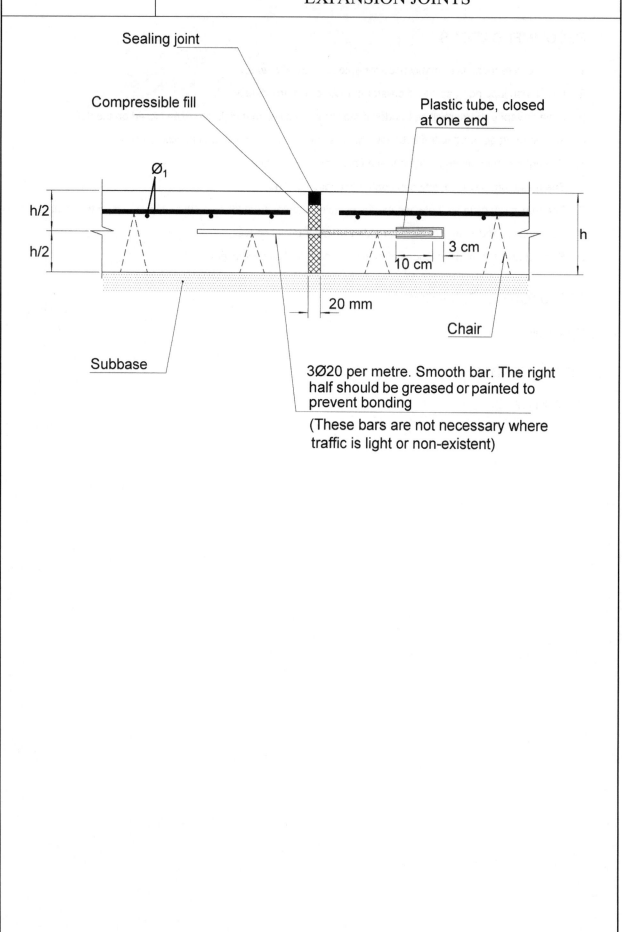

Sealing joint

Compressible fill

Plastic tube, closed at one end

$\varnothing_1$

h/2

h/2

h

3 cm

10 cm

20 mm

Chair

Subbase

3$\varnothing$20 per metre. Smooth bar. The right half should be greased or painted to prevent bonding

(These bars are not necessary where traffic is light or non-existent)

1. RECOMMENDATIONS

1. Welded wire mesh is normally used for ϕ_1 to s_1 grids (# ϕ_1 to s_1).

2. Inter-mesh lapping need be no greater than 20 cm in any direction.

3. A water vapour barrier will be needed if moisture-sensitive materials are to be stored on the floor.

4. The percentage of silt in the subbase should ensure a smooth post-compaction surface.

5. The equipment needed for the surface finish must be specified.

6. The maximum distance between expansion joints ranges from 20 m to 40 m, depending on the climate.

7. See 1.2 and 1.3 for descriptions of how to tie bars and place spacers.

2. STATUTORY LEGISLATION

None in place.

3. RECOMMENDED ALTERNATIVE CODES

See ACI (40).

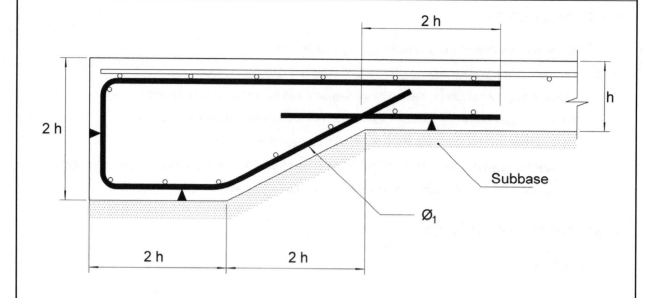

1. RECOMMENDATIONS

1. The edge should be strengthened when it is exposed to rain-induced softening the subbase.

2. In addition, the surface of the ground around the area strengthened may also be sealed.

3. See 1.2 and 1.3 for descriptions of how to tie bars and place spacers.

2. STATUTORY LEGISLATION

None in place.

3. RECOMMENDED ALTERNATIVE CODES

See ACI (40).

GALLERIES.
DUCTWAYS

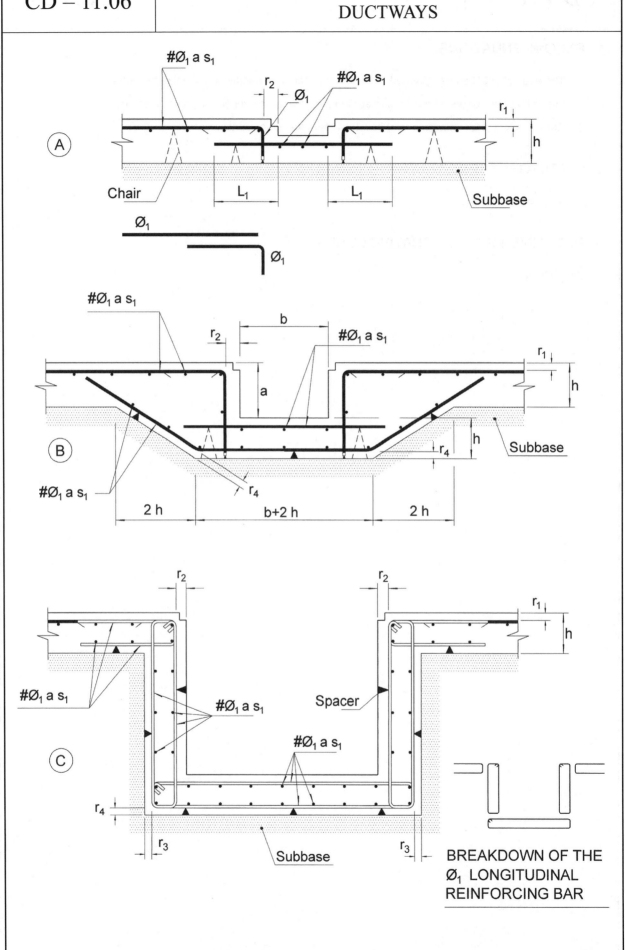

BREAKDOWN OF THE
$\emptyset_1$ LONGITUDINAL
REINFORCING BAR

1. RECOMMENDATIONS

1. The details refer to galleries and ductways to house ducts or machinery.

2. r_1 = 5 cm.

   r_2 = 5 cm. A thicker cover may be needed if required by the supports for the lids.

   r_3 = 7.5 cm.

   r_4 = 7.5 cm.

3. See CD – 11.01 for all other cover specifications.

4. See 1.2 and 1.3 for descriptions of how to tie bars and place spacers.

2. STATUTORY LEGISLATION

None in place.

3. RECOMMENDED ALTERNATIVE CODES

See ACI (40).

Group 12

Chimneys, towers and cylindrical hollow columns

CD – 12.01	CHIMNEYS, TOWERS AND CYLINDRICAL HOLLOW COLUMNS. GENERAL LAYOUT

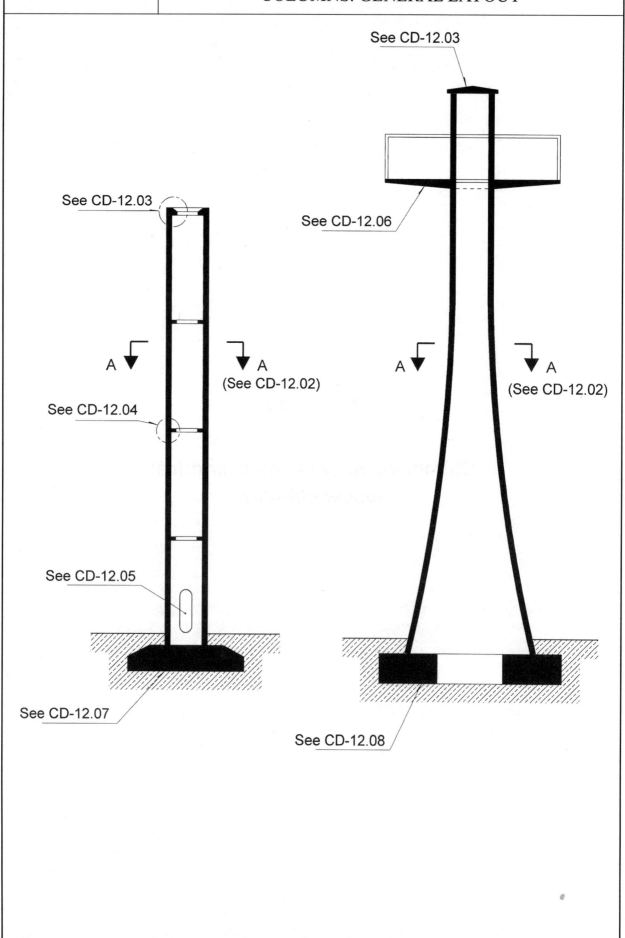

1. RECOMMENDATIONS

1. This group addresses details specific to this type of structure only. See also Group 01, Group 04 and Group 10.

2. STATUTORY LEGISLATION

None in place.

3. RECOMMENDED ALTERNATIVE CODES

See ACI 307-88 (41).

4. SPECIFIC REFERENCES

See (42).

CHIMNEYS, TOWERS AND CYLINDRICAL HOLLOW COLUMNS. GENERAL ARRANGEMENT OF REINFORCEMENT

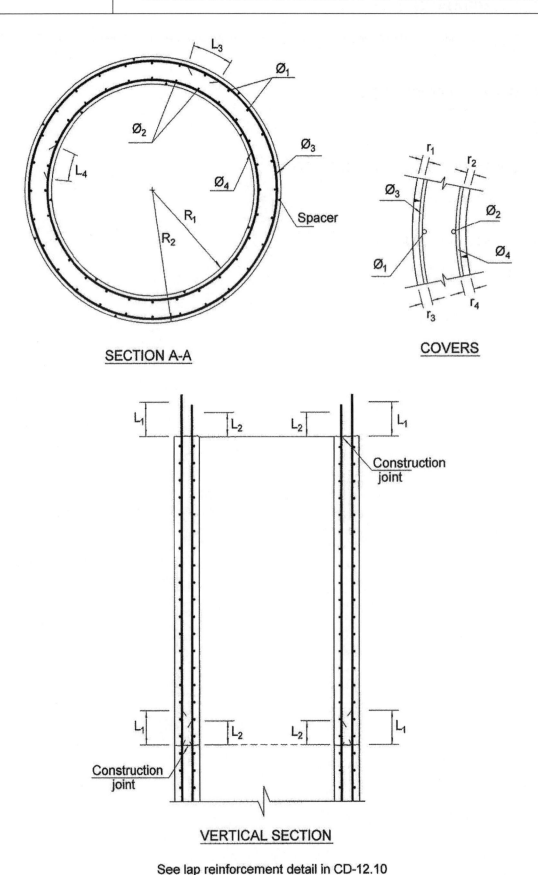

SECTION A-A

COVERS

VERTICAL SECTION

See lap reinforcement detail in CD-12.10

1. RECOMMENDATIONS

1. $r_1 = 2.5$ cm.

$$r_2 = \begin{cases} \geq 2.5 \text{ cm (in submerged columns} = 4 \text{ cm)} \\ \geq \phi_4 \end{cases}$$

   (Attention should be paid to operating temperature-related requirements.)

   $r_3 \geq \phi_1$.

   $r_4 \geq \phi_2$.

2. L_1 is the ϕ_1 bar lap length, L_2 is the ϕ_2 bar lap length, L_3 is the ϕ_3 bar lap length and L_4 is the ϕ_4 bar lap length. See CD – 12.10 for lap distribution.

3. Where the ϕ_1 and ϕ_2 bars cantilever (see CD – 12.10), the assembly should be duly stiffened during construction to prevent oscillation under wind action, which would break the bond with the concrete.

4. The horizontal joints should not be smoothed, for a rough surface such as is generated by the vibrator is needed. They should be cleaned and moistened and their surface allowed to dry before concrete pouring is resumed.

5. See 1.2 and 1.3 for descriptions of how to tie bars and place spacers.

2. STATUTORY LEGISLATION

None in place.

3. RECOMMENDED ALTERNATIVE CODES

See ACI 307-88 (41).

4. SPECIFIC REFERENCES

See (42).

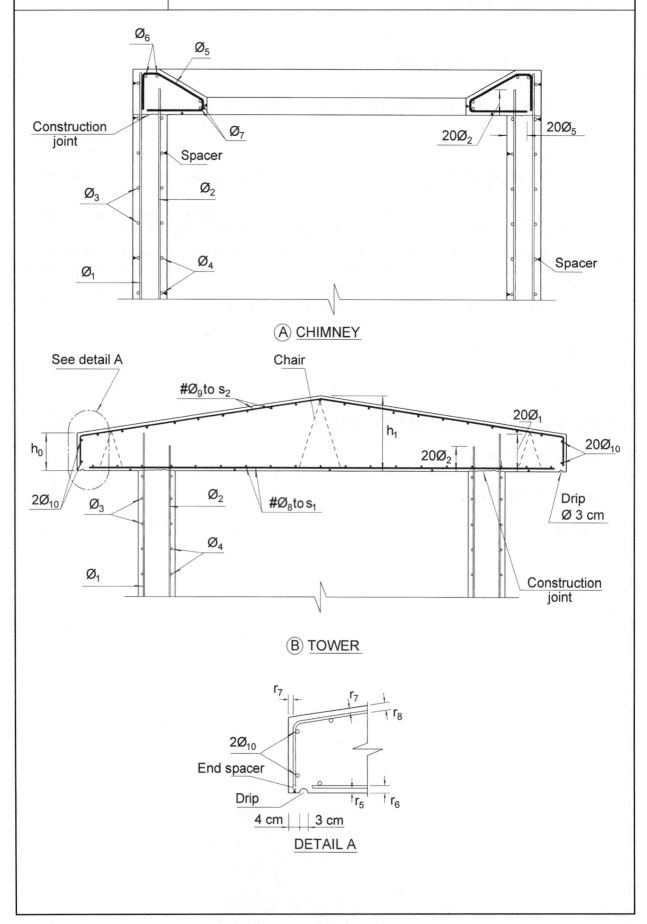

$Ø_6$

$Ø_5$

Construction
joint

$Ø_7$

Spacer

$20Ø_2$

$20Ø_5$

$Ø_3$

$Ø_2$

$Ø_1$

$Ø_4$

Spacer

Ⓐ CHIMNEY

See detail A

Chair

#$Ø_9$ to s_2

$20Ø_1$

h_1

h_0

$20Ø_2$

$20Ø_{10}$

$20Ø_{10}$

$Ø_3$

$Ø_2$

#$Ø_8$ to s_1

Drip
Ø 3 cm

$Ø_4$

$Ø_1$

Construction
joint

Ⓑ TOWER

r_7

r_7

r_8

$2Ø_{10}$

End spacer

Drip

r_5

r_6

4 cm | | 3 cm

DETAIL A

1. RECOMMENDATIONS

1. See CD – 12.02 for cover specifications.

2. The ϕ_5 bars should be covered with 3 cm of concrete.

3. $r_5 = 2.5$ cm $\geq \phi_8$.

   $r_6 \geq 2.5$ cm $+ \phi_8$.

   $r_8 = r_7 + \phi_9$.

4. The top in A is smoothed with a float or power float.

5. If the transverse slope on the slab in B is under 3 per cent, the cover over ϕ_3 should be 3 cm.

6. The slab surface in B is smoothed with a float or power float.

7. The horizontal joints should not be smoothed, for a rough surface such as is generated by the vibrator is needed. They should be cleaned and moistened and their surface allowed to dry before concrete pouring is resumed.

8. See 1.2 and 1.3 for descriptions of how to tie bars and place spacers.

2. STATUTORY LEGISLATION

None in place.

3. RECOMMENDED ALTERNATIVE CODES

See ACI 307-88 (41).

4. SPECIFIC REFERENCES

See (42).

CD – 12.04	CHIMNEYS, TOWERS AND CYLINDRICAL HOLLOW COLUMNS. BRACKET TO SUPPORT LINING

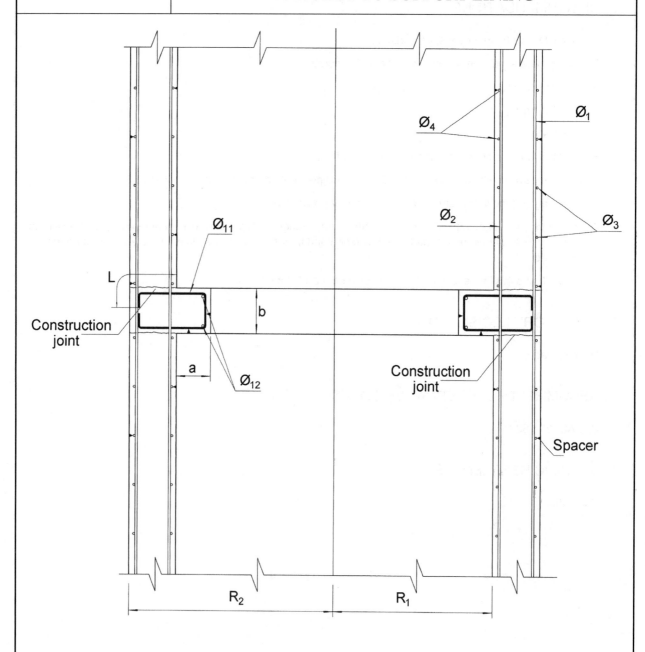

VERTICAL SECTION

1. RECOMMENDATIONS

1. See CD – 12.02 for cover specifications.

2. The cover over the ϕ_{11} bar must be at least 3 cm, but may be conditioned by the chimney operating temperature.

3. Since the ϕ_{11} bars are tied to the ϕ_1 bars, their positions must be coordinated.

4. The top of the bracket is smoothed with a float or power float.

5. See 1.2 and 1.3 for descriptions of how to tie bars and place spacers.

2. STATUTORY LEGISLATION

None in place.

3. RECOMMENDED ALTERNATIVE CODES

See ACI 307-88 (41).

4. SPECIFIC REFERENCES

See (42).

CD – 12.05	CHIMNEYS, TOWERS AND CYLINDRICAL HOLLOW COLUMNS. DUCT INLETS

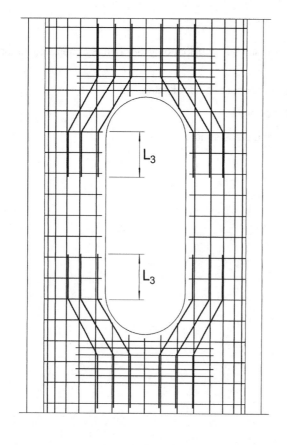

ELEVATION

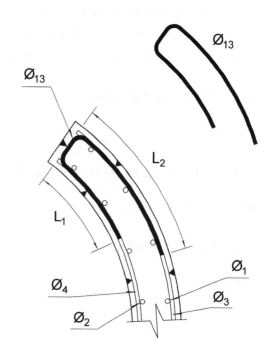

DETAIL A

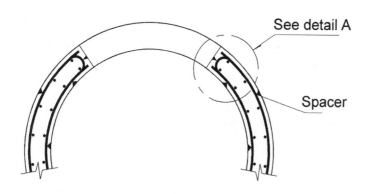

PLAN

1. RECOMMENDATIONS

1. See CD – 12.02 for cover specifications.

2. L_1 is the lap length in the larger of reinforcing bars ϕ_4 or ϕ_{13}.

3. L_2 is the lap length in the larger of reinforcing bars ϕ_3 or ϕ_{13}.

4. Diameter ϕ_{13} is equal to the larger of ϕ_3 or ϕ_4.

5. On the outer surface, L_3 is the lap length of the ϕ_1 bars, and on the inner surface, L_3 is the lap length of the ϕ_2 bars.

6. See 1.2 and 1.3 for descriptions of how to tie bars and place spacers.

2. STATUTORY LEGISLATION

None in place.

3. RECOMMENDED ALTERNATIVE CODES

See ACI 307-88(41).

4. SPECIFIC REFERENCES

See (42).

CD – 12.06	CHIMNEYS, TOWERS AND CYLINDRICAL HOLLOW COLUMNS. CIRCULAR SLAB AROUND THE TOP

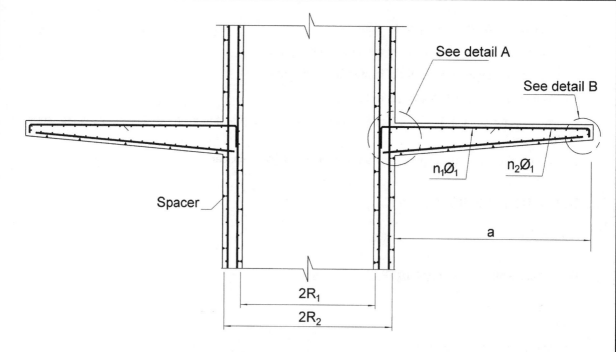

VERTICAL SECTION

(See (a)(b)(c) in CD-05.02)

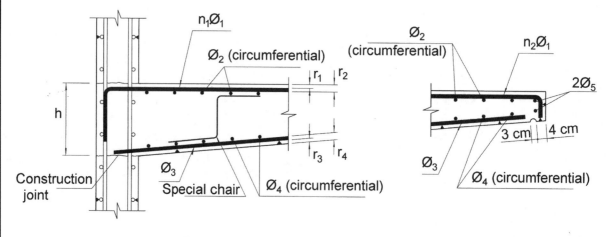

DETAIL A

DETAIL B

1. RECOMMENDATIONS

1. See CD – 12.02 for concrete cover specifications and general details.

2. $r_1 = 2.5$ cm $\geq \phi_1$.

   If the top is horizontal and exposed to rain or condensed water, $r_1 = 3$ cm.

   $r_2 \leq \phi_2$.

   $r_3 = 2.5$ cm $\geq \phi_3$.

   $r_4 \geq \phi_4$.

3. The ϕ_1 reinforcing bars are tied to the outer horizontal bars.

4. See 1.2 and 1.3 for descriptions of how to tie bars and place spacers.

2. STATUTORY LEGISLATION

None in place.

3. RECOMMENDED ALTERNATIVE CODES

See ACI 307-88 (41).

4. SPECIFIC REFERENCES

See (42).

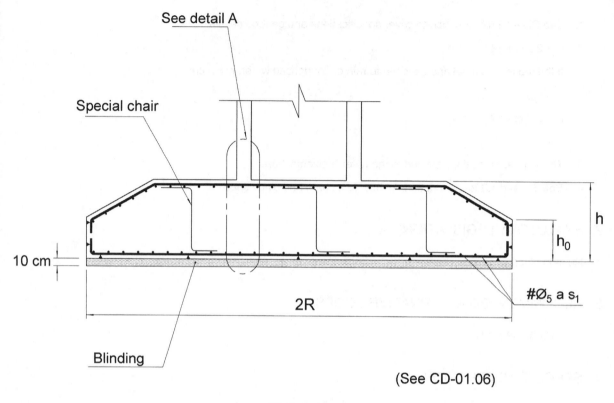

ELEVATION

(See CD-01.06)

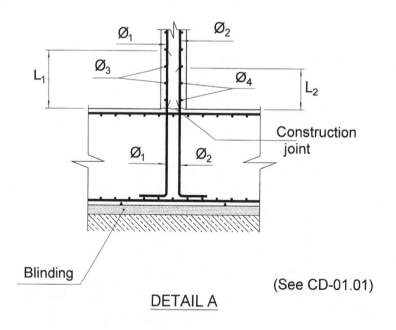

DETAIL A

(See CD-01.01)

1. RECOMMENDATIONS

1. See CD – 12.02 for general cover specifications.
2. See CD – 01.06 for cover specifications for the foundation.
3. See CD – 01.01 for the wall springing at foundation level.
4. See CD – 01.05 and CD – 01.06 for the reinforcement in the circular foundation.
5. See 1.2 and 1.3 for descriptions of how to tie bars and place spacers.

2. STATUTORY LEGISLATION

None in place.

3. RECOMMENDED ALTERNATIVE CODES

See ACI 307-88 (41).

4. SPECIFIC REFERENCES

See (42).

CHIMNEYS, TOWERS AND CYLINDRICAL HOLLOW COLUMNS. ANNULAR FOOTINGS

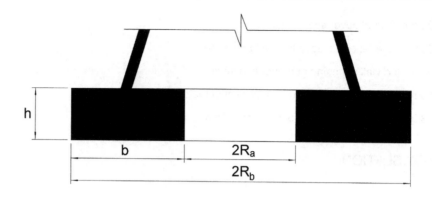

ELEVATION

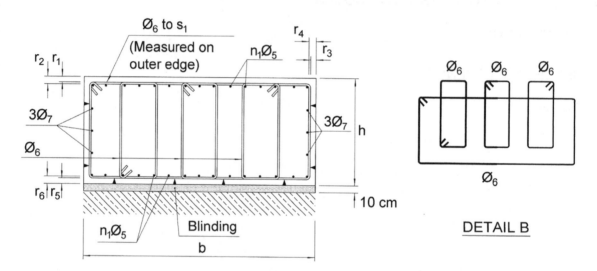

$\emptyset_6$ to s_1
(Measured on outer edge)

$n_1\emptyset_5$

$3\emptyset_7$

$\emptyset_6$

$n_1\emptyset_5$

Blinding

DETAIL A

$\emptyset_6$ $\emptyset_6$ $\emptyset_6$

$\emptyset_6$

DETAIL B

The $\emptyset_5$ reinforcing bars are placed circumferentially
The $\emptyset_6$ reinforcing bars are placed radially

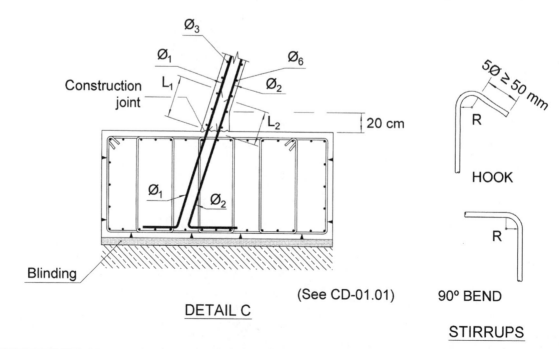

$\emptyset_3$

$\emptyset_1$ $\emptyset_6$

Construction joint L_1 $\emptyset_2$

L_2

20 cm

$\emptyset_1$ $\emptyset_2$

Blinding

DETAIL C

(See CD-01.01)

$5\emptyset \geq 50$ mm

R

HOOK

R

90º BEND

STIRRUPS

430

1. RECOMMENDATIONS

1. See CD – 12.02 for general cover specifications.

2. $r_1 = 2.5$ cm.

   $r_2 = 2.5$ cm $+ \phi_6$.

   $r_3 = 7.5$ cm if the concrete is poured directly against the soil.

   $r_4 = r_3 + \phi_6$.

   $r_5 = 2.5$ cm $\geq \phi_6$.

   $r_6 \leq \phi_5$.

3. The top is smoothed with floats or a power float.

4. The horizontal joints should not be smoothed, for a rough surface such as is generated by the vibrator is needed. They should be cleaned and moistened and their surface allowed to dry before concrete pouring is resumed.

5. See 1.2 and 1.3 for descriptions of how to tie bars and place spacers.

2. STATUTORY LEGISLATION

None in place.

3. RECOMMENDED ALTERNATIVE CODES

See ACI 307-88 (41).

4. SPECIFIC REFERENCES

See (12) and (42).

	CHIMNEYS, TOWERS AND CYLINDRICAL HOLLOW
CD – 12.09	COLUMNS. PLASTIC WATER INFLOW PIPES IN
	SUBMERGED HOLLOW COLUMNS

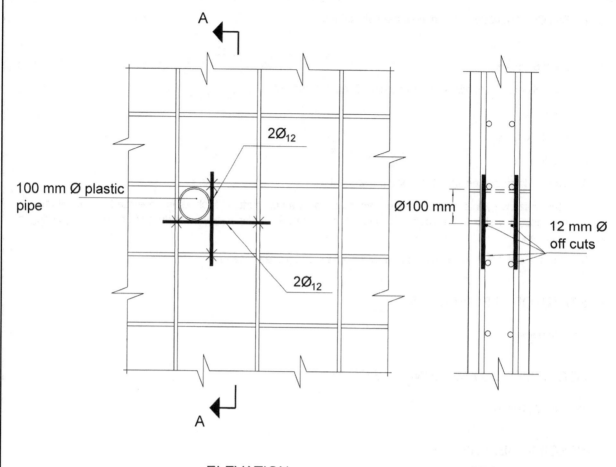

ELEVATION   SECTION A-A

1. RECOMMENDATIONS

1. See CD – 12.02 for general cover specifications.

2. The plastic pipes should be stiff and thick enough to withstand the pressure of fresh vibrated concrete without warping. They should be provisionally lidded to prevent the concrete or grout from flowing inside during concrete placement.

3. In columns in rivers or reservoirs, the pipes should preferably be placed on the downstream side to reduce the risk of clogging due to substances carried by the water.

4. See 1.2 and 1.3 for descriptions of how to tie bars and place spacers.

2. STATUTORY LEGISLATION

None in place.

3. RECOMMENDED ALTERNATIVE CODES

See ACI 307-88 (41).

4. SPECIFIC REFERENCES

See (42).

CD – 12.10	CHIMNEYS, TOWERS AND CYLINDRICAL HOLLOW COLUMNS. REINFORCEMENT LAPS

REINFORCEMENT LAP

(A) VERTICAL BARS

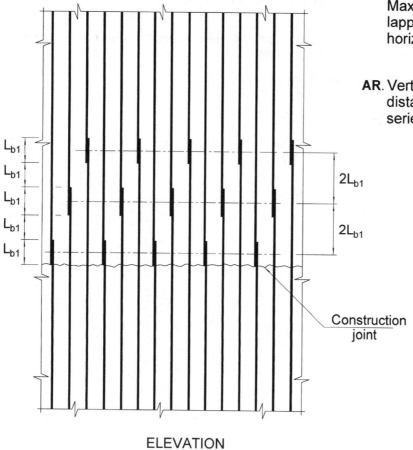

Maximum percentage of lapped bars in the same horizontal section:

33%

AR. Vertical centre-to-centre distance between lap series:

$2L_{b1}$

ELEVATION

(B) HORIZONTAL BARS

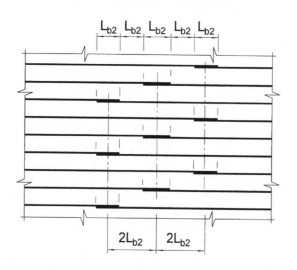

Maximum percentage of lapped bars in the same vertical section:

33%

AR. Horizontal centre-to-centre distance between lap series:

$2L_{b1}$

ELEVATION

CD – 12.10		NOTES

1. RECOMMENDATIONS

1. Cantilevered bars should be duly stiffened during construction to prevent oscillation under wind action, which would break the bond with the concrete.

2. The specified spacing refers to both the inner and outer sides.

3. Laps may concur on the two sides.

4. See Recommendation 4 in CD – 12.02 for construction joints.

5. See 1.2 and 1.3 for descriptions of how to tie bars and place spacers.

2. STATUTORY LEGISLATION

None in place.

3. RECOMMENDED ALTERNATIVE CODES

See ACI 307-88 (41).

4. SPECIFIC REFERENCES

See (42).

Group 13

Silos, caissons and rectangular hollow columns

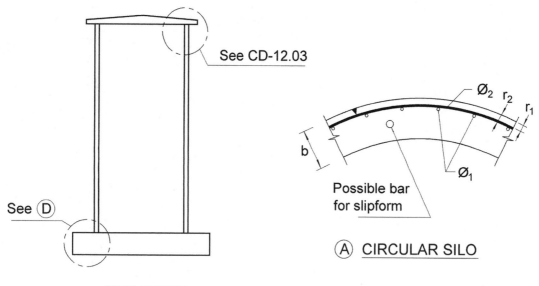

See CD-12.03

See (D)

ELEVATION

$\varnothing_2$ r_2 r_1

$\varnothing_1$

Possible bar for slipform

(A) CIRCULAR SILO

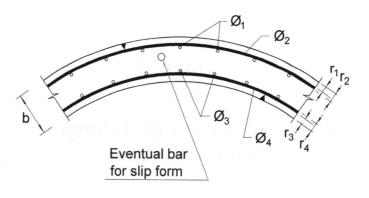

$\varnothing_1$ $\varnothing_2$

$\varnothing_3$

$\varnothing_4$

$r_1 r_2$

r_3 r_4

b

Eventual bar for slip form

(B) CIRCULAR SILO

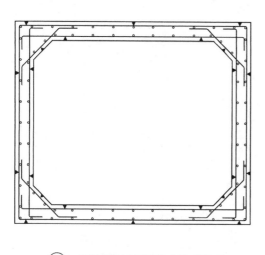

(C) RECTANGULAR SILO

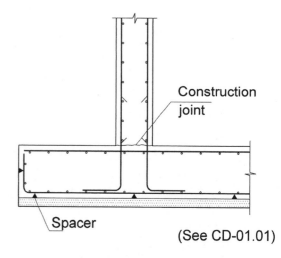

Construction joint

Spacer

(See CD-01.01)

(D) FOOTING

438

1. RECOMMENDATIONS

1. r_1 = 2.5 cm or 3 cm, depending on the case (see EC2, Table 4.1),
    but always $\geq \phi_2$.

    $r_2 \geq \phi_1$.

    r_3 = 2.5 cm or 3 cm, depending on the case (see EC2, Table 4.1),
    but always $\geq \phi_4$.

    $r_4 \geq \phi_3$.

2. See the recommendations in CD – 12.03 for the roof slab.

3. See CD – 01.01 for the wall springing at foundation level.

4. See 1.2 and 1.3 for descriptions of how to tie bars and place spacers.

2. STATUTORY LEGISLATION

See EC2 (5).

3. RECOMMENDED ALTERNATIVE CODES

See ACI 313 (43) and ACI 350.2R (44).

INTERSECTION IN CIRCULAR SILOS

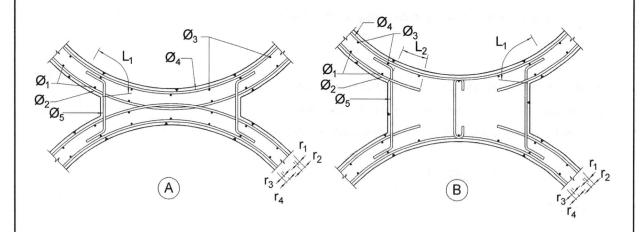

CORNERS OF RECTANGULAR SILOS

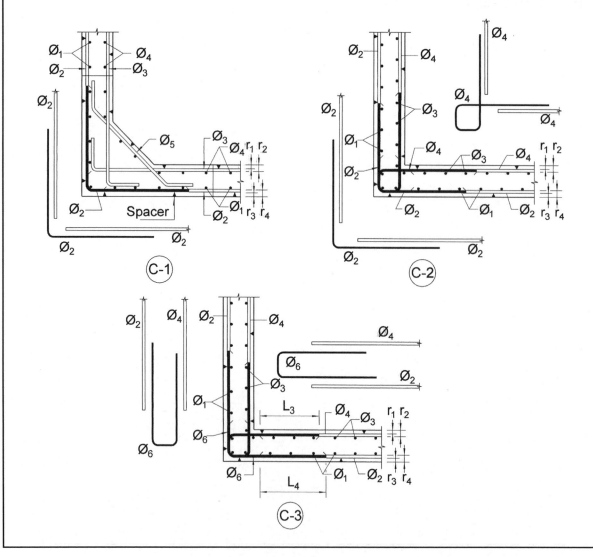

1. RECOMMENDATIONS

1. See CD – 13.01 for general cover specifications.

2. In details C-1 and C-2, the angled ϕ_2 bars overlap with the straight ϕ_2 bars along their lap length.

3. L_1 is the ϕ_5 bar lap length and L_2 is the ϕ_2 bar lap length.

   L_3 is the lap length in the larger of the reinforcing bars, ϕ_4 or ϕ_6, and L_4 is the lap length in the larger of the ϕ_2 or ϕ_6 bars.

4. Diameter ϕ_6 is equal to the larger of ϕ_2 or ϕ_4.

5. L_5 is the ϕ_6 bar lap length.

6. See 1.2 and 1.3 for descriptions of how to tie bars and place spacers.

2. STATUTORY LEGISLATION

See EC2 (5).

3. RECOMMENDED ALTERNATIVE CODES

See ACI 313 (43) and ACI 350.2R (44).

CONNECTION IN RECTANGULAR SILOS (Horizontal Sections)

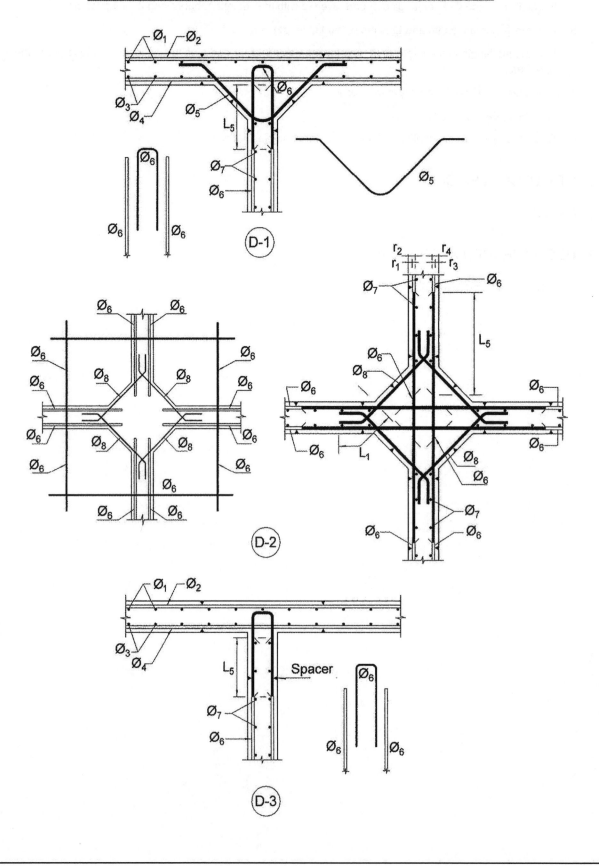

1. RECOMMENDATIONS

1. See CD – 13.01 for general cover specifications.

2. In details C-1 and C-2, the angled ϕ_2 bars overlap with the straight ϕ_2 bars along their lap length.

3. L_1 is the ϕ_5 bar lap length and L_2 is the ϕ_2 bar lap length.

   L_3 is the lap length in the larger of the reinforcing bars, ϕ_4 or ϕ_6, and L_4 is the lap length in the larger of the ϕ_2 or ϕ_6 bars.

4. Diameter ϕ_6 is equal to the larger of ϕ_2 or ϕ_4.

5. L_5 is the ϕ_6 bar lap length.

6. See 1.2 and 1.3 for descriptions of how to tie bars and place spacers.

2. STATUTORY LEGISLATION

See EC2.

3. RECOMMENDED ALTERNATIVE CODES

See ACI 313 (43) and ACI 350.2R (44).

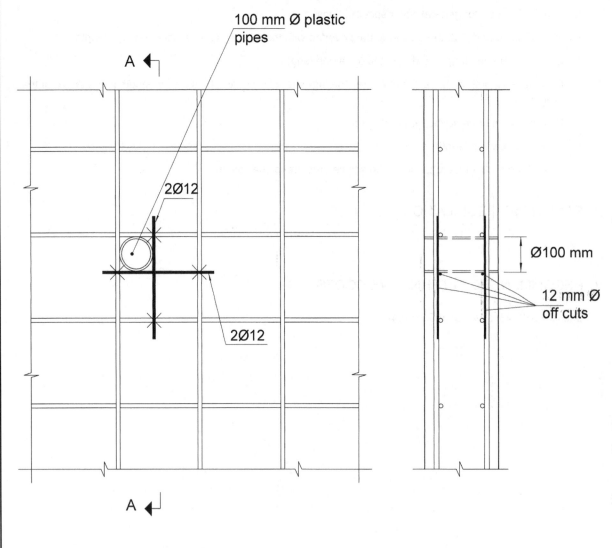

100 mm Ø plastic pipes

A

2Ø12

2Ø12

A

Ø100 mm

12 mm Ø off cuts

ELEVATION

SECTION A-A

1. RECOMMENDATIONS

1. See CD – 13.01 for general cover specifications.

2. The plastic pipes should be stiff and thick enough to withstand the pressure of fresh vibrated concrete without warping. They should be provisionally lidded to prevent the concrete or grout from flowing inside during concrete placement.

3. In columns in rivers or reservoirs, the pipes should preferably be placed on the downstream side to reduce the risk of clogging due to substances carried by the water.

4. See 1.2 and 1.3 for descriptions of how to tie bars and place spacers.

2. STATUTORY LEGISLATION

See EC2 (5).

SILOS, CAISSONS AND RECTANGULAR HOLLOW COLUMNS. HOPPERS

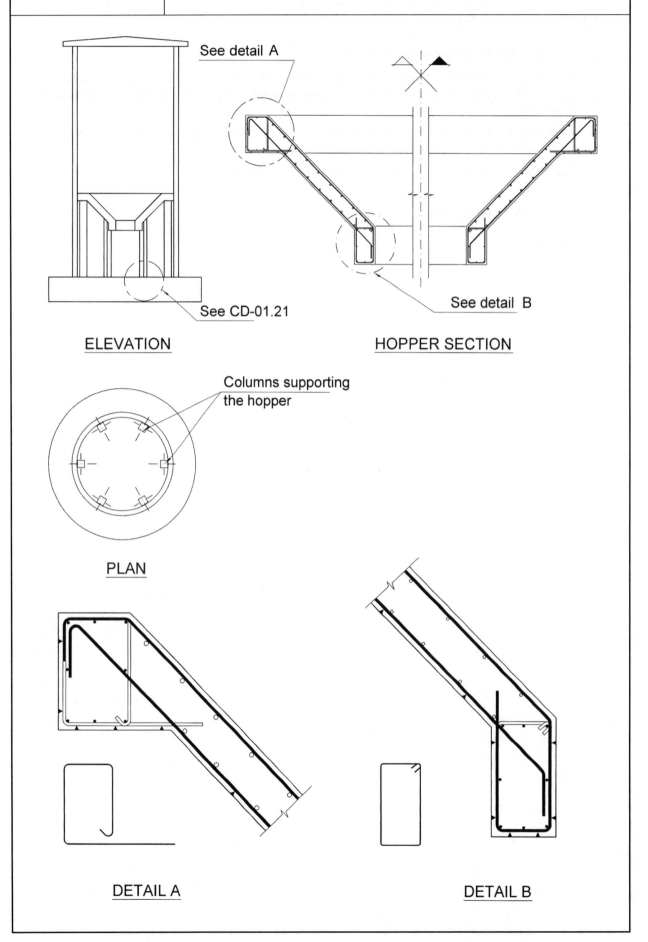

See detail A

See CD-01.21

ELEVATION

See detail B

HOPPER SECTION

Columns supporting the hopper

PLAN

DETAIL A

DETAIL B

1. RECOMMENDATIONS

1. See Group 05 for beam reinforcement cover specifications.
2. See Group 06 for slab reinforcement cover specifications.
3. See CD – 01-21 for the column springing.
4. See 1.2 and 1.3 for descriptions of how to tie bars and place spacers.

2. STATUTORY LEGISLATION

See EC2 (5).

3. RECOMMENDED ALTERNATIVE CODES

See ACI 313 (43) and ACI 350.2R (44).

CD – 13.05	SILOS, CAISSONS AND RECTANGULAR HOLLOW COLUMNS. FOUNDATIONS

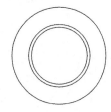

(See CD-12.07)

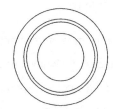

(See CD-12.08)

(A) INDIVIDUAL FOOTING WITH CIRCULAR SLAB

(B) INDIVIDUAL FOOTING WITH ANNULAR SLAB

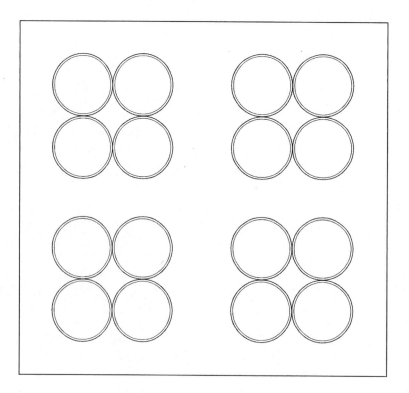

(See CD-01.21)

(C) GROUP OF SILOS WITH COMMON SLAB FOUNDATION

1. RECOMMENDATIONS

1. See CD – 01.05, CD – 01.06, CD – 12.07 and the respective recommendations for circular slab foundations.

2. See CD – 12.08 and the respective recommendations for annular foundations.

3. See CD – 01.21 and the respective recommendations for foundation slabs.

2. STATUTORY LEGISLATION

See EC2 (5).

3. RECOMMENDED ALTERNATIVE CODES

See ACI 313 (43) and ACI 350.2R (44).

CD – 13.06	SILOS, CAISSONS AND RECTANGULAR HOLLOW COLUMNS. REINFORCEMENT LAPS

REINFORCEMENT LAP

Ⓐ VERTICAL BARS

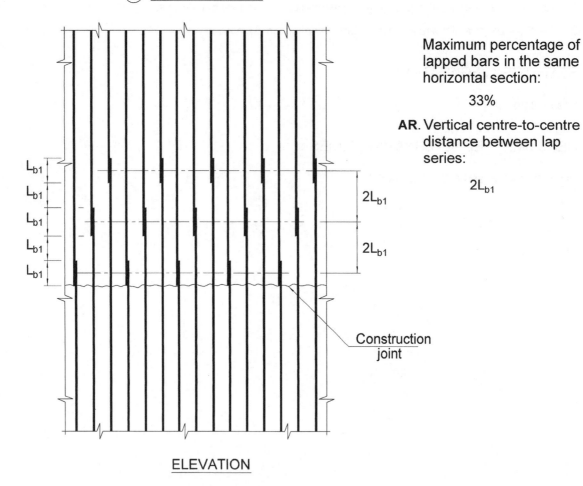

Maximum percentage of lapped bars in the same horizontal section:

33%

AR. Vertical centre-to-centre distance between lap series:

$2L_{b1}$

Construction joint

ELEVATION

Ⓑ HORIZONTAL BARS

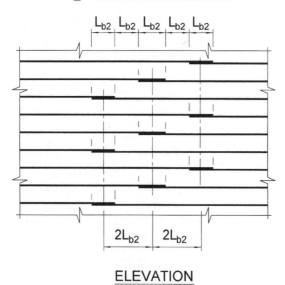

Maximum percentage of lapped bars in the same vertical section:

33%

AR. Horizontal centre-to-centre distance between lap series:

$2L_{b1}$

ELEVATION

1. RECOMMENDATIONS

1. Cantilevered bars should be duly stiffened during construction to prevent oscillation under wind action, which would break the bond with the concrete.

2. The spacing specified refers to both the inner and outer sides.

3. Laps may concur on the two sides.

4. See Recommendation 4 in CD – 12.02 for construction joints.

5. See 1.2 and 1.3 for descriptions of how to tie bars and place spacers.

2. STATUTORY LEGISLATION

See EC2 (5).

3. RECOMMENDED ALTERNATIVE CODES

See ACI 313 (43) and 350.2R (44).

Group 14

Reservoirs, tanks and swimming pools

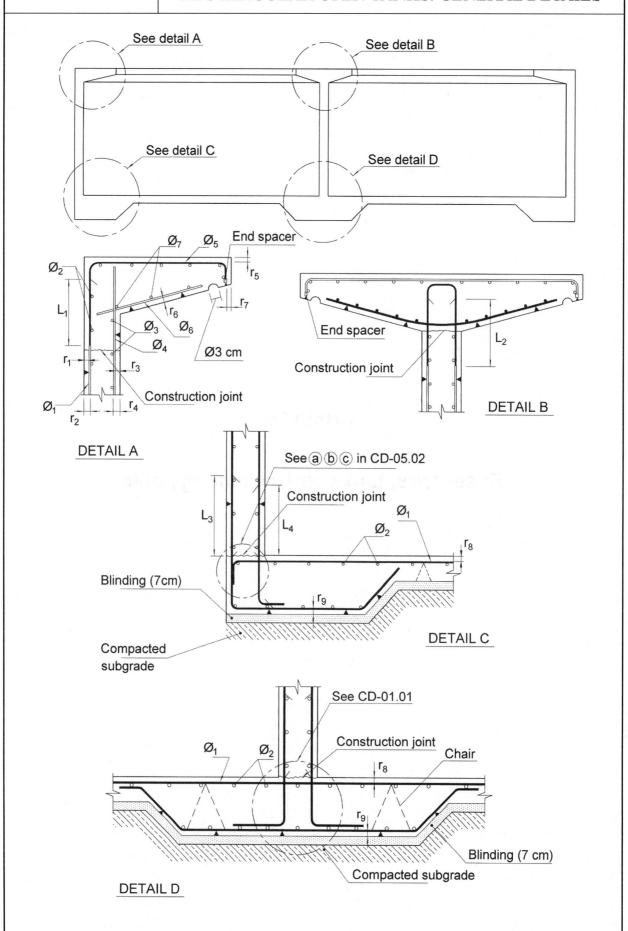

See detail A

See detail B

See detail C

See detail D

DETAIL A

DETAIL B

DETAIL C

DETAIL D

See (a) (b) (c) in CD-05.02

See CD-01.01

End spacer

Construction joint

Blinding (7cm)

Compacted subgrade

Chair

1. RECOMMENDATIONS

1. Covers:

   $r_1 = 7.5$ cm if the concrete is poured directly against the soil.

   $r_2 \geq \phi_2$.

   $r_3 = 2.5$ cm $\geq \phi_4$.

   $r_4 \leq \phi_3$.

   $r_5 = 3$ cm $\geq \phi_5$.

   $r_6 = 3$ cm.

   $r_7 = 3$ cm.

2. L_1 is the lap length of the thicker of the ϕ_1 or ϕ_5 bars.

3. L_2 is the lap length of the vertical bars.

4. $r_8 = 5$ cm.

   $r_9 = 3$ cm $\geq \phi_1$.

5. The tops are smoothed with a float or power float. The bottom is smoothed with a vibrating screed. This is why a 5 cm cover is needed. Otherwise, the vibrating screed would cause the coarse aggregate to rise up over the transverse reinforcement, making it difficult to smooth the surface.

6. The blinding is floated or smoothed with a power float.

7. A rough surface such as is generated by vibration is needed in the construction joints. These joints should be cleaned and moistened and their surface allowed to dry before concrete pouring is resumed.

8. See CD – 14.08 for the use of waterstops where watertightness is desired.

9. See 1.2 and 1.3 for descriptions of how to tie bars and place spacers.

2. STATUTORY LEGISLATION

See EC2 (5).

3. RECOMMENDED ALTERNATIVE CODES

See ACI 318-08 (22).

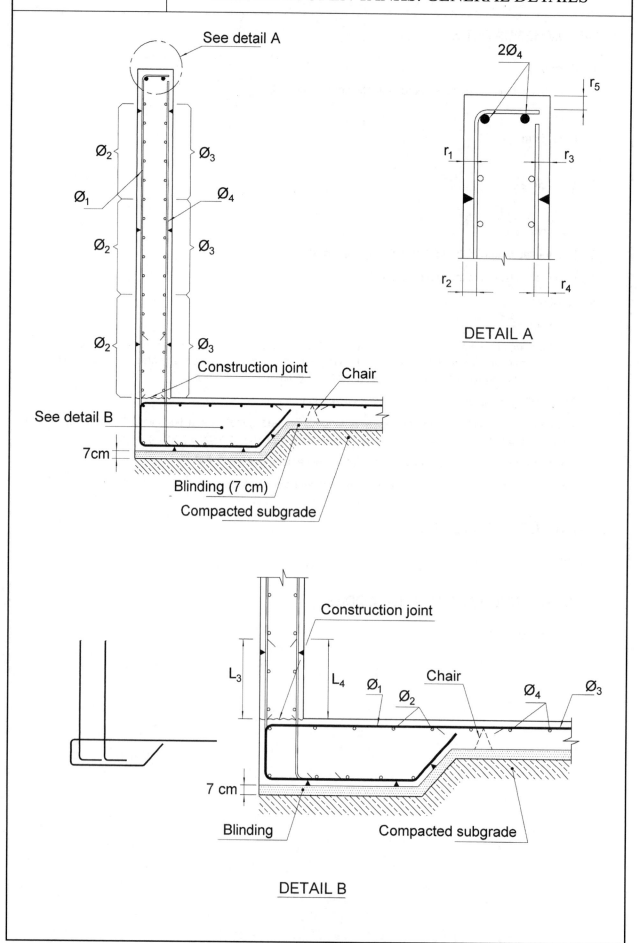

DETAIL A

DETAIL B

CD – 14.02	NOTES

1. RECOMMENDATIONS

1. See CD – 14.01 for cover specifications.

2. L_3 is the ϕ_1 bar lap length.

3. L_4 is the ϕ_4 bar lap length.

4. See wall crown detail in CD – 02.03. for the value of ϕ_4 (detail A), which depends on the height of the tank.

5. The top is smoothed with a float or power float. The bottom is smoothed with a vibrating screed. This is why a 5 cm cover is needed. Otherwise, the vibrating screed would cause the coarse aggregate to rise up over the transverse reinforcement, making it difficult to smooth the surface.

6. A rough surface such as is generated by vibration is needed in the construction joints. These joints should be cleaned and moistened and their surface allowed to dry before concrete pouring is resumed.

7. See CD – 14.08 for the use of waterstops where watertightness is desired.

8. See 1.2 and 1.3 for descriptions of how to tie bars and place spacers.

2. STATUTORY LEGISLATION

See EC2 (5).

3. RECOMMENDED ALTERNATIVE CODES

See ACI 318-08 (22).

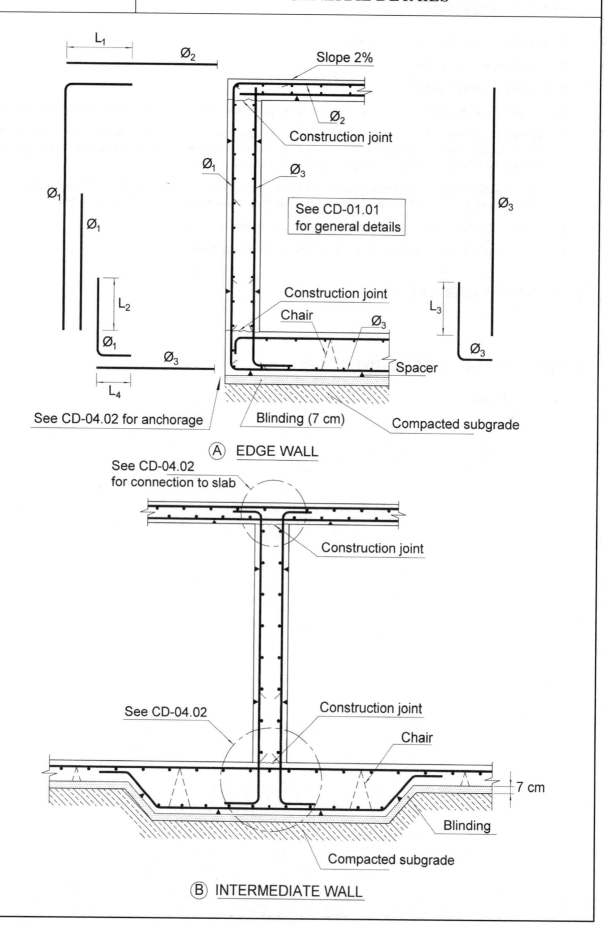

Slope 2%

$\varnothing_2$

Construction joint

$\varnothing_1$

$\varnothing_3$

See CD-01.01 for general details

L_1

$\varnothing_2$

$\varnothing_1$

$\varnothing_1$

L_2

$\varnothing_1$

$\varnothing_3$

L_4

See CD-04.02 for anchorage

Construction joint

Chair

$\varnothing_3$

L_3

$\varnothing_3$

Spacer

Blinding (7 cm)

Compacted subgrade

(A) EDGE WALL

See CD-04.02 for connection to slab

Construction joint

See CD-04.02

Construction joint

Chair

7 cm

Blinding

Compacted subgrade

(B) INTERMEDIATE WALL

1. RECOMMENDATIONS

1. See CD – 14.01 for cover specifications.

2. L_1 is the lap length of the thicker of the ϕ_1 or ϕ_2 bars.

3. L_2 is the ϕ_1 bar lap length.

4. L_3 is the ϕ_3 bar lap length.

5. L_4 is the lap length of the thicker of the ϕ_1 or ϕ_3 bars.

6. The tops are smoothed with a float or power float. The bottom is smoothed with a vibrating screed. This is why a 5 cm cover is needed. Otherwise, the vibrating screed would cause the coarse aggregate to rise up over the transverse reinforcement, making it difficult to smooth the surface.

7. A rough surface such as is generated by vibration is needed in the construction joints. These joints should be cleaned and moistened and their surface allowed to dry before concrete pouring is resumed.

8. See CD – 14.08 for the use of waterstops where watertightness is desired.

9. See 1.2 and 1.3 for descriptions of how to tie bars and place spacers.

2. STATUTORY LEGISLATION

See EC2 (5).

3. RECOMMENDED ALTERNATIVE CODES

See ACI 318-08 (22).

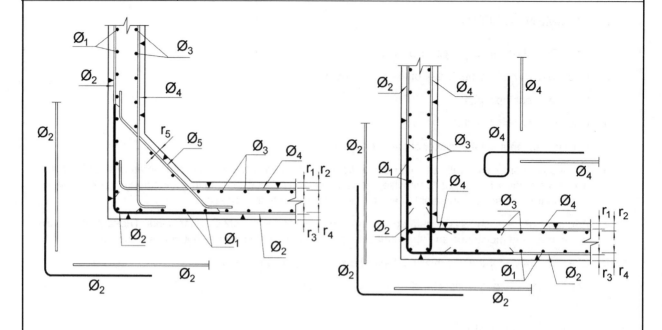

A

B

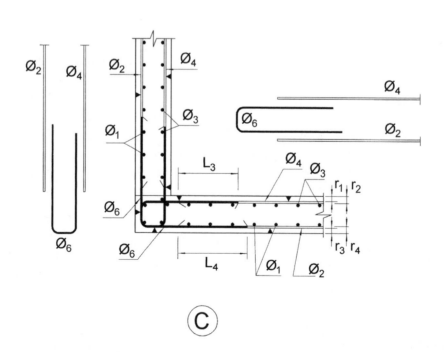

C

1. RECOMMENDATIONS

1. See CD – 14.01 for cover specifications.

2. In details A and B, the ϕ_2 angled bar overlaps with the ϕ_2 straight bars along their lap length.

3. L_1 is the ϕ_8 bar lap length.

4. L_2 is the ϕ_6 bar lap length.

5. L_3 is the lap length in the larger of the ϕ_4 or ϕ_6 reinforcing bars; L_4 is the lap length in the larger of the ϕ_2 or ϕ_6 bars.

6. Diameter ϕ_6 is equal to the larger of ϕ_2 or ϕ_4.

7. See 1.2 and 1.3 for descriptions of how to tie bars and place spacers.

2. STATUTORY LEGISLATION

See EC2 (5).

3. RECOMMENDED ALTERNATIVE CODES

See ACI 318-08 (22).

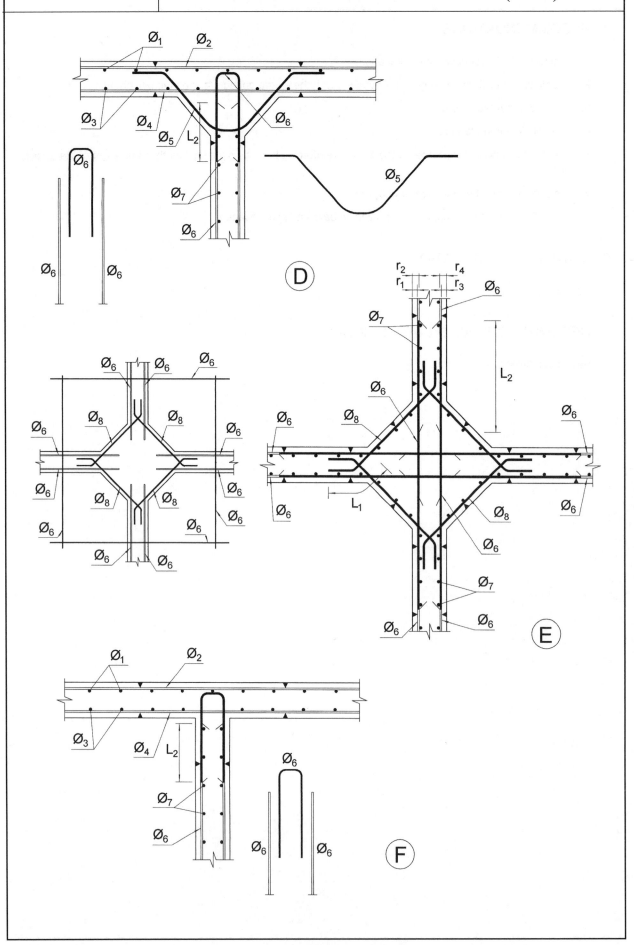

1. RECOMMENDATIONS

1. See CD – 14.01 for cover specifications.

2. In details A and B, the ϕ_2 angled bar overlaps with the ϕ_2 straight bars along their lap length.

3. L_1 is the ϕ_8 bar lap length.

4. L_2 is the ϕ_6 bar lap length.

5. L_3 is the lap length in the larger of the ϕ_4 or ϕ_6 reinforcing bars; L_4 is the lap length in the larger of the ϕ_2 or ϕ_6 bars.

6. Diameter ϕ_6 is equal to the larger of ϕ_2 or ϕ_4.

7. See 1.2 and 1.3 for descriptions of how to tie bars and place spacers.

2. STATUTORY LEGISLATION

See EC2 (5).

3. RECOMMENDED ALTERNATIVE CODES

See ACI 318-08 (22)

CD – 14.05	RESERVOIRS, TANKS AND SWIMMING POOLS. JOINTS AND BEARINGS IN WALLS

- See CD-02.09 for contraction joints in walls
- See CD-02.10 for expansion joints in walls

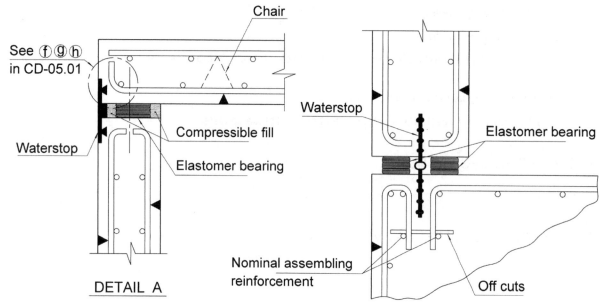

See (f)(g)(h) in CD-05.01

Chair

Waterstop

Compressible fill

Elastomer bearing

Waterstop

Elastomer bearing

Nominal assembling reinforcement

Off cuts

DETAIL A
ROOF – WALL SLIDE BEARING

DETAIL B
WALL – FOUNDATION SLIDE BEARING

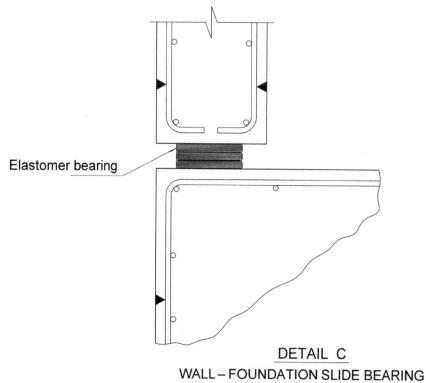

Elastomer bearing

DETAIL C
WALL – FOUNDATION SLIDE BEARING

1. RECOMMENDATIONS

1. See CD – 09.02 and CD – 09.03 for elastomer bearing placement.
2. See CD – 02.09 and CD – 02.10 for waterstop placement.
3. Detail C may only be used when the weight of the wall can ensure the effectiveness of the seal.

2. STATUTORY LEGISLATION

See EC2 (5).

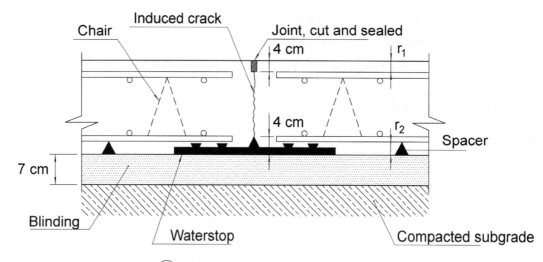

(A) CONTRACTION JOINT

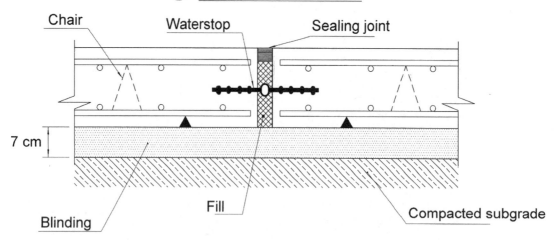

(B) EXPANSION JOINT (VARIATION 1)

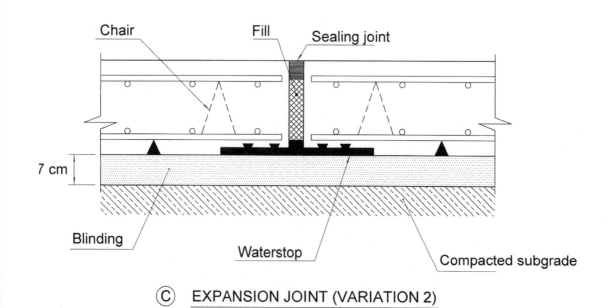

(C) EXPANSION JOINT (VARIATION 2)

1. RECOMMENDATIONS

1. Covers:

   r_1 = 5 cm. (Since the slab is to be finished with a vibrating screed, if a thinner cover were used, the gravel would rise up over the transverse wires to the detriment of the finish.)

   r_2 = 3 cm $\geq \phi$ (diameter of the bottom bar).

2. The blinding is floated or smoothed with a power float.

3. The area around the waterstops must be vigorously and meticulously vibrated.

4. The watertight band is difficult to position correctly with expansion joint type B.

2. STATUTORY LEGISLATION

See EC2 (5).

CD – 14.07	RESERVOIRS, TANKS AND SWIMMING POOLS. JOINTS IN WALLS

- See CD-02.09 for contraction joints
- See CD-02.10 for expansion joints

DISTRIBUTION OF CONTRACTION JOINTS

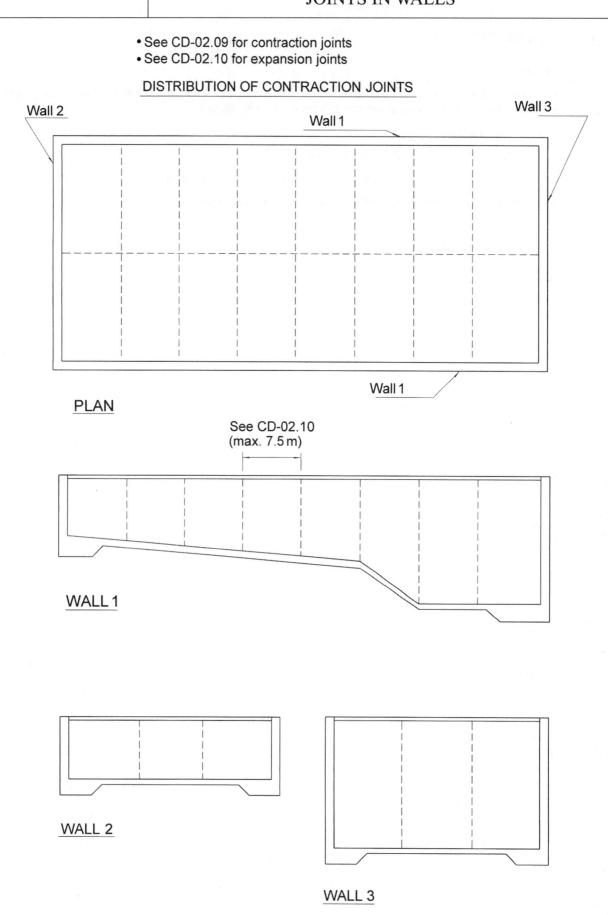

PLAN

See CD-02.10
(max. 7.5 m)

WALL 1

WALL 2

WALL 3

1. RECOMMENDATIONS

1. See CD – 14.06.

2. STATUTORY LEGISLATION

None in place.

RESERVOIRS, TANKS AND SWIMMING POOLS. SPECIFIC DETAILS FOR IMPROVING CONSTRUCTION JOINT WATERTIGHTNESS

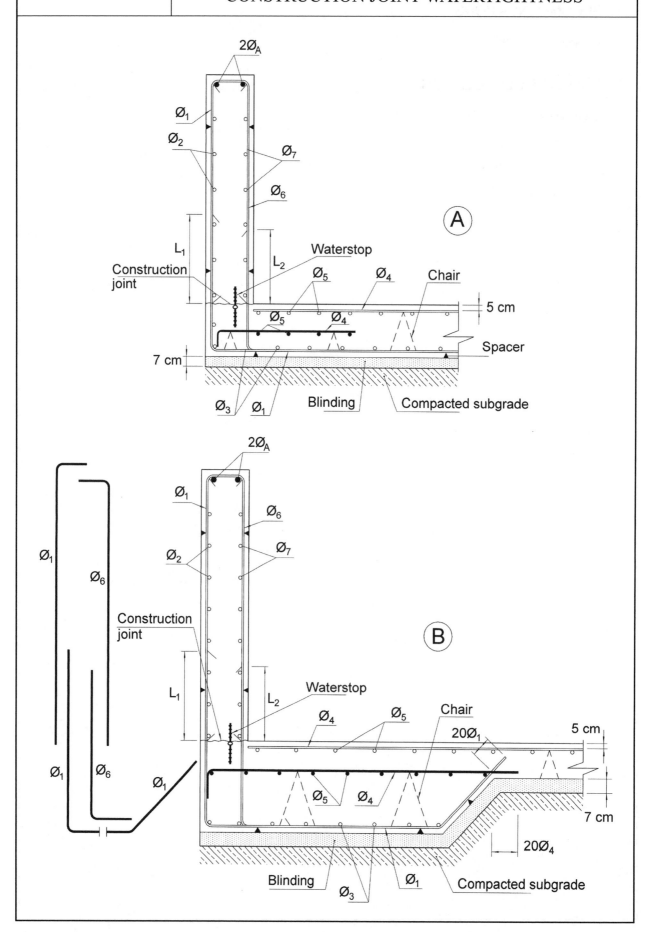

A

$2\emptyset_A$

$\emptyset_1$

$\emptyset_2$

$\emptyset_7$

$\emptyset_6$

L_1

L_2

Construction joint

Waterstop

$\emptyset_5$

$\emptyset_4$

Chair

5 cm

$\emptyset_5$

$\emptyset_4$

Spacer

7 cm

$\emptyset_3$

$\emptyset_1$

Blinding

Compacted subgrade

B

$2\emptyset_A$

$\emptyset_1$

$\emptyset_6$

$\emptyset_2$

$\emptyset_7$

Construction joint

L_1

L_2

Waterstop

$\emptyset_4$

$\emptyset_5$

Chair

$20\emptyset_1$

5 cm

$\emptyset_1$

$\emptyset_6$

$\emptyset_1$

$\emptyset_6$

$\emptyset_1$

$\emptyset_5$

$\emptyset_4$

7 cm

$20\emptyset_4$

Blinding

$\emptyset_3$

$\emptyset_1$

Compacted subgrade

1. RECOMMENDATIONS

1. See CD – 14.01 for cover and lapping specifications.

2. The area around the waterstops must be vigorously and meticulously vibrated.

3. See CD – 02.09 and CD – 02.10 on securing the watertight band during concrete pouring.

4. The blinding is floated or smoothed with a power float.

5. A rough surface such as is generated by vibration is needed in the construction joints. These joints should be cleaned and moistened and their surface allowed to dry before concrete pouring is resumed.

6. See 1.2 and 1.3 for descriptions of how to tie bars and place spacers.

2. STATUTORY LEGISLATION

See EC2 (5).

Group 15

Special construction details
for earthquake zones

1. **Eurocode 8 (EN 1998-1):** *Design of structures for earthquake resistance – Part 1: General rules, seismic actions and rules for buildings* (hereinafter EC8) (4) is the one that is relevant to Group 15.

   EC8 (Part 1) is supplemented by the following Parts:

   - EN 1998-2 contains specifications for bridges.
   - EN 1998-3 contains specifications for seismic assessment and retrofitting of existing buildings.
   - EN 1998-4 contains specifications relevant to silos, tanks and pipelines.
   - EN 1998-5 contains specifications relevant to foundations, retaining structures and geotechnical aspects.
   - EN 1998-6 contains specifications relevant to towers, masts and chimneys.

2. SEISMIC ZONES

For the purposes of EC8, the national authorities will have to subdivide their territory into seismic zones.

For most applications of EC8, the hazard is described in terms of a single parameter, which is the peak acceleration value A_{SR}.

In cases of *low seismicity*, simplified calculation procedures may be used for certain structures. This information is included in each country's National Annex.

In cases of *very low seismicity*, application of the specifications in EC8 is not required. This information is also included in each country's National Annex.

3. SEISMIC MEMBERS

EC8 makes a distinction between:

- *Primary seismic members.* Members considered to form part of the structural system that withstands the seismic action.
- *Secondary seismic members.* Members that are not considered to form part of the structural system that withstands the seismic action and whose strength and rigidity is diminished by these effects.

4. CRITICAL REGIONS

This is the zone of a primary seismic member that is subject to the most adverse combination of action effects possible and where plastic hinges could form.

5. MATERIALS

- *Steel reinforcement.* In primary seismic elements, the steel reinforcement must be class B or C, as specified in EC2, Table C.1.
- With the exception of closes, stirrups and cross-ties, only corrugated bars should be used as reinforcing in critical regions of primary seismic elements.
- Wire meshes may be used if they have ribbed wires and are class B or C.
- *Concrete.* Concrete lower than C 16/20 may not be used in primary seismic elements.

1. RECOMMENDATIONS

1. Some countries of the European Union have not published their National Annexes yet.

2. Some countries of the European Union have territorial zones located on other continents.

3. **AR.** The requirement of C 16/20 concrete appears to be very modest. A lower limit of C 20/25 would be preferable.

2. STATUTORY LEGISLATION

See EC2 (5) and EC8 (4).

1. BEAMS

1.1 Eccentricity of joints: the value of e (Figures 1 and 2) between the axes of the columns and the beams must be limited to:

$$e \leq \frac{b_c}{4}$$

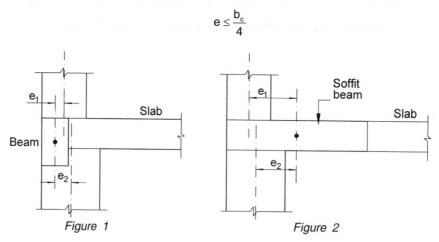

Figure 1   Figure 2

where b_c is the longest dimension of the cross-section of the column perpendicular to the beam axis.

AR. It is understood that this limitation must be fulfilled on the two pillars that are joined to the beam (e_1 and e_2 in Figures 1 and 2).

1.2 To take into account the favourable effect of the column compression on the bond of the horizontal bars passing the joints (Figure 3), the width b_w of a primary seismic beam must satisfy the following condition:

$$b_w \leq \min. \left\{ b_c + h_w ; 2b_c \right\}$$

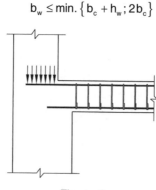

Figure 3

where b_c has already been defined in the previous clause

         h_w is the depth of the beam.

2. COLUMNS

2.1 Unless $\theta \leq 1$, the dimensions of the cross-section of primary seismic columns must not be less than one-tenth of the greatest distance between the points of inflection of the deformed axis of the column and the ends of the column.

3. DUCTILE WALLS

The thickness of the web, b_{wo}, in meters, must fulfil the following condition:

$$b_{wo} \geq \max. \left\{ 0.15, h_s / 20 \right\}$$

where h_s is the free floor height, in metres.

1. RECOMMENDATIONS

1. The angle θ is the inter-storey drift resistivity coefficient. See EC-8 (4).

2. STATUTORY LEGISLATION

See EC2 (5) and EC8 (4).

3. RECOMMENDED ALTERNATIVE CODES

See Chapter 21 in ACI 318-08 (22).

1. BEAMS

1.1 The regions of a primary seismic beam at a distance $\ell_{cr} = h_w$ (where h_w is the depth of the beam) from the end section where the beam joins to a beam or column joint, as well as on both sides of any section that is susceptible to yielding in the seismic design situation, must be considered to be critical regions.

1.2 Unless more detailed studies of the local ductility requirements are done, they could, in a simplified manner, be considered to be fulfilled on both flanges of the beam, provided that the following requirements are met.

- In the compressed reinforcing zone, there is at least half of the reinforcing installed in the tension zone. (This is apart from the compression reinforcing needed to satisfy the U.L.S.).

- The reinforcement ratio of the tension zone, ρ, does not reach the value

$$\rho_{max} = \rho' + \frac{0.0018}{\mu_\phi \, \varepsilon_{sy,d}} \cdot \frac{f_{cd}}{f_{yd}}$$

where the reinforcement ratios of the tension zone and compression zone, ρ and ρ' are both normalised to bd, where:

- b   is the width of the compression flange of the beam.
- $\varepsilon_{sy,d}$ is the design value of steel strain at yield
- μ_ϕ is the curvature ductility factor.

- Along the entire length of the beam, the reinforcement ratio of the tension zone, ρ, must not be less than the following value:

$$\rho_{min} = 0.5 \left(\frac{f_{ctm}}{f_{yk}} \right)$$

- Within the critical regions of primary seismic beams, the stirrups must fulfil the following requirements:

   (a)   the diameter must not be less than 6 mm;
   (b)   the spacing s, of stirrups, in mm, must not exceed the following value:

$$s = \min. \left\{ h_w / 4; \; 24 \, d_{bw}; \; 225; \; 8 d_{bL} \right\}$$

where

        d_{bL} is the diameter of the narrowest longitudinal bar
        h_w is the beam depth in mm
        d_{bw} is the stirrup diameter in mm;

   (c)   the first stirrup will not be more than 50 mm from the end surface on the beam (Figure 4).

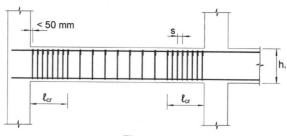

Figure 4

2. COLUMNS. DETAILING OF PRIMARY SEISMIC COLUMNS FOR LOCAL DUCTILITY

2.1 The total longitudinal reinforcement ratio, ρ_1, must be between 0.01 and 0.04. In members with symmetrical cross-sections, the reinforcing must also be symmetrical ($\rho = \rho'$).

2.2 At least one intermediate bar must be installed on each side of the piece to ensure the integrity of the beam–column joints.

2.3 The regions at a distance ℓ_{cr} from the ends of a primary seismic column must be considered to be critical regions.

2.4 In the absence of more precise calculations, the length of the critical region ℓ_{cr} (in metres) can be calculated based on the following equation:

$$\ell_{cr} = \max. \left\{ h_c; \; \ell_{c1}/6; \; 0.45 \right\}$$

1. RECOMMENDATIONS

1. In most cases, condition 1.2 prevents the use of soffit beams in seismic zones.

2. STATUTORY LEGISLATION

See EC2 (5) and EC8 (4).

3. RECOMMENDED ALTERNATIVE CODES

See Chapter 21 in ACI 318-08 (22).

where

h_c is the longer dimension of the column's cross-section (in metres)

ℓ_{c1} is the clear length of the column (in metres).

2.5   If $\ell_{c1}/h_c < 3$, the entire height of the column must be considered to be critical and reinforced accordingly.

2.6   In confinement of columns, a minimum value of ω_{wd} equal to 0.08 must be provided in the critical zone at the base of the primary seismic columns. The value of ω_{wd} is given by the equation

$$\omega_{wd} = \frac{\text{volume of confining hoops}}{\text{volume of concrete core}} \cdot \frac{f_{yd}}{f_{cd}}$$

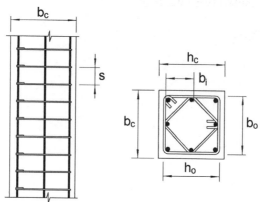

Figure 5

2.7   The spacing, s, of the hoops (in mm) must not exceed the value

$$s = \min.\left\{ b_o / 2;\ 175;\ 8d_{bL} \right\}$$

where

b_o (in mm) is the minimum dimension of the concrete core

d_{bL} is the diameter of the narrowest longitudinal bar.

2.8   The distance between consecutive longitudinal bars engaged by hoops or cross-ties must not exceed 200 mm.

3. DUCTILE WALLS

3.1   The height of the critical region, above the base of the wall, can be estimated by applying the equation:

$$h_{cr} = \max.\left\{ \ell_w,\ h_w / 6 \right\}$$

where

ℓ_w is the length of the wall cross-section

h_w is the height of the wall

but

$$h_{cr} \leq \begin{cases} 2\ell_w \\ \\ \begin{cases} h_s & \text{for } n \leq 6 \text{ storeys} \\ 2h_s & \text{for } n \geq 7 \text{ storeys} \end{cases} \end{cases}$$

where h_s is the open height between floors.

3.2   Premature web shear cracking of walls shall be prevented by providing a minimum ratio of web reinforcement: $\rho_{h,min} = \rho_{v,min} = 0.002$.

3.3   The web reinforcement should be provided in the two grids of bars, one at each face of the wall. The grids should be connected by cross-ties spaced about 500 mm apart.

3.4   Web reinforcement must have a diameter of not less than 8 mm, but not greater than one-eighth of the width of the wall. Spacing must not exceed 250 mm or 25 times the diameter of the bars, whichever is less.

1. RECOMMENDATIONS

1. In most cases, condition 1.2 prevents the use of soffit beams in seismic zones.

2. STATUTORY LEGISLATION

See EC2 (5) and EC8 (4).

3. RECOMMENDED ALTERNATIVE CODES

See Chapter 21 in ACI 318-08 (22).

1. COUPLING ELEMENTS OF COUPLED WALLS

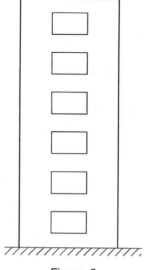

Figure 6

1.1 Wall coupling using just the floor slabs is not effective in the case of seismic zones (Figure 6).

1.2 To ensure against prevailing flexure mode of failure, it is necessary that

$$\frac{\ell}{h} \geq 3$$

1.3 When diagonal reinforcing is used, it should be arranged in column-like elements with side lengths at least equal to $0.5\ b_w$; its anchorage length should be 50 per cent greater than the value required by EC2 (Figure 7).

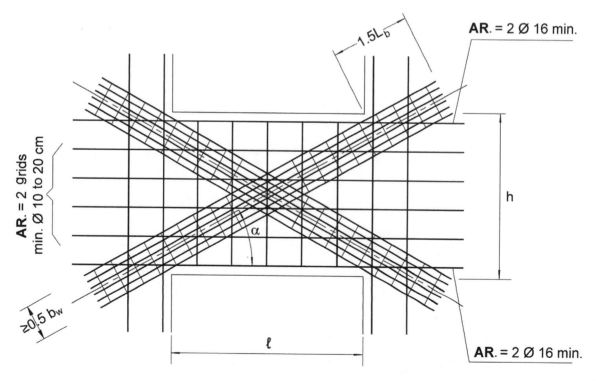

Figure 7

Hoops should be provided around these column-like element to prevents buckling of the longitudinal bars.

1. RECOMMENDATIONS

None.

2. STATUTORY LEGISLATION

See EC2 (5) and EC8 (4).

3. RECOMMENDED ALTERNATIVE CODES

See Chapter 21 in ACI 318-08 (22).

1. COLUMNS

1.1 When designing the length of the lap of the main reinforcing bars of a column, $A_{s,req}/A_{s,prov}$ shall be assumed to be 1.

1.2 If, in the seismic design situation, the axial force on a column is tensile, the anchorage lengths must be increased by 50 per cent with respect to the values specified in EC2.

2. BEAMS

2.1 The part of the longitudinal reinforcing of the angle beam at the joints to anchor it must always be inside the corresponding column hoops.

2.2 If it is not possible to fulfil the anchorage lengths specified in EC8, 5.6.2.2. P (4) on the outside of the beam–column joint, one of the solutions shown in Figure 8 may be adopted.

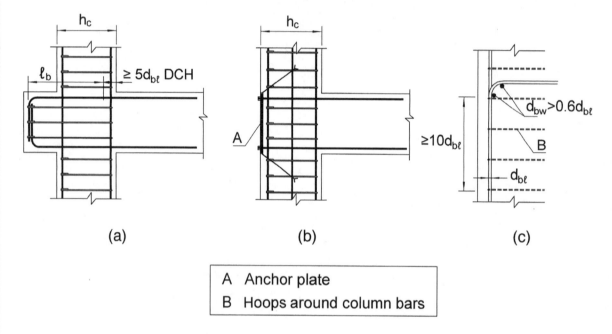

(a) (b) (c)

| A | Anchor plate |
| B | Hoops around column bars |

Figure 8

1. RECOMMENDATIONS

1. Specification 1.1 is included to guarantee that the bars reach their yield point in the failure of the structure.

2. STATUTORY LEGISLATION

See EC2 (5) and EC8 (4).

3. RECOMMENDED ALTERNATIVE CODES

See Chapter 21 in ACI 318-08 (22).

1.　Welded laps should not be located inside critical regions of elements.

2.　Mechanical couplers can be used in columns and walls if they have been tested and shown to have the necessary ductility.

3.　In addition to fulfilling the specifications in EC2, the transverse reinforcement provided within the lap length must comply with the following additional requirements.

- If the anchored and the continuing bar are in a plane parallel to the transverse reinforcement, the sum of the areas of all of the lapping bars must be used in the calculation of the transverse reinforcement.

- If both bars are within a normal plane to the transverse reinforcement, the transverse reinforcement will be calculated based on the area of the thickest lapping bar.

- The spacing, s, of the transverse reinforcement in the lap zone (in mm) must not exceed the value

$$s = \min. \left\{ h / 4; \ 100 \right\}$$

where h is the smaller dimension of the cross-section in mm.

1. RECOMMENDATIONS

1.   Note that the rules for transverse reinforcement in laps are more stringent.

2. STATUTORY LEGISLATION

See EC2 (5) and EC8 (4).

3. RECOMMENDED ALTERNATIVE CODES

See Chapter 21 in ACI 318-08 (22).

1. The short columns between the upper side of the foundation and the lower side of the tie-beams are not allowed in seismic zones.

2. The tie-beams and foundation beams must have a minimum width of 0.25 m and a minimum depth of 0.4 m for buildings of up to three storeys and 0.5 m for four or more storeys. (These values may be modified by the National Annex.)

3. The tie-beams must have a steel ratio along their full length of 0.4 per cent. (This value may be modified by the National Annex.)

1. RECOMMENDATIONS

1. Remember that the ground slabs (Group 11) do not serve as tie elements because they have separation joints around the pillars (See CD – 11.02).

2. STATUTORY LEGISLATION

See EC2 (5) and EC8 (4).

3. RECOMMENDED ALTERNATIVE CODES

See Chapter 21 in ACI 318-08 (22).

4. SPECIFIC REFERENCES

See Chapter 3 in (12).

References

(1)   EN 10080. *Steel for the reinforcement of concrete.*

(2)   EN ISO 17760. *Permitted welding process for reinforcement.*

(3)   EN ISO 3766:2003. *Construction drawings. Simplified representation of concrete reinforcement.*

(4)   EUROCODE 8 (EN 1998.1). *Design of structures for earthquake resistance – Part 1: General rules, seismic actions and rules for buildings.* December 2004.

(5)   EUROCODE 2: *Design of concrete structures – Part 1-1; General rules and rules for buildings.* December 2004.

(6)   CONCRETE SOCIETY. *Spacers for reinforced concrete.* Concrete Society. 1989. Report CS 101. Camberley, Surrey.

(7)   COMITÉ EURO INTERNATIONAL DU BÉTON. (CEB). Bulletin N 201. *Spacers, chairs and tying of steel reinforcement.* Lausanne. 1990.

(8)   BS 7973:2001. *Spacers and chairs for steel reinforcement and their specifications.* London. 2001.

(9)   NIKYRY, P. *Anchorage of reinforcement in concrete structures.* Proceedings of International Conference Bond in Concrete. Riga. 1992.

(10) EUROCODE 2: *Design of Concrete Structures – Part 3: Concrete Foundations.* 1998.

(11) EN ISO 17660-1. *Welding of reinforcing steel.*

(12) CALAVERA, J. *Cálculo de estructuras de cimentación.* (Foundation concrete design). 4 Edición. 2000. INTEMAC. Madrid.

(13) PARK, R; PAULAY, T. *Reinforced Concrete Structures.* John Wiley & Son. New York. 1975.

(14) CALAVERA, J. *Proyecto y Cálculo de Estructuras de Hormigón.* (Structural concrete design). 2 Edición. INTEMAC. Madrid. 2008.

(15) AMERICAN CONCRETE INSTITUTE (ACI). *ACI DETAILING MANUAL.* Farmington Hills. Michigan. 2004.

(16) CONCRETE REINFORCING STEEL INSTITUTE. *Placing reinforcing bars.* 8th Edition. Schaumburg, Illinois. 2005.

(17) THE CONCRETE SOCIETY. *Standard Method of Detailing Structural Concrete.* 3rd Edition. 2006. Camberley, Surrey.

(18) CALAVERA, J. *Manual de Detalles Constructivos en Obras de Hormigón Armado.* (Manual for Detailing Reinforced Concrete Structures). INTEMAC. Madrid. 1993.

(19) BANGASH, M.Y.H. *Structural detailing in concrete.* 2nd Edition. Thomas Telford. London. 2003.

(20) LEONHARDT, F. *Vorlesungn über massivbau.* (Structural Concrete). Springer-Verlag. Berlin. 1979.

(21) SCHLAICH, J; WEISCHEDE, D. *Ein praktisches verfahren zum methodischen bemessen und konstruieren im stahlbetonbau.* (A practical method for the design and detailing of structural concrete). Bulletin d'Information N 150, Comité Euro-International du Béton. París. Mars, 1982.

(22) ACI 318-08. *Buildings code requirements for structural concrete and commentary.* American Concrete Institute. Farmington Hills. Michigan. 2008.

(23) REYNOLDS, C.E.; STEEDMAN, J.C.; TARELFALL, A.J. *Reinforced Concrete Designer's Handbook*. 11ᵗʰ Edition. Taylor & Francis. London 2008.

(24) EN 1536. *Execution of special geotechnical work. Bored piles.*

(25) DUNHAM, C.W. *Foundations of structures*. McGraw-Hill. New York. 1962.

(26) EUROCODE 7. *Geotechnical design-Part 1: General rules.*

(27) CALAVERA, J. *Muros de Contención y Muros de Sótano*. (Retaining walls and basement walls). 3 Edición. INTEMAC. Madrid 2000.

(28) ACI 336-2R. *Suggested Analysis and Design Procedures for Combined Footings and Mats*. American Concrete Institute. Farmington Hills. Michigan. 2002.

(29) ACI. *Cast-in-Place Walls*. 2nd edition. Farmington Hills. Michigan. 2000.

(30) *Precast prestressed hollow core floors*. FIP Recommendations. Thomas Telford. London. 1988.

(31) Draft pr EN 1168. *Floor of precast prestressed hollow core elements*. August 1993.

(32) pr EN 1168-1. *Precast concrete products – Hollow core slabs for floors*. August 2002.

(33) ASSAP. Association of Manufactures of Prestressed Hollow Core Floors. *The Hollow Core Floor. Design and Applications*. Verona. 2002.

(34) EN 1337-1. *Structural bearings. Part 1: General design rules*. April 2001.

(35) Bearings EN 1337-3. *Structural bearing. Part 3: Elastomeric bearings*. November 2005.

(36) LEONHARDT, F. and MÖNNIG, E. *Sonferfälle der Bemessung im Stahlbetonbau*. (Customised solutions in reinforced concrete structural engineering) Springer-Verlag. Berlin. 1974.

(37) LEONHARDT, F. and REYMANN, H. *Betongelenke, Versuchsbericht und Vorschläge zur Bemessung und konstruktiven Ausbildung*. (Concrete joints, test report and structural engineering and design proposals). DafStb. H 175. Berlin. Ernst v. Sohn. 1965.

(38) MATTOCK, A.H.; CHEN, K.C.; SOONGSWANG, K. *The behaviour of reinforced concrete corbels*. Journal of the Prestressed Concrete Institute. March-April. Chicago pp. 55–69. 1976.

(39) NAGRODZKA-GODYCKA, K. *Reinforced Concrete Corbels and Dapped – end Beams*. Cuaderno de INTEMAC N 69. Madrid. 1ᵉʳ trimestre de 2008.

(40) ACI. SLABS ON GRADE. American Concrete Institute. Detroit. 1983.

(41) ACI 307-88. *Standard Practice for the Design and Construction of Cast-in-Place Reinforced Concrete Chimneys*. American Concrete Institute. Detroit. 1988.

(42) PINFOLD, G.M.: *Reinforced Concrete Chimneys and Towers*. Viewpoint Publications. London. 1975.

(43) ACI 313. *Standard Practice for Design and Construction of Concrete Silos and Stacking Tubes for Storing Granular Materials*. Farmington Hills. Michigan. 1997

(44) ACI 350.2R. *Concrete Structures for Containment of Hazardous Materials*. Farmington Hills. Michigan. 2004.

(45) CEB Application Manual on *Concrete Reinforcement Technology*. Georgi Publishing Company. Saint-Saphorin (Switzerland). 1983.

(46) CALAVERA, J. *Cálculo, Construcción, Patología y Rehabilitación de Forjados de Edificación*. (Design, Construction, Pathology and Strengthening of Slabs in Buildings). 5 Edición. INTEMAC Madrid. 2002.

Index